Biology and Chemistry of Eucaryotic Cell Surfaces

MIAMI WINTER SYMPOSIA

---

1. *W. J. Whelan and J. Schultz, editors:* HOMOLOGIES IN ENZYMES AND METABOLIC PATHWAYS *and* METABOLIC ALTERATIONS IN CANCER,* 1970
2. *D. W. Ribbons, J. F. Woessner, Jr., and J. Schultz, editors:* NUCLEIC ACID-PROTEIN INTERACTIONS *and* NUCLEIC ACID SYNTHESIS IN VIRAL INFECTION,* 1971
3. *J. F. Woessner, Jr. and F. Huijing, editors:* THE MOLECULAR BASIS OF BIOLOGICAL TRANSPORT, 1972
4. *J. Schultz and B. F. Cameron, editors:* THE MOLECULAR BASIS OF ELECTRON TRANSPORT, 1972
5. *F. Huijing and E. Y. C. Lee, editors:* PROTEIN PHOSPHORYLATION IN CONTROL MECHANISMS, 1973
6. *J. Schultz and H. G. Gratzner, editors:* THE ROLE OF CYCLIC NUCLEOTIDES IN CARCINOGENESIS, 1973
7. *E. Y. C. Lee and E. E. Smith, editors:* BIOLOGY AND CHEMISTRY OF EUCARYOTIC CELL SURFACES, 1974
8. *J. Schultz and R. Block, editors:* MEMBRANE TRANSFORMATION IN NEOPLASIA, 1974

**Published by North-Holland Publishing Company, Amsterdam, The Netherlands.*

MIAMI WINTER SYMPOSIA–VOLUME 7

Biology and Chemistry of Eucaryotic Cell Surfaces

edited by

E. Y. C. Lee

E. E. Smith

DEPARTMENT OF BIOCHEMISTRY
UNIVERSITY OF MIAMI SCHOOL OF MEDICINE
MIAMI, FLORIDA

Proceedings of the Miami Winter Symposia, January 14-15, 1974 organized by the Department of Biochemistry, University of Miami School of Medicine, Miami, Florida

ACADEMIC PRESS New York and London 1974
A Subsidiary of Harcourt Brace Jovanovich, Publishers

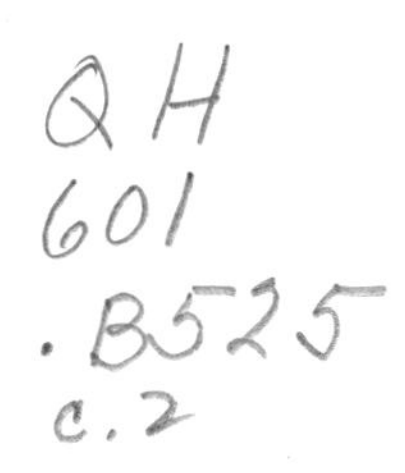

ACADEMIC PRESS, INC.
111 Fifth Avenue, New York, New York 10003

United Kingdom Edition published by
ACADEMIC PRESS, INC. (LONDON) LTD.
24/28 Oval Road, London NW1

Library of Congress Cataloging in Publication Data
Main entry under title:

Biology and chemistry of eucaryotic cell surfaces.

(Miami winter symposia, v. 7)
"Proceedings of the Miami winter symposia, January 14-15, 1974, organized by the Department of Biochemistry, University of Miami School of Medicine, Miami, Florida."
1. Cell membranes–Congresses. 2. Cytochemistry–Congresses. I. Lee, Ernest Yee Chung, ed. II. Smith, Eric Ernest, Date ed. III. Miami, University of, Coral Cables, Fla. Dept. of Biochemistry. IV. Series. [DNLM: 1. Cell membrane–Congresses. 2. Cell nucleus–Congresses. W3M1202 v. 7 1974 / WH595 S989b 1974]
QH 601.B525 574.8′75 74-10683
ISBN 0–12–441550–4

PRINTED IN THE UNITED STATES OF AMERICA

CONTENTS

Free Communications

SPEAKERS, CHAIRMEN, AND DISCUSSANTS

Ankel, H., The Medical College of Wisconsin, Milwaukee, Wisconsin

Assaf, S. A., University of Texas Medical School, Houston, Texas

Basu, S., University of Notre Dame, Notre Dame, Indiana

Behrens, N. H., Instituto de Investigaciones Bioquimicas, Buenos Aires, Argentina

Bernacki, R., Roswell Park Memorial Institute, Buffalo, New York

Bhattacharyya, S., Duke University Medical Center, Durham, North Carolina

Boone, C., National Institutes of Health, Bethesda, Maryland

Bretthauer, R. K., University of Notre Dame, Notre Dame, Indiana

Christopher, C. W., Harvard Medical School, Boston, Massachusetts

Czech, M. P., Duke University Medical Center, Durham, North Carolina

Daniel, J., Papanicolaou Cancer Institute, University of Miami, Miami, Florida

DeLuca, L., National Institutes of Health, Bethesda, Maryland

Elbein, A. D., University of Texas Health Science Center, San Antonio, Texas

Fullerton, W. W., Merck Institute for Therapeutic Research, West Point, Pennsylvania

Gardner, D. A., Miles Laboratories, Inc., Elkhart, Indiana

Gerisch, G., Friedrich-Miescher-Laboratorium des Max Planck Gesellschaft, Tübingen, Germany

Ginsburg, V., National Institute of Arthritis and Metabolic Disorders, Bethesda, Maryland

Glick, M. C., Children's Hospital of Philadelphia, Philadelphia, Pennsylvania

Goldstein, J., New York Blood Center, New York, New York

Hakomori, S., University of Washington, Seattle, Washington

Heath, E. C., University of Pittsburgh School of Medicine, Pittsburgh, Pennsylvania

Hempling, H. G., Medical University of S. Carolina, Charleston, South Carolina

Hochstadt, J., Worcester Foundation for Experimental Biology, Shrewsbury, Massachusetts

Horowitz, M., New York Medical College, Valhalla, New York

Huijing, F., (Session Chairman), University of Miami School of Medicine, Miami, Florida

Jamieson, G. A., The American National Red Cross Blood Research Laboratory, Bethesda, Maryland

Jeanloz, R. W., Harvard Medical School, Boston, Massachusetts

Kalos, J., Columbia University, New York, New York

Keston, A., Institute for Medical Research and Studies, New York, New York

Koch, G., Roche Institute of Molecular Biology, Nutley, New Jersey

Kornfeld, S. A., Washington University, St. Louis, Missouri

Leloir, L. F., (Session Chairman), Instituto de Investigaciones Bioquimicas, Buenos Aires, Argentina

Longton, R., National Naval Medical Center, Bethesda, Maryland

Lubin, M., Dartmouth Medical School, Hanover, New Hampshire

Lynen, F., (Session Chairman), Max-Planck Institut für Biochemie, Munich, Germany

Marchesi, V. T., Yale University, New Haven, Connecticut

Montes de Gomez, V., National University of Colombia, Bogota, Colombia

Morrison, M., St. Jude Children's Research Hospital, Memphis, Tennessee

Nathenson, S. G., Albert Einstein College of Medicine, New York, New York

Nicolson, G. L., The Salk Institute for Biological Studies, San Diego, California

Ostrand-Rosenberg, S., California Institute of Technology, Pasadena, California

Pickart, L. R., University of California, San Francisco, California

Poretz, R. D., Rutgers University, New Brunswick, New Jersey

Quash, G., University of the West Indies, Kingston, Jamaica

Rabinowitz, M., National Institutes of Health, Bethesda, Maryland

Reddi, A. H., University of Chicago, Chicago, Illinois

Reinhold, V. N., Arthur D. Little, Inc., Cambridge, Massachusetts

Rieber, M., Center of Microbiology and Cell Biology, Caracas, Venezuela

Robbins, P. W., Massachusetts Institute of Technology, Cambridge, Massachusetts

Roseman, S., John Hopkins University, Baltimore, Maryland

Rosenberg, S., (see S. Ostrand-Rosenberg)

Roth, J., Biocenter, Basel, Switzerland

Schenkein, I., New York University Medical Center, New York, New York

Srere, P. A., Veterans Administration Hospital, Dallas, Texas

Stanford, H. K., (Session Chairman), President, University of Miami, Coral Gables, Florida

Sutherland, E. W., (Session Chairman), University of Miami School of Medicine, Miami, Florida

Vaheri, A., University of Helsinki, Helsinki, Finland

Warren, L., University of Pennsylvania, Philadelphia, Pennsylvania

Whelan, W. J., Chairman, Department of Biochemistry, University of Miami School of Medicine, Miami, Florida

Wolff, J., National Institutes of Health, Bethesda, Maryland

Wood, H. G., (Session Chairman), Case Western Reserve University, Cleveland, Ohio

PREFACE

In January 1969, the Department of Biochemistry of the University of Miami and the University-affiliated Papanicolaou Cancer Research Institute joined in sponsoring and presenting two symposia on biochemical topics. Those symposia have now developed into an annual event, now in its sixth year. In 1970 the two symposia were published as the first volume of a continuing series entitled the "Miami Winter Symposia." In 1972 we initiated the publication of the proceedings of the two symposia as separate volumes in the series.

The major emphasis in the selection of the topics for our symposia has been to identify the frontier areas in which progress in biochemistry is leading toward an understanding of the molecular bases of biological phenomena. We follow a pattern in which a common theme is dealt with, in our symposium, on a fundamental basis, and then in the Papanicolaou Cancer Research Institute Symposium, as it relates to an understanding of malignant processes. This volume, the seventh in the series, contains the proceedings of the Biochemistry Department's Symposium on the "Biology and Chemistry of Eucaryotic Cell Surfaces," and will be published simultaneously with the proceedings of the Papanicolaou Cancer Research Institute's symposium on "Membrane Transformations in Neoplasia."

In 1975 the topics for the symposium will be on a theme in basic immunology followed by a symposium on cancer immunology, taking place during 13-17 January.

Associated with the symposia is a featured lecture, the Feodor Lynen Lecture, named in honor of the Department of Biochemistry's distinguished Visiting Professor. Past speakers were Drs. George Wald, Arthur Kornberg, Harland G. Wood, Earl W. Sutherland, Jr., and this year we heard from an old friend of the Department, Dr. Luis F. Leloir. These lectures are of an autobiographical nature, devoted to the life histories of these great scientists. They are greatly enjoyed by their audience, and for ourselves, remain a source of inspiration. The Lynen lecturer for 1975 will be Dr. Gerald M. Edelman.

Our arrangement with the speakers and publishers is to ensure publication as rapidly as possible. We thank the secretarial staff for their help in bringing this about. To bring forward as much of the recent work as possible, we now include a session of short communications in the meeting, the abstracts of which are included in this volume. We thank the speakers for submitting their manuscripts promptly and the participants whose interest and discussions provided the interactions that bring a symposium to life.

The financial assistance of the University of Miami Departments of Anesthesiology and Medicine, the Howard Hughes Medical Institute, Dermatology Foundation of Miami, Eli Lilly and Company, Hoffmann-La Roche Inc., MC/B Manufacturing Chemists and the International Union of Biochemistry, is gratefully acknowledged. The Union provided funds for the participation of Latin-American biochemists. This symposium volume also constitutes volume 62 of the IUB series.

E. Y. C. Lee
E. E. Smith

THE FIFTH FEODOR LYNEN LECTURE

"I HATE TO BORE PEOPLE WITH MY RECOLLECTIONS"

L.F. LELOIR
Instituto de Investigaciones Bioquímicas
"Fundación Campomar", y Facultad de Ciencias
Exactas y Naturales

When Bill Whelan invited me to give the Lynen Lecture, which is of an autobiographical nature, I accepted with reluctance because I don't think that anything in my life is worth telling and much less, entertaining. Furthermore, I have already written an historical review called "Polysaccharide synthesis seen from Buenos Aires" (1) and it is difficult to write two biographies without having led a double life.

My first difficulty with this lecture was that I could not find a title. I spent such a long time thinking, that Whelan phoned me and it was then that I said "I hate boring people with my recollections". The long distance call was rather noisy and Whelan took that phrase as the title. Since I did not find a better one it was left as it is.

My life covers the period of biochemistry in which most of the vitamins, hormones, enzymes and coenzymes were discovered and in which most of the reactions of intermediary metabolism were unravelled. I was born in 1906, the same year in which Harden and Young (2) published their paper on the coenzyme of yeast fermentation and which initiated the studies on cofactors in which we became involved many years later.

After fairly normal studies in the primary and secondary school, I studied medicine, in Buenos Aires. I was an average student, interested in scientific subjects but had no occasion of thinking on research. This happened while I was working in the hospital and I heard that Dr. Houssay was doing very interesting studies on carbohydrate metabolism. They were described to me as revolutionary findings and turned out to be in fact of great interest and eventually led to the award of the Nobel Prize to Houssay in 1947. What Houssay had found was related to the role of the pituitary. If the pancreas is removed from an animal it becomes diabetic. This had been discovered by von Mering and Minkowsky in 1886. The new finding was that if the pituitary was also removed then the animals did not become diabetic. The animals without pancreas and pituitary became known as Houssay's dogs in the physiological literature.

While I was studying medicine I had no idea of what I wanted to do nor which was the field I was more apt for. In relation to this I remember discussing the problem with some of my colleagues at the hospital and one of them told me "you are not very intelligent but maybe you can be successful because you are persevering". About two years elapsed before I was introduced to Dr. Houssay and started working at the Institute of Physiology. It was there where I did my thesis work on the Role of the Adrenals in Carbohydrate Metabolism. Houssay helped me a lot, not only did he do the thinking but he also carried out the adrenalectomies on the dogs. He was a really extraordinary person. As a scientist he was self-made and he developed himself under completely adverse conditions. He had one teacher of great quality but whom he never met. This was the famous French physiologist Claude Bernard. It was by reading the book: "Introduction a l'etude de la Medicine Experimentale" that Houssay became a scientist. Besides the laboratory work, the teaching of thousands of students and the organization of research he carried out a lasting campaign for the advancement of science. A very great part of research in Argentina developed thanks to his efforts. He was president of our National Research Council for about 15 years and he did a splendid job. He was an untiring worker and used to say "I have no time to be ill". It has been for me a great privilege to be in close contact for many years with such an outstanding personality as Dr. Houssay.

During my thesis work I realized that it would be interesting to understand Physiology more deeply and this is what led me to study Biochemistry. This was similar to the previous change when I wanted to understand medicine more deeply and began studying Physiology. Perhaps it was fortunate that I did not continue trying to understand things more deeply because I might have ended up in Nuclear Physics or in Philosophy.

Cambridge. After some consultations I decided to improve my knowledge of Biochemistry by going to the Biochemical Laboratory of Cambridge. At the time (1935) it was one of the leading world centers in biochemical research. The head of the laboratory was Sir Frederick Gowland Hopkins a pioneer of studies on vitamins, the discoverer of trytophan and one of the first great British biochemists. He was awarded the Nobel Prize in 1929 together with Eijkman for "his discovery of growth stimulating vitamins".

First I worked with Malcolm Dixon on the action of cyanide and pyrophosphate on dehydrogenases (3), then with Norman L. Edson on the utilization of acetoacetate and β-hydroxybutyrate by tissue slices (4). Afterwards I worked with David E. Green on β-hydroxybutyrate dehydrogenase (5).

Cambridge was a great experience. The laboratory was different from the German laboratories of that time where there was a Herr Professor and all the others worked directly under his directions. In Cambridge the groups worked independently and there were many outstanding investigators; Marjorie Stephenson, one of the pioneers in Biochemical Microbiology, Dorothy Needham, who worked on muscle, Robin Hill who became so well known from the Hill effect, Norman Pirie who at the time had succeeded in crystallyzing tobacco mosaic virus. This he did simultaneously with Wendel Stanley who received the Nobel Prize for this work. After one year in Cambridge I thought I had acquired some rudimentary knowledge of biochemical research and decided to return to Buenos Aires.

Fatty acids. In those times it was generally believed that the oxidation of fatty acids would only take place in intact cells. This belief was based on the fact that cell homogenates were completely inactive. With J.M. Muñoz (6) we decided to try to obtain a soluble preparation which would oxidize fatty acids.

We used to measure the disappearance of volatile fatty acid on incubation with liver homogenates. The analytical procedure consisted in distilling the acids and oxidizing them with chromic acid. When the acid disappeared the reaction mixture remained brown and if there was no fatty acid consumption the final mixture was green. In most experiments nothing happened and the final mixture was green. I remember having the feeling that our faces also turned green after many of these failures. We used to work with guinea pig liver homogenates and fractionated them in a centrifuge which was cooled by wrapping an inner automobile tire filled with freezing mixture around it. At that time there were no commercial refrigerated centrifuges. We had no guinea pigs nearby and had to send our assistant to fetch them from rather far away. He usually took a basket but one day he forgot to take it and instead used a large paper bag. He boarded a bus with the guinea pigs but these were clever enough to find a way of escaping; they wetted the paper so that it became soft and then worked their way out. The bus was crowded so that the result was that the ladies screamed and the men complained. Our assistant had to get down and leave the guinea pigs behind.

Finally after innumerable failures we overcame these and many other difficulties and obtained an enzyme preparation which oxidized fatty acids when complemented with adenylic acid or adenosine triphosphate, a C_4 dicarboxilic acid, cytochrome C and magnesium ions. I remember that one of the things that intrigued us was that activity disappeared very rapidly on leaving the enzyme mixture in water. Now that we know that the activity is mitochondrial this does not seem surprising. The separation of mitochondria by differential centrifugation in sucrose solutions was developed by Claude years later and the mechanism of fatty acid oxidation was clarified by several groups of workers, including Fedor Lynen.

Angiotensin. Circumstances led me to change my line of research several times in my life. I doubt if this is desirable but in some cases it is inevitable.

In the Institute of Physiology of the Faculty of Medicine of Buenos Aires, Juan Carlos Fasciolo had been working on the mechanism of arterial hypertension produced by constriction of the renal artery. He had reached the conclusion that a pressor substance was involved. With J.B. Muñoz and Eduardo Braun Menendez we decided to collaborate in this project and in a rather short time we made several findings of considerable importance and furthermore we had a good time. This period as well as a few others in which I worked with pleasant and enthusiastic people were the most enjoyable experiences in my career.

It was known that kidney contains a protein, which when injected into animals, produces an increase in blood pressure. Renin is a protein and is thermolabile. On the other hand the substance which appeared in the experiments with isolated kidneys was thermostable. This led us to do experiments in which extracts of kidney were incubated under various conditions and then tested for thermostable pressor substance. The results were always negative. One day Braun Menendez came along and proposed to incubate partially purified renin with blood plasma and then to test it. I argued that I had incubated crude extracts which certainly contained renin and it did not work, therefore testing renin seemed hopeless. However, to make him happy I did the experiment as he proposed. We obtained a beautiful effect and became very excited. We learned afterwards that the negative results with crude extracts were due to the presence of an enzyme which destroyed our pressor substance. In a short time we found out that the renin acted on another protein present in blood to yield the pressor substance which we named hypertensin. The blood protein we called hypertensinogen and the enzyme which inactivated the pressor substance was named hypertensinase. To our dismay Page and Helmer published similar findings practically simultaneously. The pressor substance which we called hypertensin (7) was called angiotonin by Page and Helmer. Each group tried to impose its name until Braun Menendez and Page solomonically proposed the name angiotensin which is now used universally.

The hypertension team lasted about one year but the amount of work it carried out is really surprising (8). We found out that angiotensin is a polypeptide but with the methods available it was practically impossible to obtain more information. A lot of work has been done in the field since then. Now it is known that renin hydrolyzes off a decapeptide, angiotensin I, from angiotensinogen. Angiotensin I is practically inactive but by the action of a converting enzyme is transformed into an active octapeptide, angiotensin II. This substance increases blood pressure and also stimulates the secretion of aldosterone from the adrenal gland.

The work on hypertension was interrupted by extraneous reasons and the team disintegrated. In one of our periodical political upheavals Dr. Houssay was dismissed from his post at the University. We had to leave the Institute of Physiology and were left without a laboratory.

St. Louis and New York. In compensation for the scientific misfortune I had the luck of getting happily married and I decided to spend some time doing research in the U.S.A. I was interested in going to Dr. Cori's laboratory in Washington University in St. Louis which was one of the Mecca's of workers in carbohydrate metabolism. I don't recall the circumstances too well but according to Dr. Cori I presented myself unannounced and when he asked me when I would start working I answered "right now". The laboratory was very pleasant with the two Coris, Colowick, Arda Green, Slein, Burger, Taylor and others. There were few people because all this occurred during the 1939-45 war. The Cori's had recently published a big paper on the crystallization of phosphorylase and on some of its kinetic properties. While I was in Washington University I worked with Ed Hunter on citric acid formation (9).

After a time we decided to go to New York where I again met David Green with whom I had worked at Cambridge. We worked on transaminases and were able to separate two of them (10). David Green was always full of ideas and projects and working with him was most interesting. He was also very critical and often became iconoclastic. It was at that time that paper chromatography was developed by Consden, Gordon and Martin. I remember very well that Green showed me the paper in the Biochemical Journal and said "have a look at this, it seems rather interesting". Since it dealt with aminoacid separation I was not interested and did not appreciate the importance the new method would acquire.

We lived fairly near the College of Physicians and Surgeons of Columbia University. My wife was still perfecting her cooking abilities and one day gave me a piece of roasted liver. It looked so repulsive that while she went out of the room I rapidly put it in an old envelope and threw it out of the door. I was happy not to have hurt her feelings. On the following morning the post arrived and my wife sorted the letters. One of them was fatter and she stared at it and said "look at what Dr. So and So sends you; it must be something for the laboratory". I nearly fainted when I discovered that the roasted liver had come back with the post.

After about one year and a half in the U.S.A. we thought that it would be possible to work again in Buenos Aires.

Fundación Campomar. Dr. Houssay had been reinstated as Professor so that many of us returned to the Institute of Physiology. However, this did not last long. A few months later Houssay was again dismissed from the University and the Institute of Physiology was disintegrated again. An unexpected event came to our aid. Jaime Campomar, a well known textile manufacturer, decided to create a private biochemical research center. He asked Dr. Houssay who could take care of organizing the new center. Dr. Houssay suggested my name although, I think, he was not very convinced that I could do a good job. Presumably he did not find a better candidate.

One of the good things I did was to obtain the collaboration of Ranwell Caputto who had just returned from a Fellowship in the Biochemical Laboratory in Cambridge. We also enrolled Raul Trucco who had experience in Bacteriology because at that time we were interested in fatty acid oxidation by bacteria. At first we worked in the Institute of Biology and Experimental Medicine which was a private institution where Houssay and coworkers had taken refuge. After a short stay there we rented an old, adjoining, small house and conditions began to improve slowly. We had three small laboratories, a library where I took my private books, and a little store room. Dr. Cardini, who had been dismissed from the University of Tucumán joined us as well as A.C. Paladini who came with the first fellowship of the Fundación Campomar.

The installation of a new laboratory is always an amusing enterprise. Furthermore we were all young and enthusiastic. The Rockefeller Foundation provided us with a refrigerated centrifuge and Dr. Houssay loaned us a spectrophotometer. The first research project, fatty acid oxidation in bacteria, failed and then we went on to study lactose synthesis. As the preliminary trials were unsuccessful we thought we might get some information on the synthesis by studying lactose utilization. For this we selected a yeast that grows on lactose: *Saccharomyces fragilis*. We grew it on milk serum which was cheaper than lactose and then dried it. For this purpose we used to extend the yeast paste on the bottom of inverted precipitation beakers. Since we had only a few we selected those that were cracked for this operation. The lady who washed the glassware came one day complaining that we should not dry the yeast in that way because it cracked the flasks. I mention this because it is a kind of reasoning that is quite common and which even scientists use inadvertently.

The extracts of the dried yeast were found to contain an enzyme which phosphorylated galactose to a product which turned out to be galactose-1-phosphate. The fate of the latter was unknown at the time but it did not take long to find out that the galactose-1-phosphate was transformed to glucose-1-phosphate and the latter to glucose-6-phosphate. The change from galactose to glucose requires an inversion of the hydroxyl group at position 4. The transformation of glucose-1-phosphate to glucose-6-phosphate was known to be carried out by the enzyme phosphoglucomutase but there was no information on a cofactor requirement.

The reaction was measured by following the increase in reducing power with a copper reagent. We tried to purify the enzymes required in the process and soon found that addition of a heated extract (a Kochsaft) produced a large activation. We then wanted to find out what was the substance in the heated extract which produced the activation. According to my recollection Dr. Caputto was testing the action on glucose-1-phosphate while I was using galactose-1-phosphate. The results were very confusing until we realized that we were dealing not with one but with two thermostable substances. Once we realized this we proceeded to isolate first the cofactor of the glucose-1-phosphate - glucose-6-phosphate transformation. Our job was made easier because we had a theory on the identity and mode of action of the active substance. The theory came from the fact that the active substance had some properties similar to fructose diphosphate. According to the theory our substance was glucose-1-6-diphosphate and it acted by transferring its 1-phosphate to position 6 of glucose-1-phosphate. Thus the product of the reaction was glucose-6-phosphate and a new molecule of glucose-1-6-diphosphate. By one of these chances that rarely happen in research, the theory turned out to be correct. By using the classical isolation methods of that time we were able to isolate the substance and prove that it was glucose-1-6-diphosphate (11).

Then we turned our attention to the isolation of the other substance, that is the one that accelerated the galactose-1-phosphate → glucose-1-phosphate transformation. We used a combination of precipitation of the mercury salt and adsorption on charcoal. Our concentrates were found to adsorb light at about 260 nm and at first it was thought to be an adenine nucleotide.

Sometimes one forgets how primitive our knowledge on nucleotides was at that time. The only free nucleotides known to be present in tissues were adenosine triphosphate, adenosine diphosphate, adenosine monophosphate and inosinic acid. Uridine monophosphate was only known as the 3' phosphate obtained by hydrolysis of nucleic acid. The 5' phosphate had not been isolated and of course neither uridine diphosphate nor uridine triphosphate was known. Finally we obtained fairly pure compound and had no great difficulty in determining its structure as UDP-Glucose (12).

In the course of the isolation of UDP-Glucose we detected, with Cabib and Cardini, other similar compounds which were later separated and identified. One of them was GDP-Mannose (13) which is now well known as a mannose donor and precursor of GDP-Rhamnose. The other nucleotide which we identified was UDP-Acetylglucosamine (14). This led us to studies on hexosamines such as the enzymatic synthesis of glucosamine phosphate from hexose phosphate and glutamine (15), the decomposition of glucosamine-6-phosphate into fructose-6-phosphate and ammonia (16) and on the formation of what we thought was acetylgalactosamine (17) and that as shown by Comb and Roseman (18) turned out to be acetylmannosamine. Another nucleotide: UDP-Acetylgalactosamine, was identified by Pontis (19) from bovine liver in our laboratory.

At the time that we were working with UDP-Glucose, Ted Park together with Marvin Johnson were studying the action of penicillin on staphylococcus (20). They found that a compound containing acid labile phosphate accumulated in the presence of the antibiotic. From the mixtures Park isolated a compound which contained uridine and an unidentified sugar. The structure of this unidentified sugar, now called acetylmuramic acid, was elucidated many years later. We had more luck than Park in that our compound was made up of well known components. The muramic acid containing compound discovered by Park turned out to be of great importance in the biosynthesis of bacterial cell walls. The mechanism of this process has been studied by several workers and specially by Jack Strominger.

The mechanism by which UDP-Glucose catalyzes the transformation of galactose-1-phosphate into glucose-1-phosphate became more understandable when it was found that on incubation with the enzyme, part of the glucose was converted into galactose (21). We now know that the formation of galactose requires an oxidation at position 4 of the glucose residue of UDP-Glucose. The studies on galactose metabolism were developed further by several workers, especially by Hermann Kalckar and his group. As to the role of sugar nucleotides as sugar donors, it became apparent when Dutton and Storey (22) found that glucuronides are formed from UDP-Glucuronic acid. However I think that I did not appreciate at the time that this was a general phenomenon. The role of UDP-Glucose as glucose donor was first suggested by Calvin, Buchanan and others (23) to explain the formation of sucrose. The evidence they had at the time was not very convincing but the conclusion, we know now, was correct.

In our laboratory we were measuring the disappearance of UDP-Glucose under different conditions. We found that glucose-6-phosphate greatly increased the disappearance of UDP-Glucose in the presence of yeast enzymes (24). The changes were soon traced to the formation of trehalose-6-phosphate. Soon afterwards the synthesis of sucrose (25) and sucrose phosphate (26) with UDP-Glucose and plant enzymes was obtained and a series of other transfer reactions were discovered by various workers (27). The laboratory of Dr. Hassid deserves special mention for studies on saccharide synthesis in plants.

The synthesis of glycogen from UDP-Glucose, was a finding of some interest particularly because it was universally accepted that both synthesis and degradation were catalyzed by phosphorylase. The direction in which the reaction takes place at any moment was believed to be dictated by local concentration of glucose-1-phosphate and inorganic phosphate. Some inconsistencies of this theory were pointed out by Sutherland (28). He reasoned that it was difficult to understand why the activation of phosphorylase should always produce degradation of glycogen and that if it was also involved in synthesis it should lead to glycogen formation if the equilibrium of the reaction was favorable. The suggestion that UDP-Glucose might be involved in glycogen synthesis was formulated by Niemeyer (29) but it was the detection of glycogen synthetase that finally settled the problem. The activity of the enzyme was found to be very low. Here again we tried the old trick that was used by Harden and Young in the days when I was born. We added a heated extract of liver and obtained a great increase in glycogen formation from UDP-Glucose (30). The substance responsible for the effect was found to be acid stable, alkali labile and retained by an anion exchange resin. Several known substances with these properties were then tested. One of them, glucose-6-phosphate was found to be active and furthermore many others were without effect. That is, the effect of glucose-6-phosphate was quite specific. This led to the idea that glucose-6-phosphate might have a regulatory effect on glycogen synthetase while adenylic acid would regulate phosphorylase. Dr. Belocopitow (31) carried out some experiments on the effect of adrenalin. He found that adrenalin produced a decrease of glycogen synthetase activity in rat diaphragm. The field which was developed rapidly by Larner, Villar Palasi and many others is now a full chapter of the enzyme regulating mechanisms.

Besides the studies on glycogen formation in which the transfer of glucose was measured with radioactive tracers, other experiments were carried out in order to obtain "in vitro" a product similar to the native glycogen present in liver.

Glycogen isolated by procedures which avoid degradation has molecular weights ranging from 10 to 1.000 million daltons. Samples prepared "in vitro" with purified synthetase, and branching enzymes gave preparations of the same molecular weight as native glycogen and that had the same lability to acid and alkali and a similar aspect when examined with the electron microscope (32). Glycogen prepared with phosphorylase and glucose-1-phosphate had a similar molecular weight but differed in several other properties.

The explanation of the properties of the high molecular weight glycogen remain obscure but may be related to the fact that the molecules seem to grow on a protein core (33), on which subparticles are joined by acid labile linkages.

Experiments designed to inform on the subcellular distribution of glycogen synthetase showed that on fractional centrifugation it sedimented together with the glycogen (34). This finding was important in connection with the biosynthesis of starch. The bonds in glycogen and in starch are α-1→4 with α-1→6 links in the branch point. The difference between the two resides in that starch consists of two components amylose and amylopectin.

The first is a linear chain while the second is ramified but has longer chains than glycogen. After working on glycogen synthesis we naturally became interested in that of starch. It seemed obvious that the precursor should be UDP-Glucose. However experiments with crude plant extracts gave negative results until it was reasoned that if glycogen synthetase activity goes with glycogen then starch synthetase should be found in starch.

Experiments with María Fekete and Cardini (35) showed that bean starch incubated with UDP-Glucose gave rise to the formation of uridine diphosphate and to transfer of the glucose. We had the impression that the activity was rather low for a polysaccharide which is formed at rather high rate in plants. For this reason it was considered worth while to test some other substrates besides UDP-Glucose. For instance we thought of a sugar nucleotide such as UDP-Maltose which would act as donor of maltose and also of nucleotides of glucose with different bases. At the time Dr. Eduardo Recondo who was trained as an organic chemist, came to our laboratory. Furthermore appropriate synthetic methods had been developed by Khorana. One of the first nucleotides that was synthesized was ADP-Glucose and it was tested for starch formation. Surprisingly this compound turned out to act as glucose donor much more efficiently than UDP-Glucose (36). The reaction was actually about ten times faster. The observation was interesting and indicated that ADP-Glucose could be the glucose donor for starch synthesis. This idea was strengthened by Espada's (37) finding of a specific enzyme which leads to the synthesis of ADP-Glucose from adenosine triphosphate and glucose-1-phosphate.. At present it is accepted that ADP-Glucose is a natural substrate for starch biosynthesis and for bacterial glycogen. Many reactions leading to polysaccharide synthesis with bacterial plant or animal enzymes have been studied. Our beliefs on the mechanism of these have been changing for many years. The study of saccharide biosynthesis actually started before I was born when Croft Hill, in 1898, incubated concentrated glucose with yeast enzymes and obtained an α-linked disaccharide. Years later Zemplen and Bourquelot also used glucose but an enzyme of plant origin and obtained β-linked disaccharides. Bourquelot also obtained disaccharides from galactose or mannose solutions. From these facts it was deduced that reversal of hydrolysis was a possible mechanism for polysaccharide synthesis. Many years later in 1939 the Cori's obtained the synthesis of glycogen from glucose-1-phosphate with the enzyme phosphorylase. After that reversal of phosphorylase was believed to be the mechanism of polysaccharide synthesis. Another route was discovered in 1941 by Hehre. This was the synthesis of dextran from sucrose with a bacterial enzyme. A similar process in which the fructose moiety of sucrose is transferred has also been described. The next advancement in this field was the discovery of the role of nucleotide sugars. Several transfer reactions leading to the synthesis of polysaccharides such as chitin, callose, cellulose and many others have been studied (38). In some cases the sugar nucleotides are not the direct donors and lipid intermediates are involved. This is one of the most important recent developments in the field.

Dolichol. In the last years we have been concerned with the role of dolichol phosphate in sugar transfer in animal tissues. This has turned out to be a difficult but very exciting problem. Studies on bacteria by Robbins, Strominger, Osborn, Lennarz and others have shown that polyprenoid alcohols are involved in the synthesis of bacterial cell wall constituents.

Marcelo Dankert from our laboratory was a member of one of the teams, with Phil Robbins and A. Wright. When he returned to our laboratory he told us about the function of the polyprenoid alcohols and we became convinced that this was a very important breakthrough in carbohydrate metabolism.

We had at hand radioactive UDP-Glucose and some liver extracts so we incubated them together and then measured the radioactivity of the butanol soluble substances. There were only a few counts over the control but after changing the conditions the system was found to form a radioactive lipid reproducibly. The next step forward was when, with Nicolás Behrens (39) we found a stimulation by a lipid fraction of liver. After fractionation and a study of the properties we concluded that our stimulating lipid might be dolichol phosphate. The latter was prepared by synthesis and found to be the same as the natural compound. Incubation of various sugar nucleotides with dolichol phosphate and liver microsomes showed that dolichol derivatives are formed from UDP-Glucose, GDP-Mannose and UDP-Acetylglucosamine but apparently not with UDP-Galactose or UDP-Acetylgalactosamine (40). As to UDP-Glucuronic acid no definite conclusions were reached. The mannose containing compound has been studied carefully by Hemming and coworkers who have concluded that it really is a derivative of dolichol phosphate. As to the compound formed from UDP-Acetylglucosamine it appears to have a pyrophosphate and not a phosphate residue.

The glucose containing compound, dolichol-monophosphate-glucose, on incubation with liver microsomes is transformed so that the glucose moiety is transferred to an endogenous acceptor yielding a compound which appears to be Dolichol-P-P-oligosaccharide (41). Incubation of this compound with microsomes and manganese ions results in a transfer of the oligosaccharide to protein (42). Incubation of dolichol-monophosphate-mannose with microsomes gives rise to a transfer of mannose to an acceptor giving a lipid-oligosaccharide. The latter can then act as donor in transfer of the oligosaccharide to protein (43). The acetylglucosamine containing compound incubated again with UDP-Acetylglucosamine gives rise to the formation of a substance with two acetylglucosamine residues joined β-1,4 (44). This is a finding of considerable interest because disaccharide formed from two acetylglucosamines, that is N,N'-diacetylchitobiose, is found as the innermost residue in the oligosaccharide of several glycoproteins such as: thyroglobulin, aspergillus amylase, ovalbumin and probably many others.

All the studies we have carried out up to now with Behrens, Carminatti, Parodi, Staneloni and others, have been done following the radioactivity and not by regular analytical methods. Therefore many of the conclusions remain to be confirmed. The picture which emerges from the experiments is as follows. The first step would be a reaction between dolichol-monophosphate and UDP-Acetylglucosamine to give dolichol-diphosphate-acetylglucosamine according to the following equation:

$$DMP + UDP\text{-}GlcNAc \rightarrow DDP\text{-}GlcNAc\ UMP$$

The reaction product would then react with another UDP-Acetylglucosamine molecule yielding a N,N'-diacetylchitobiose containing lipid.

$$DDP\text{-}GlcNAc + UDP\text{-}GlcNAc \rightarrow DPP(GlcNAc)_2 + UDP$$

The product would in turn act as acceptor either directly from GDP-Mannose or from DMP-Mannose. Several mannose residues linked α or β would be thus added followed by some acetylglucosamine so that the product would be a dolichol diphosphate linked oligosaccharide. This oligosaccharide would then be transferred to the acceptor protein and the glycoprotein would thus be completed.

For some glycoprotein glucose would be added from dolichol-monophosphate-glucose before the transfer to protein.

The fact that mannose is involved and some experiments in which lability to alkali was measured indicate that the glycoproteins in question are of the asparagine type. Therefore it seems that the oligosaccharide is built up joined to dolichol phosphate and then transferred to an asparagine residue in a protein. Probably other residues of N-acetylglucosamine, of galactose and of neuraminic acid may be added after the oligosaccharide is transferred to protein.

It seems that this pathway of synthesis is not general for all the glycoproteins. Several cases have been studied rather carefully and no evidence for the intermediate formation of lipid intermediates was found. Such is the case with collagen or glomerular basement membrane which has a glucosyl α-1$\rightarrow$2-galactosyl residue β-linked to hydroxylysine. Two transferases have been detected, one for glucose and another for galactose.

As to the glycoprotein in which the oligosaccharide is linked to a hydroxyaminoacid there are several which have been studied in order to clarify their biosynthesis. Ovine submaxillary mucin has a disaccharide formed by neuraminic acid and galactosamine. The linkage region of chondroitin sulfate contains glucuronic acid, two galactoses and xylose linked to serine. In none of these cases does it seem that lipid intermediates are involved. It seems therefore that the role of the dolichol intermediates is limited to the glycoproteins in which the oligosaccharide is joined to asparagine.

About 40 years have now elapsed since I first started to do research. These have been years of fairly hard work but with many agreeable moments. Research has many aspects which make it an attractive venture. One of them is the intellectual pleasure of discovering previously unknown facts. There are also the human aspects which are worth mentioning. Some of the most pleasant periods in my career were those in which I could work with people that were enthusiastic and clever and had a good sense of humour. The discussion of research problems with these people has always been a most stimulating experience.

The less agreeable part of research, which is the routine work and the sensation of failure which accompanies most of the experiments, is more than compensated by the agreeable aspects which include meeting and sometimes winning the friendship of people of superior intellect and having the occasion of travelling and visiting different parts of the world. The balance is clearly positive.

REFERENCES

(1) L.F. Leloir, in: Biochemistry of the Glycosidic Bond, ed. R. Piras and H. Pontis (Academic Press, New York and London, 1972) p.1.

(2) A. Harden and N.J. Young. Proc. Roy. Soc. B. 78 (1906) 369.

(3) L.F. Leloir and M. Dixon. Enzymologia. 2 (1937) 81.

(4) N.L. Edson and L.F. Leloir. Biochem. J. 30 (1936) 2319.

(5) D.E. Green, J.G. Dewan and L.F. Leloir. Biochem. J. 31 (1937) 934.

(6) J.M. Muñoz and L.F. Leloir. J. Biol. Chem. 147 (1943) 355.

(7) J.M. Muñoz, E. Braun Menendez, J.C. Fasciolo and L.F. Leloir. Am. J. Med. Sci. 200 (1940) 608.

(8) L.F. Leloir. Special publications of the N.Y. Acad. Sci. III, (1946) 60-76.

(9) F.E. Hunter and L.F. Leloir. J. Biol. Chem. 159 (1945) 295.

(10) D.E. Green, L.F. Leloir and V. Nocito. J. Biol. Chem. 161 (1945) 559.

(11) C.E. Cardini, A.C. Paladini, R. Caputto, L.F. Leloir and R. Trucco. Arch. Biochem. Biophys. 22 (1949) 87.

(12) R. Caputto, L.F. Leloir, C.E. Cardini and A.C. Paladini. J. Biol. Chem. 184 (1950) 333.

(13) E. Cabib and L.F. Leloir. J. Biol. Chem. 206 (1954) 779.

(14) E. Cabib, L.F. Leloir and C.E. Cardini. J. Biol. Chem. 203 (1955) 1055.

(15) L.F. Leloir and C.E. Cardini. Biochem. Biophys. Acta. 12 (1953) 15.

(16) L.F. Leloir and C.E. Cardini. Biochem. Biophys. Acta. 20 (1956) 33.

(17) C.E. Cardini and L.F. Leloir. J. Biol. Chem. 225 (1957) 317.

(18) D.G. Comb and S. Roseman. Biochem. Biophys. Acta. 29 (1958) 653.

(19) H.G. Pontis. J. Biol. Chem. 216 (1955) 195.

(20) J.T. Park and M.J. Johnson. J. Biol. Chem. 179 (1949) 585.

(21) L.F. Leloir. Arch. Biochem. Biophys. 33 (1951) 186.

(22) G.J. Dutton and D.E. Storey. Biochem. J. 53 (1953) XXXVII.

(23) J.G. Buchanan, J.A. Bassham, A.A. Benson, D.F. Bradley, M. Calvin, L.L. Daus, M. Goodman, P.M. Hayes, V.H. Lynch, L.T. Norris and A.T. Wilson. A Symposium on Phosphorus Metabolism, Vol. II, ed. Williams D. McElroy and Bentley Glass (The Johns Hopkins Press, Baltimore, 1952).

(24) L.F. Leloir and E. Cabib. J. Amer. Chem. Soc. 75 (1953) 5445.

(25) C.E. Cardini, L.F. Leloir and J. Chiriboga. J. Biol. Chem. 214 (1955) 149.

(26) L.F. Leloir and C.E. Cardini. J. Biol. Chem. 214 (1955) 157.

(27) L.F. Leloir and C.E. Cardini. J. Amer. Chem. Soc. 79 (1957) 6340.

(28) E.W. Sutherland. Ann. N.Y. Acad. Sci. 54 (1951b) 693.

(29) H. Niemeyer. Metabolismo de los hidratos de carbono en el hígado (Universidad de Chile, Santiago, 1955) p. 148.

(30) L.F. Leloir, J.M. Olavarría, S. Goldemberg and H. Carminatti. Arch. Biochem. Biophys. 81 (1959) 508.

(31) E. Belocopitow. Arch. Biochem. Biophys. 93 (1961)

458.

(32) A.J. Parodi, J. Mordoh, C.R. Krisman and L.F. Leloir. Arch. Biochem. Biophys. 132 (1969) 111.

(33) C.R. Krisman. Ann. N.Y. Acad. Sci. 210 (1973) 81.

(34) L.F. Leloir and S.H. Goldemberg. J. Biol. Chem. 235 (1960). 919.

(35) M.A.R. de Fekete, L.F. Leloir and C.E. Cardini. Nature, London. 187 (1960) 918.

(36) E.F. Recondo and L.F. Leloir. Biochem. Biophys. Res. Commun. 6 (1961) 85.

(37) J. Espada. J. Biol. Chem. 237 (1962) 3577.

(38) H. Nikaido and W.Z. Hassid. Advan. Carbohydrate Chem. Biochem. 26 (1971) 351.

(39) N.H. Behrens and L.F. Leloir. Proc. Natl. Acad. Sci. USA. 66 (1970) 153.

(40) N.H. Behrens, A.J. Parodi, L.F. Leloir and C.R. Krisman. Arch. Biochem. Biophys. 143 (1971) 375.

(41) N.H. Behrens, A.J. Parodi and L.F. Leloir. Proc. Natl. Acad. Sci. USA. 68 (1971) 2857.

(42) A.J. Parodi, N.H. Behrens, L.F. Leloir and H. Carminatti. Proc. Natl. Acad. Sci. USA. 69 (1972) 3268.

(43) N.H. Behrens, H. Carminatti, R.J. Staneloni, L.F. Leloir and A.I. Cantarella. Proc. Natl. Acad. Sci. USA (in press).

(44) L.F. Leloir, R.J. Staneloni, H. Carminatti and N.H. Behrens. Biochem. Biophys. Res. Commun. 52 (1973) 1285.

STOCHASTIC STUDIES ON CELL SURFACE STICKINESS

P.A. SRERE and M. MILAM
Pre-Clinical Science Unit, V. A. Hospital and Department of Biochemistry, University of Texas Health Science Center, Dallas, Texas

INTRODUCTION

The cell surface (cell periphery, cell membrane) is that part of a cell that isolates the cell from and at the same time serves as a communicator with the surrounding environment. It is biochemically complex and only recently have we gained any insight into its structure and function. The properties of the cell surface are believed to be important in controlling most aspects of a cell's growth, differentiation, and morphogenesis.

One of the properties of cell surfaces, the ability to stick to each other or to certain materials (known as cell adhesion), has been the subject of intense research. One reason for this interest in the adhesive properties of cells is that most normal cells cannot be grown in culture unless they are adherent to a substratum. When normal cells are transformed to become tumorigenic, they often also become modified so that they are less adhesive and can be grown in suspension culture. The study of cell adhesion has potential as a relatively simple approach in the study of a difference between normal and tumor cells.

Cell adhesion is believed to be involved in cell aggregation (1), cell mobility (2) and intercellular communication (3). It is also believed that intercellular adhesiveness is stronger in normal than in malignant cells (4) and that malignant cells may be less sensitive to "contact inhibition" than normal cells (5) because of differences in cellular adhesiveness (6).

The mechanisms of attachment have not been elucidated for any single system and each appears to be complex. The various adhesion processes are clearly not identical but there has been a tendency to consider them the result of a single adhesive mechanism. A number of models have been suggested to account for adhesion and aggregation. Detailed descriptions can be found in the reviews by Weiss (4) and Curtis (7). Of the many proposals to explain these phenomena, until recently two models were accorded most credence.

Curtis and Weiss and their collaborators believe that cell-cell and cell-solid support contacts occur at distances of the order of 100 Å or more and that adhesion can be explained by analogy to lyophobic colloid formation. Pethica (8), on the other hand, favors the long postulated idea of ionic bridging involving metal or metal-protein bridges either intercellularly or between cells and solid supports. Pethica (8) extended the bridging hypothesis to include the participation of micropseudopods in making initial cell-cell or cell-surface contacts. The bridging theory has been discussed recently by Steinberg (9), Moscona (10), Lilien (11), and Gingell *et al.* (12). The works of Gingell and Palmer (13), Jones *et al.* (14) and Kemp and Jones (15) demonstrate the presence and activity of contractile protein at the cell surface and these authors have suggested that contractile events are involved in adhesion. Microscopic studies have led to the conclusion that microvillus formation may be involved in adhesion. In addition microtubule formation within the cell has also been implicated in the adhesive process (16).

Cell To Surface Adhesion

Cells adhere to a large variety of solid substrates including plastics, different types of glass, teflon, stainless steel, and various absorbed protein layers (Weiss (17), Rosenberg (18), Nordling *et al.* (19)). Since adhesion must involve close molecular contact between some component(s) of the cell surface and the substratum one would have thought that a systematic variation of chemical composition of the substratum coupled with chemical modification of the cell surface would yield definitive information. However, there are few detailed studies correlating physico-chemical surface properties of solid supports with their efficacy for cell adhesion. Weiss and Blumenson (20) have reported differences in the adhesion of sarcoma cells to glass and teflon and have related their experimental observations to wetting ability of glass and teflon. They observed differences only in the absence of serum but none in the presence of serum.

Curtis (7) has reviewed the binding of cells to substrata of varying composition and Rappaport (21) has reported the variation of adhesion with varying composition of glass used as substrata. The limited number of surfaces tested in each of these experiments did not allow any generalizations to be drawn from these results.

We have pursued two separate lines of experimentation in our study of cell adhesion. One approach was to study the physical and physical-chemical aspects of the cell surface adhesive process. We first established a quantitative kinetic assay for cell to surface adhesion, and then tested the adhesion of cells to a series of substrata with widely differing properties; glass, polystyrene, stainless steel, aluminum, polycarbonate, polyethylene, TPX and polypropylene. The adhesion of HTC cells and three types of BHK cells to these surfaces was measured (22,23). Generally the ability to adhere to the various surfaces was in the same order regardless of cell type, i.e. cells adhered best to glass and treated polyethylene and least well to TPX and polypropylene. A rough correlation existed in how well cells adhered to a surface and the contact angle (wetability) for that surface. These results indicated that the forces involved in biological adhesion are analogous to those postulated in physical adhesion, i.e., short-range forces involving molecular contacts (22,23).

Our second approach was to characterize the adhesion process biochemically. The presence of free sulfhydryl groups on the surfaces of mammalian cells has been reported. Grassetti (24) demonstrated their presence on Ehrlich ascites tumor cells while Garvin (25) showed that treatment of whole blood with iodoacetamide resulted in decreased adhesiveness of lymphocytes. No other studies have been reported on specific chemical groups necessary for adhesion. We first showed that different types of sulfhydryl binding reagents with the exception of DTNB, inhibited the adhesion of rat hepatoma cells (HTC) and baby hamster kidney cells (BHK-13-21s and BHK-py) to various substrata. These reagents also inhibited the strengthening of cell attachment (26,27). The effects of SH binding reagents on cell adhesion is quite specific since various classes of reagents (mercurials, alkylating agents, and arsenicals) inhibit adhesion. Moreover, the effect does not reflect general cell death since the cells were still able to exclude trypan blue (a test for viability). Other metabolic poisons (F^-, N_3^- and cycloheximide) did not inhibit adhesion. We have also found that attached cells bind less N-ethylmaleimide than free cells and the inhibition of adhesiveness was reduced.

A divalent metal requirement for maintenance of intercellular adhesion originally was shown by the dissociation of frog embryos in Ca^{2+} free medium. The disruption of cell aggregates by EDTA is attributed to the metal requirement for formation and stabilization of cell aggregates although other factors may also be involved (Gingell *et al.* (12)). In general, there is cell specificity with respect to which metal is most effective in promoting adhesiveness (Steinberg (9), Armstrong (28), Armstrong and Jones (29)).

The exact role of divalent metals in adhesive events is unclear. The lyophobic colloid theory predicts a metal requirement for maintenance of proper cell surface charge density. However, Armstrong (28) has shown that Mg^{2+}, Ca^{2+}, Sr^{2+}, and Ba^{2+} have identical effects on surface charge but dissimilar effects on cell adhesiveness. Weiss (30), Gingell and Garrod (31), and Gingell *et al.* (12) have shown that the average effects exerted by metals on cell surface properties do not conform to the predictions of the lyophobic colloid theory; however, localized effects on the cell surface have not been studied. The participation of metal ions in a direct or indirect bridge is in agreement with the data available in the literature.

In addition to SH groups and divalent cations, other chemical groups have been indirectly shown to participate in adhesion. For instance, neuraminidase which cleaves sialic acid moieties has been shown to inhibit adhesiveness of some cells (Kemp (32)) and l-glutamine is required for aggregation of mouse teratoma cells (Oppenheimer (33)). Vicker and Edwards (34), on the other hand, have reported that neuraminidase treatment of BHK-21-C13 cells increase their adhesiveness. Some proteolytic enzymes, especially trypsin, are known to break intercellular and cell-surface adhesive bonds. This observation is consistent with the participation of protein molecule bridges in adhesion and aggregation. Weiss (35), Rosenberg (36), Daniel (37), and others have reported evidence suggesting the presence of extracellular material between cells and the surfaces to which they are adherent, but this material has not been isolated or directly shown to participate in adhesion.

A further difficulty in the understanding of cell to surface adhesion derives from the fact that both the cell surface and the support (or other cell) to which it eventually adheres carry a negative charge (4). From a theoretical standpoint therefore, the close approximation of the two surfaces is not at all favored. Pethica (8) calculated that it is more difficult for a negatively charged sphere with the radius of most tissue culture cells to approach a negatively charged surface than it would be for a microvillus of the cell which has a much smaller radius of curvature to approach the surface. In such a case the rate of attachment and flattening of cells would depend on the rate that cell surface was brought in contact with the substrate.

We thought if one could artificially flatten the cell against the substrate, then one could test the hypothesis that the rate limiting step of adhesion is in the mechanism of bringing the two surfaces together. Centrifugation of cells in a specifically constructed carrier rapidly produced a flattening effect and brought about a ten fold increase in the first order rate constants for adhesion for both BHK-12-21s and BHK-py cells (38). This does not give any insight into the reason for slowness of approximation of the two surfaces but indicates that adhesion sites are probably already present on the cell surface.

This model is an elaboration of the one presented by Taylor (39) of cell flattening and consistent with the postulate of Pethica (8) concerning cell microvilli formation. Weiss and Harlos (40) currently have considered two mechanisms to explain the delay in adhesion of cells 1) the time for protrusion of probes and 2) the extrusion of an extracellular glue, but cannot favor one mechanism over the other, although this latter mechanism is not favored by our experiments with centrifuged cells.

Several other observations have been reported concerning cell adhesion. Cyclic AMP has been reported to increase the adhesion of fibroblasts to a substratum (41). We examined this by using both BHK-py and BHK-21-13s cells and found that cyclic AMP decreases the detachability of BHK-py cells but does not affect the rate of adhesion of either cell line (42).

Recently Ballard and Tomkins (43) showed that hepatoma cells require serum for adhesion, but that in the absence of serum, cells could be induced by glucocorticoids to synthesize a factor which permits adhesion to occur. In this case, a cellular product is directly implicated in the adhesive process. In contrast we find that BHK cells stick much better in the absence of serum than in its presence.

As mentioned above, adhered cells have been detached routinely by treatment with low concentrations of trypsin. Such treatment does not affect the adhesive properties of the cells. More extensive treatment of the cells with trypsin causes a progressive loss in the adhesive ability of the cells (27). Incubation of the trypsin-treated cells in the absence of trypsin results in a recovery of adhesive ability. The recovery of cell to surface adhesion can be inhibited by cycloheximide but not by chloramphenicol and only partially inhibited by actinomycin D.

Our data on adhesion of cells to high energy surfaces coupled with the ability of cells to adhere to surfaces rapidly in a centrifugal field indicate 1) that close molecular contacts must be made and that the adhesive sites are already present on the cell surface. In addition, the involvement of SH groups as well as the loss of adhesiveness by trypsin treatment and the cycloheximide inhibitable recovery of adhesiveness lends support to the involvement of proteins at surface adhesive sites.

Cell To Cell Adhesion

Our first studies were concerned primarily with cell to surface adhesion which is just one aspect of cellular stickiness. In our attempts to study the effects of various hormones on cell adhesion we started studies on two other types of cell adhesion; cell to cell aggregation and lectin induced cell aggregation.

Roseman (44) has recently emphasized the idea that there were at least two types of cell adhesion, nonspecific and specific. It is interesting to note that this represents one of the first clear proposals that "cell adhesion" may be an umbrella term covering several quite different cellular capabilities and molecular events. Specific adhesion is the ability of like cells to aggregate and is believed to involve aggregation factors and cell receptors (Pessac and Defendi (45)).

There are two distinct mechanisms postulated for the process of cell aggregation. Trinkhaus and Lenz (46) studied the process of sorting out in a mixed cell population and concluded that no chemical mediator was responsible for the observed cellular aggregation and segregation. In a more recent study Merrell and Glaser (47) showed that specific recognition sites existed on plasma membranes isolated from embryonic chick neural, retinal and cerebellum cells. No evidence for the existence of an aggregation factor was found. That this process was not a general one for all cells was indicated by the fact that no similar recognition sites could be found on liver cell plasma membrane.

Another body of evidence has existed for a number of years showing the existence of aggregation factors (formation of large aggregates) necessary for the re-aggregation of certain cell types (Hausman and Moscona (48)). The isolation and study of these factors has continued. Pessac and Defendi (45) have shown the need for both the aggregation factors and receptor sites for the complete aggregation process. Burger and his colleagues (49) have shown that a carbohydrate group on the aggregation factor was probably involved in the aggregation process.

The involvement of cell surface carbohydrates (probably as glycoproteins) in the aggregation process has been indicated by many different experimental approaches. Thus neuraminidase treatment of BHK cells has increased their adhesiveness to each other (Vicker and Edwards (34)). Roseman (50) has proposed that aggregation occurs via sialic acid residues on one cell binding surface to sialyl transferases on another. Roseman and his coworkers (51) have also shown that BHK cells will aggregate onto Sephadex beads which contain specific carbohydrate residues. Studies on induced aggregation with concanavalin A, a carbohydrate binding protein, also indicate the involvement of carbohydrate residues in cell aggregation (52,53).

We have attempted to look for similarities between cell to surface, cell to cell and lectin induced cell to cell adhesion and the results of these experiments are presented here.

Effect of Hormones

A number of workers have indicated that hormone treatment of various tissue culture cells affected the adhesive ability of the cells (43,23). The action of many hormones is believed to be mediated by changes in cellular cyclic AMP concentration, so that many experiments have attempted to relate the cyclic AMP levels of cells to their ability to adhere either to surface or to each other. As noted earlier, Pastan and his coworkers (41) had shown that cyclic AMP affected the adhesiveness of BHK cells while we (42) later showed that this was an effect on the strength of adhesion rather than the rate of adhesion. Although not considered to act through cAMP, insulin has been reported to stimulate growth of cells and counteract some effects of cAMP.

It is believed that increased cAMP concentrations are typical of transformed cells (54) and lead to the differences in properties of transformed cells (or dividing normal cells) and normal cells.

After studying the effect of dibutyryl cAMP on cell to surface adhesions, we looked for another means of altering intracellular cAMP concentration. We tried to assess the effect of glucagon and theophylline on cell adhesion (Table 1 and Fig. 1).

TABLE 1

Effect of glucagon, insulin and theophylline on agglutination, adhesion and strength of adhesion of BHK-21-13b cells

Addition	Agglutination	% Adhered	% Detached
Control (3)	0	79.3 ± 3	32.0 ± 4
Glucagon (3) (1 μg/ml)	++	75.7 ± 3	39.8 ± 3
Insulin (3) (1 u/ml)	++	77.7 ± 3	34.8 ± 5
Theophylline (3) (1 mM)	+	67.7 ± 3	38.5 ± 5

Cells were grown with the indicated additives in suspension culture at 37°C for 24 hrs. The cells were collected by centrifugation and resuspended in 0.1 ml of the same incubation medium. For agglutination measurements 25λ were placed in 1 cm circles on glass cover slips, and rocked on the aliquot mixer for 5 min at room temp. The cover slips were inverted over the well of a slide and viewed microscopically. Preparation with only a few small (3-5 cells) clumps visible were scored 0. Large aggregates with no or very few single cells were scored ++++. Intermediate aggregations were estimated as +, ++, or +++. Adhesion measurements were carried out in Falcon polystyrene flasks. 1 x 10^6 cells were suspended in 5 ml of adhesion salts + 5% fetal calf serum and placed in each flask. The flasks were incubated at 35°C for 30 min. The flasks were placed on a New Brunswick reciprocal shaker for 1 min at 100 rpm. The free cells were removed with a pipette and 5 ml of detachment medium + 0.1 ml of 25% human serum albumin were added to each flask. After 3 min at 37°C, the flasks were placed on the shaker for 1 min at 100 rpm. The detached cells were removed with a pipette, and 5 ml of detachment medium + 0.1 ml of 25% human serum albumin were added. The flasks were incubated at 37°C for

10 min. All the cells were detached. After collection by centrifugation each pellet was suspended in 2 ml adhesion medium (5% FCS) and turbidity measured at 640 nm.

Fig. 1a shows control cells i.e. cells with no additions which show a uniform field of cells having only few small clumps of cells.

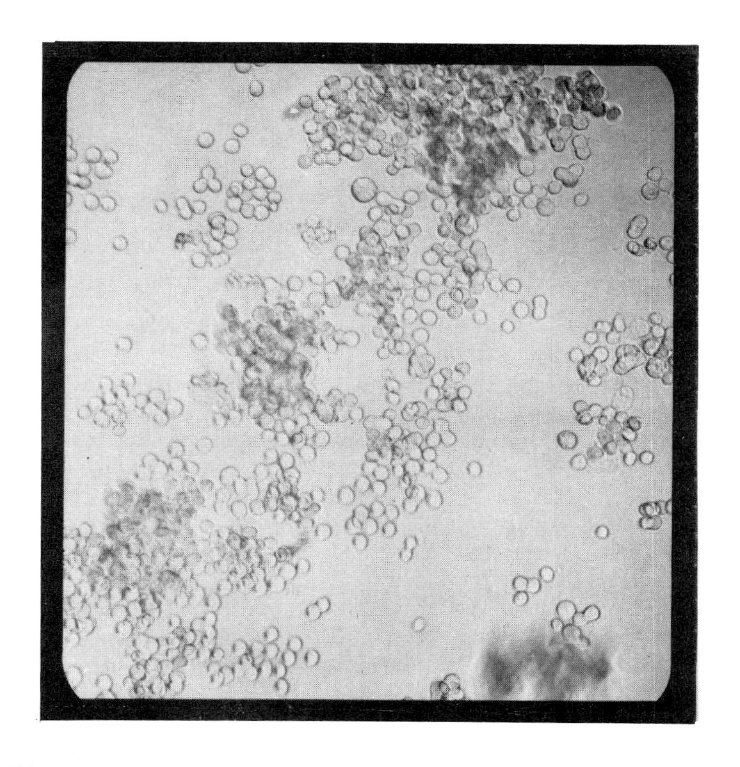

Fig. 1b shows cells grown in the presence of 1 mM theophylline for 24 hrs. In this case there are large aggregates of cells with only a few small aggregates. The agglutinability of cells grown in the presence of theophylline, glucagon or insulin was shown to be time dependent and also dependent on the concentration of the material added. The lowest concentrations at which agglutination could be directed was about 10^{-7} M for both insulin and glucagon. Preliminary data indicate that cAMP levels are increased in the cells grown in the presence of glucagon and theophylline. In every case it has also been observed that cell growth is inhibited (10-20%) in the presence of these materials.

There are several apparent differences and similarities between the data reported here and data from other laboratories. First, Puck (55) reported that dibutyryl cAMP induced association in CHO-K1 cells. He also reported that the CHO-K1 cells treated with dibutyryl cAMP adhered more firmly than untreated cells. In the present experiments it can be seen that our treatment had no effect on the stickiness or the detachability of these cells to a surface. Further a number of workers (56,57,58) have seen that insulin stimulates cell growth and in many systems cAMP levels decrease after insulin treatment. We have tried combinations of insulin and glucagon and always observed additive effects on agglutination.

These hormone effects clearly showed separate mechanisms for cell to surface adhesion and for cell to cell adhesion. Although our technique of measuring agglutination did not have the quantitative precision of our cell to surface adhesion measurements the technique did allow us to test for similarities between these two kinds of adhesion.

We had shown previously that cell to surface adhesion could be inhibited by a variety of SH reagents. When similar concentrations were tried on cells agglutinated by glucagon treatment, no inhibition of agglutination was seen. In fact $HgCl_2$ always stimulated agglutination slightly (Table 2). EDTA at concentrations used to detach cells from surfaces had no apparent effect of agglutination (Table 2 and Table 3).

TABLE 2

Effect of EDTA on $HgCl_2$ on glucagon-induced agglutination of BHK-21-13b cells

Incubation	Treatment	% Agglutinated
Control	None	0
	EDTA	0
	$HgCl_2$	+
Glucagon	None	++
	EDTA	++
	$HgCl_2$	+++

Cells were incubated in spinner flasks for 24 hrs ± glucagon (1 μg/ml). Aliquots of 5 x 10^5 cells were removed and collected by centrifugation. Controls were incubated in 5 ml adhesion salts + 0.025% bovine serum albumin, $HgCl_2$-treated cells in 5 ml adhesion salts + 0.025% BSA and 0.1 mM $HgCl_2$, and EDTA-treated cells in adhesion salts minus Mg^{++} + 0.25% BSA and 5 mM EDTA for 10 min at room temp. Agglutination was measured as described in Table 1.

TABLE 3

Effect of EDTA on adhesion to Falcon polystyrene and agglutination by ConA of BHK-21-13b cells

Treatment	% Adhered	Agglutination
None	76	++
5 mM EDTA	20	++

Adhesion and agglutination tests were performed as described previously in the presence of adhesion salts minus Mg^{++} containing 5 mM EDTA and 5% fetal calf serum (adhesion) or 0.025% bovine serum albumin (agglutination).

NANase and trypsin have been shown to increase agglutination of cells. We have tested the effect of trypsin and NANase on BHK-21-13b cells and on the same cells made agglutinable by glucagon treatment. Trypsin induces agglutination in control cells and enhances it in glucagon grown cells but no effect of NANase was observed (Table 4).

TABLE 4

Effect of trypsin and neuraminidase on agglutination of BHK-21-13b cells by glucagon

Incubation	Addition	Agglutination
Control	None	0
Control	Trypsin	++
Control	Neuraminidase	0
Glucagon	None	++
Glucagon	Trypsin	++++
Glucagon	Neuraminidase	++

Cells were incubated for 24 hrs in spinner flasks ± glucagon (1 μg/ml). Aliquots of each flask were removed, and the cells were washed once with adhesion medium. Cells were treated with 0.4% trypsin at 30°C for 10 min, with 0.39 units of neuraminidase for 30 min at 37°C or with adhesion medium for 30 min at 37°C (control). Agglutination tests were performed as described previously.

Since trypsin and EDTA are often used to detach cells from surfaces or to disaggregate them we have tried both these agents on glucagon induced agglutination and untreated cells (Table 5) as before no effect of EDTA was observed and a slight stimulation by trypsin was obtained. As stated earlier, NANase had no effect on adhesion to cells to surfaces (Table 6).

TABLE 5

Effect of trypsin and EDTA on BHK-21-13b cells agglutinated by glucagon

Incubation	Treatment	Agglutination
Control	None	0
	EDTA	0
	Trypsin	+
Glucagon	None	++
	EDTA	++
	Trypsin	+++

Cells were grown for 24 hrs ± glucagon (1 μg/ml). 5×10^6 cells from each flask were removed, collected by centrifugation, and suspended in 0.1 ml of adhesion salts less Mg^{++} + 0.025% BSA. 20λ was placed on a 1 cm circle on a cover slip and the cover slips were placed on the aliquot mixer for 5 min. Then 20λ of 5 mM EDTA, 20λ of 0.25% trypsin or 20λ adhesion salts less Mg^{++} + 0.025% BSA were added as indicated. The mixing continued for another 2 min and then the cells were examined microscopically.

TABLE 6

Effect of neuraminidase on adhesion of BHK-21-13s cells

Treatment	Nanomoles NANA cleaved/5 x 10^7 cells	Shaker RPM	% Attached	% Detached
None	0	100	75	35
		300	72	76
Neuraminidase	175	100	69	38
		300	67	77

5×10^7 cells were treated with 0.078 units of *Clostridium* neuraminidase in 2.5 ml of adhesion medium containing 0.25% BSA for 30 min at 37°C. NANA released was measured by the method of Warren. Cell adhesion was measured in Falcon polystyrene flasks containing 1×10^6 cells in 5 ml of adhesion medium containing 5% Fetal Calf Serum (FCS) for 30 min at 30°C. Strength of cell adhesion was measured by placing flasks on a reciprocal shaker for 1 min at the indicated speeds.

These experiments emphasize the differences between the adhesion of cells to surfaces and cells to each other.

Lectin Induced Agglutination

It has been reported by a number of laboratories (52, 53) that certain plant proteins (called lectins) have the ability to agglutinate certain cells grown in culture. Transformed cells have been shown to be more sensitive to this induced agglutination than are normal cells. However, it has been shown that trypsin treatment can cause even normal cells to become agglutinated by lectin. It has also been shown that normal cells during the M phase of their growth cycle are agglutinable when lectins are added.

Recently Grinnell (59) has shown that ConA increases the adhesiveness of BHK cells to a surface. We have repeated those experiments and they are shown in Table 7 which shows that ConA does affect the strength of attachment of cells to a surface. The specificity of this reaction is shown in that trypsinized ConA is not active in this test (Table 8).

TABLE 7

Effect of ConA on strength of adhesion of BHK-21-13b cells

2×10^6 cells in 5 ml of adhesion medium were placed in each of 8 Falcon polystyrene flasks. The flasks were incubated at 30°C for 30 min. Free cells were suspended on the reciprocal shaker for 1 min at 100 rpm. The cells were removed with a pipet and 2 ml of adhesion salts ± 80 μg ConA were added to each flask. After 5 min at room temperature, the supernatant was discarded, and 5 ml of detachment medium + 0.1 ml of human serum albumin were added to each flask. The flasks were incubated for 5 or 10 min at 30°C and then placed on the reciprocal shaker for 1 min at 250 rpm. The detached cells were removed with a pipet, and the remaining bound cells were removed with detachment medium and HSA in 10 min at 30°C.

Addition	Min of Detachment	% Bound	% Detached	Average
None	5	86	28	34
None	5	89	40	
ConA	5	87	4	3
ConA	5	87	2	
None	10	90	74	76
None	10	89	77	
ConA	10	86	5	4
ConA	10	86	3	

TABLE 8

Effect of monovalent ConA on strength of adhesion of BHK-21-13b cells

Treatment	% Bound	Shaker Speed	% Detached
Control	76	100	38
ConA (trypsin)	67	100	44
Control	76	300	52
ConA (trypsin)	69	300	54

ConA (1 mg/ml) was treated with trypsin (mg for mg) for 5 hrs at 37°C. The reaction was stopped by the addition of a neutralizing amount of soybean trypsin inhibitor.

Cells were treated as indicated with trypsinized ConA of a concentration of 250 μg/ml in adhesion salts + 0.025% bovine serum albumin (BSA) for 20 min at 37°C. Control cells were incubated in adhesion salts + 0.025% BSA for the same time period.

Cell adhesion was measured as previously described for 30 min at 30°C. After the free cells were decanted, 5 ml of detachment medium and 0.1 ml of human serum albumin were added to each flask. The flasks were incubated at 37°C for 3 min. Then the flasks were placed on the reciprocal shaker at the indicated speed for 1 min. The detached cells were decanted, and the remaining bound cells were detached as previously described.

When ConA is used to agglutinate cells neither EDTA, trypsin or NANase is effective in disrupting the aggregates (Table 9). A small effect of monovalent ConA can be seen when tried with glucagon induced agglutination (Table 10).

TABLE 9

Effect of EDTA, trypsin, and neuraminidase on BHK-21-13b cells agglutinated by ConA

Treatment	% Agglutination	Addition	% Agglutination
None	0	None	0
None	0	5 mM EDTA	0
None	0	0.25% Trypsin	+
None	0	0.78 u/ml Neuraminidase	0
ConA	++++	None	++++
ConA	++++	5 mM EDTA	++++
ConA	++++	0.25% Trypsin	++++
ConA	++++	0.78 u/ml Neuraminidase	++++

Agglutination was measured as described previously in the presence or absence of 0.1 mg/ml ConA. To each cell suspension on the cover slip was added an equal amount of the stated material or adhesion salts in the case of the controls. The cells were returned to the aliquot mixer for another 5 min and then the cells were examined microscopically.

TABLE 10

Effect of monovalent ConA on agglutination of BHK-21-13b cells by glucagon

Incubation	Treatment	% Agglutination
Control	None	0
	Monovalent ConA	0
	Trypsin + Trypsin Inhibitor	0
	Trypsin + Trypsin Inhibitor + ConA	++++
Glucagon	None	++
	Monovalent ConA	+
	Trypsin + Trypsin Inhibitor	++
	Trypsin + Trypsin Inhibitor + ConA	++++

Monovalent ConA was prepared by treating 1 mg/ml of whole ConA with trypsin mg for mg for 5 hrs at 37°C. The trypsin was neutralized with an equvalent amount of trypsin inhibitor. Trypsin + trypsin inhibitor was prepared by mixing 1 mg of trypsin + 1 mg trypsin inhibitor in 1 ml adhesion salts and incubating 5 hrs at 37°C. Trypsin + trypsin inhibitor + ConA was prepared by mixing 1 mg ConA with 1 mg trypsin with 1 mg trypsin inhibitor in 1 ml of adhesion salts and incubated for 5 hrs at 37°C.

10^6 cells in 0.75 ml of adhesion salts + 0.025% BSA were mixed with 0.25 ml of the designated mixture and incubated for 20 min at 37°C. Agglutination was measured as described previously.

Finally, similar to our results with glucagon induced aggregation we find that ConA induced aggregation is insensitive to SH reagents (Table 11).

TABLE 11

Effect of sulfhydryl reagents on adhesion to Falcon polystyrene and agglutination by ConA of BHK-21-13b cells

Treatment	% Adhered	Agglutination
None	76	++
$HgCl_2$ (0.1 mM)	11	++
NEM (0.1 mM)	21	++
pMB (0.4 mM)	51	++
$NaAsO_2$ (1.0 mM)	58	++

3 x 10^6 cells were placed in 5 ml of adhesion medium (5% fetal calf serum) in each tube. Additions were made as indicated. After 10 min at room temperature, the cells were collected by centrifugation and resuspended in 0.6 ml of adhesion salts + 0.025% bovine serum albumin. To each of two Falcon polystyrene flasks already containing 4.8 ml of adhesion medium (5% FCS) was placed 0.2 ml of the cell suspension. The flasks were incubated for 30 min at 30°C and cell adhesion was measured as described previously. Of the remaining 0.2 ml cell suspension, 0.1 ml was transferred to a tube containing 10λ of 1 mg/ml Concanavalin A. Agglutination was measured as described previously on 25λ of the original cell suspension and on the ConA containing suspensions.

DISCUSSION

The results presented here emphasize the individuality of at least two kinds of adhesive events (Table 12).

TABLE 12

Influence of various agents on three kinds of cell adhesion

Effector	Cell→surface	Cell→cell	Cell→lectin→cell
EDTA	↓	0	0
Trypsin	↓	↑	↑
NANase	0	0	0
SH reagents	↓	0	0
ConA	↑	-	↑
ConA (trypsin)	0	-	↓
Cyclic AMP*	↑±	↑	?

*This presumes an adequate solution to the anomalous insulin effects (see text).

The process of cell to surface adhesion although not completely characterized, involves the bringing two surfaces close to each other. Several kinds of interactions may be involved; divalent metal bridges and protein-surface contacts are likely whereas carbohydrate moieties do not seem to be crucial. Cell to surface adhesion is relatively insensitive to cyclic AMP concentrations but extremely sensitive to a serum factor and sulfhydryl reagents.

A second case of adhesion, cell to cell adhesion resulting in large aggregates was also studied and no loss of adhesiveness was observed by treatment with SH reagents EDTA, or trypsin. This type of adhesion may be related to the cyclic AMP content of the cell although insulin gave anomalous results in this respect. Data from other laboratories would seem to indicate that certain carbohydrate moieties were involved in this type of adhesion.

Carbohydrate residues certainly are involved in lectin induced aggregation. This process shares many properties with the cell to cell aggregation phenomena in that it is insensitive to EDTA, trypsin and SH reagents.

Roseman stated that it was doubtful that the various methods of measuring adhesion measure the same molecular event. Our data support this conclusion.

REFERENCES

(1) J.E. Lilien, in: *Current Topics in Developmental Biology*, Vol. 4, eds. A.A. Moscona and A. Monroy (Academic Press, New York, 1969) p. 169.

(2) M.H. Gail and Ch.W. Boone, *Exp. Cell Res.* 70 (1972) 33.

(3) N.B. Gilula, O.R. Reeves and A. Steinbach, *Nature* 235 (1972) 262.

(4) L. Weiss, *The Cell Periphery, Metastasis and Other Contact Phenomena* (Wiley & Sons, New York, 1967) p. 294.

(5) M. Abercrombie and E.J. Ambrose, *Cancer Res.* 22 (1962) 525.

(6) J.P. Trinkaus, T. Betchaku, and L.S. Krulikowski, *Exp. Cell Res.* 64 (1971) 291.

(7) A.S.C. Curtis, *The Cell Surface; Its Molecular Role in Morphogenesis* (Academic Press, New York, 1967).

(8) B.A. Pethica, *Exp. Cell Res. Suppl.* 8 (1961) 123.

(9) M.S. Steinberg, in: *Biological Interactions in Normal and Neoplastic Growth*, eds. M.J. Brennan and W.I. Simpson (Little, Brown and Co., Boston, 1962).

(10) A.A. Moscona, *Developmental Biology* 18 (1968) 250.

(11) J.E. Lilien, *Current Topics in Developmental Biology* 4 (1969) 169.

(12) D. Gingell, D.R. Garrod and J.F. Palmer, in: *Calcium and Cellular Function*, ed. A.W. Cuthbert (Wm. Clowes and Sons, Ltd., London, 1970) p. 59.

(13) D. Gingell and J.F. Palmer, *Nature* 217 (1968) 98.

(14) B.M. Jones, R.B. Kemp and U. Grüschel-Stewart, *Nature* 226 (1970) 261.

(15) R.B. Kemp and B.M. Jones *Exp. Cell Res.* 63 (1970) 293.

(16) J.A. Witkowski and W.D. Brighton,*Exp. Cell Res.* 70 (1972) 41.

(17) L. Weiss, *Exp. Cell Res.* 17 (1969) 508.

(18) M.D. Rosenberg, *Proc. Nat. Acad. Sci. USA* 48 (1962) 1342.

(19) S. Nordling, K. Penttinen and E. Saxen, *Exp. Cell Res.* 37 (1965) 161.

(20) L. Weiss and L.E. Blumenson, *J. Cell Physiol.* 70 (1967) 23.

(21) C. Rappaport, in: *The Chemistry of Biosurfaces,* Vol. 2, ed. M.L. Hair (Marcel Dekker, New York, 1972) p. 449.

(22) F. Grinnell, M. Milam and P.A. Srere, *Arch. Biochem. Biophys.* 153 (1972) 193.

(23) F. Grinnell, M. Milam and P.A. Srere,*Biochem. Med.* 7 (1973) 87.

(24) D.R. Grassetti, *Nature* 228 (1970) 282.

(25) E.J. Garvin, *J. Exp. Med.* 114 (1961) 51.

(26) F. Grinnell and P.A. Srere, *J. Cell Physiol.* 78 (1971) 153.

(27) F. Grinnell, M. Milam and P.A. Srere, *J. Cell Biol.* 56 (1973) 659.

(28) P.B. Armstrong, *J. Exp. Zool.* 163 (1966) 99.

(29) P.B. Armstrong and D.P. Jones, *J. Exp. Zool.* 167 (1968) 275.

(30) L. Weiss, *Exp. Cell Res.* 51 (1968) 609.

(31) D. Gingell and D.R. Garrod, *Nature* 221 (1969) 192.

(32) R.B. Kemp, *Nature* 218 (1968) 1255.

(33) S.B. Oppenheimer, M. Edidin, C.W. Orr and S. Roseman, *Proc. Nat. Acad. Sci. USA* 63 (1969) 1969.

(34) M.G. Vicker and J.G. Edwards, *J. Cell Sci.* 10 (1972) 759.

(35) L. Weiss, *Exp. Cell Res. Suppl.* 8 (1961) 141.

(36) M.D. Rosenberg, *Biophys. J.* 1 (1969) 137.

(37) M.R. Daniel, *Exp. Cell Res.* 46 (1967) 191.

(38) M. Milam, F. Grinnell and P.A. Srere, *Nature New Biol.* 244 (1973) 83.

(39) A.C. Taylor, *Exp. Cell Res. Suppl.* 8 (1961) 154.

(40) L. Weiss and J.P. Harlos, *J. Theor. Biol.* 37 (1972) 169.

(41) G.S. Johnson and I. Pastan, *Nature New Biol.* 236 (1972) 247.

(42) F. Grinnell, M. Milam and P.A. Srere, *Nature New Biol.* 241 (1973) 82.

(43) P.L. Ballard and G.M. Tomkins, *J. Cell Biol.* 47

(1970) 222.

(44) B.T. Walther, R. Öhman and S. Roseman, *Proc. Nat. Acad. Sci. USA* 70 (1973) 1569.

(45) B. Pessac and V. Defendi, *Nature New Biol.* 238 (1972) 13.

(46) J.P. Trinkaus and J.P. Lentz, *Developmental Biol.* 9 (1964) 115.

(47) R. Merrell and L. Glaser, *Proc. Nat. Acad. Sci. USA* 70 (1973) 2794.

(48) R.E. Hausman and A.A. Moscona, *Proc. Nat. Acad. Sci. USA* 70 (1973) 3111.

(49) G. Weinbaum and M.M. Burger, *Nature* 244 (1973) 509.

(50) S. Roseman, *Chem. Phys. Lipids* 5 (1970) 270.

(51) S. Chipowsky, Y.C. Lee and S. Roseman, *Proc. Nat. Acad. Sci. USA* 70 (1973) 2309.

(52) M. Inbar and L. Sachs, *Proc. Nat. Acad. Sci. USA* 63 (1969) 1419.

(53) M.M. Burger and K.D. Noonan, *Nature* 228 (1970) 512.

(54) J.R. Sheppard, *Nature New Biol.* 236 (1972) 14.

(55) T. Puck, in: *The Role of Cyclic Nucleotides in Carcinogenesis*, Vol. 6, eds. J. Schultz and H.G. Gratzner (Academic Press, New York, 1973) p. 283.

(56) L. Jimenez de Asua, E.S. Surian, M.M. Flawia and H.N. Torres, *Proc. Nat. Acad. Sci. USA* 70 (1973) 1388.

(57) A. Vaheri, E. Ruoslahti, T. Hovi and S. Nordling, *J. Cell. Physiol.* 81 (1973) 355.

(58) S.M. Naseem and V.P. Hollander, *Cancer Res.* 33 (1973) 2909.

(59) F. Grinnell, *J. Cell Biol.* 58 (1973) 602.

DISCUSSION

M. RABINOWITZ: Several years ago Belkin and Hardy demonstrated that different sulfhydryl reagents caused different types of blebing in suspended cells. Did you examine your cells for blebing or any damage to the cell surface?

P.A. SRERE: We always looked at our cells to see if they will exclude trypan blue, and none of the cells took up trypan blue. We never report results from damaged cells. Under certain conditions, we do see blebing, but none of the results were with cells of that type.

M. RABINOWITZ: I was surprised at the high concentrations of mercuric chloride which you claim did not affect cell growth over a period of 4 days - do you have any explanation for this?

P.A. SRERE: No, I don't, except that much of the mercuric chloride at these concentrations will be bound by sulfhydro groups in the medium and on the cell surface.

M. RABINOWITZ: Under growth conditions, then, you add materials which lower the effective concentration of added sulfhydryl reagents, whereas in your adhesion studies, you did not add these materials?

P.A. SRERE: No, any time we replaced the growth medium we would put the same amount of mercuric chloride in. At the concentrations of the mercuric chloride that inhibited cell-to-surface adhesion growth was not inhibited.

J. HOCHSTADT: I was most interested in your centrifugation experiments. After the centrifugation, did you follow cell viability and later growth? Would you recommend this as a way to increase plating efficiency; the technique might be very useful for selection and isolation of specific mutants.

P.A. SRERE: The cells were viable afterwards and would go on and divide. We always got complete cell recovery, that is one thing we always looked for.

J. WOLFF: You mentioned that the repulsive effect of surface charges during adhesion may be diminished by the formation of microvilli. Is it possible to change the ionic strength at constant osmolarity to see if adhesion reactions are changed? Secondly, do you know if there has been any confirmation of a report that cells fed with unsaturated fatty acid exhibit large changes in adhesiveness?

P.A. SRERE: I do not know the answer to your second question. In answer to your first question, we have looked at changes in ionic strength and up to a certain level there are small effects which usually level off above a certain ionic strength. However, we have to repeat all the kinetics in the absence of serum.

RECEPTORS FOR INTERCELLULAR SIGNALS IN AGGREGATING CELLS OF THE SLIME MOLD, *DICTYOSTELIUM DISCOIDEUM*

G. GERISCH, H. BEUG, D. MALCHOW, H. SCHWARZ and A.v.STEIN
Friedrich-Miescher-Laboratorium
der Max-Planck-Gesellschaft,74 Tuebingen, Germany

Abstract: Cell aggregation in *Dictyostelium discoideum* is the result of cell-to-cell adhesion, chemotaxis, and wave-like transmission of periodic stimuli which trigger the release of a chemotactic factor. The two latter reactions are apparently mediated by cyclic AMP. Cell surface constituents presumably responsible for cyclic-AMP reception and cell-to-cell adhesion are regulated during development in accord with changes in cell behavior. These constituents are (1) Cyclic-AMP binding sites, (2) a cyclic-AMP phosphodiesterase, and (3) one class of target sites of univalent antibody fragments (Fab) that block cell adhesion. The target sites of blocking Fab fragments, referred to as contact sites, are not identical with the bulk of cell surface antigens present at aggregating cells. Two independent systems of such sites can be distinguished by differences in immunospecificity, developmental regulation and pattern of cell assembly. Non-aggregating mutants were used in combination with immuno-absorption assays for the analysis of one system. Electronmicroscopic data indicate that contact sites are distributed in a fine pattern between other surface constituents, and suggest that they extend into the intercellular space. The expression of the developmentally regulated cell surface sites is controlled by a common genetic system.

MECHANISMS OF CELL AGGREGATION

The multicellular state of the cellular slime mold, Dictyostelium discoideum, arises by aggregation of previously independent amoeboid cells. Exhaustion of food bacteria, or removal of growth-medium in the case of an axenic strain (1), induces these cells to differentiate within a couple of hours into aggregation-competent cells (2, 3). These cells are characterized by their abilities to associate into stream-like assemblies by cell-to-cell adhesion, with a preference for adhesion at the tips of the elongated cells; to respond chemotactically to cyclic AMP with much greater sensitivity than in the growth-phase (4); and to propagate chemotactic pulses from cell to cell in a wave-like pattern (5, 6, 7), probably again by means of cyclic AMP as a transmitter (8). In this paper we discuss molecular changes of the plasma membrane as an expression of cell differentiation to aggregation-competence, and as a possible basis of the ability for cell-to-cell adhesion and cyclic-AMP reception.

CYCLIC-AMP BINDING SITES AND PHOSPHODIESTERASE

The chemotactic response to cyclic AMP suggests the existence at the cell surface of binding sites functioning as cyclic-AMP receptors. The detection of binding was complicated by a phosphodiesterase exposed in living cells to external cyclic AMP, hydrolysing up to 10^8 molecules per cell per minute (3). This phosphodiesterase activity is characteristic for aggregation-competent cells, growth-phase cells exhibiting only low activity (3, 9). Based on the findings that cyclic GMP is well hydrolysed (10) but chemotactically about 1000 fold less active than cyclic AMP (11), binding to living cells was measured in presence of excess cyclic GMP in order to compete with cyclic-AMP hydrolysis (12). Scatchard plots of binding data obtained under these conditions revealed a number of about 5×10^5 cyclic-AMP binding sites per aggregation-competent cell. Similar to cell-bound phosphodiesterase activity, the ability to bind cyclic AMP increases in the course of cell differentiation (fig.1), which is in accord with the increased chemotactic sensitivity of aggregation-competent cells (4).

The association of the binding sites with an effector-hydrolysing enzyme relates the cyclic-AMP receptor system to the synaptic acetylcholine/cholinesterase system. Like the cholinesterase, the phosphodiesterase functions in terminating the signal. Termination is also in the slime mold system a prerequisite for the transmission of periodic pulses.

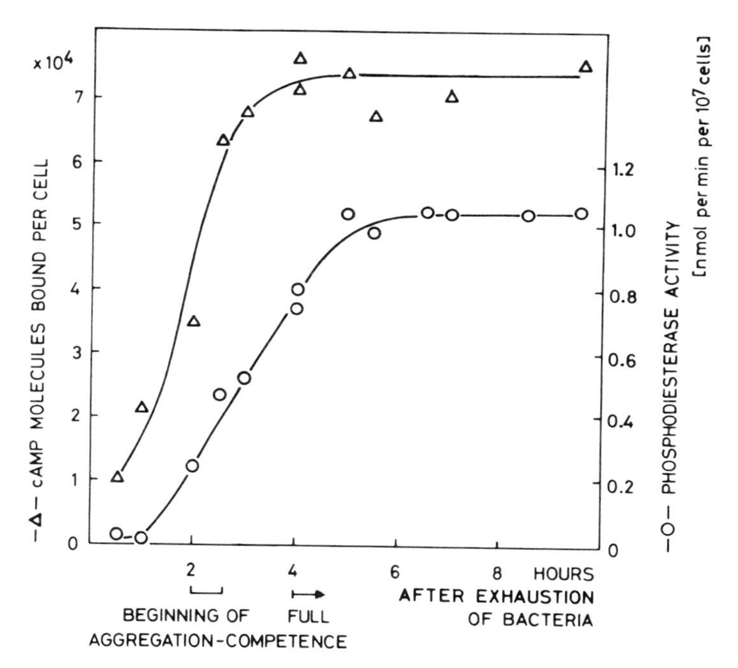

Fig. 1. Developmental regulation of cell-bound cyclic-AMP phosphodiesterase (O) and of cyclic-AMP binding activity (Δ) in D.discoideum M2. Phosphodiesterase activity was measured using living cells as described previously (3). Binding was assayed by simultaneously adding under intense shaking 5×10^{-4}M cyclic GMP and 7×10^{-8}M ^{3}H-cyclic AMP to cells suspended in 15 mM imidazol-HCl buffer, pH 6.0. The cells were sedimented at 5 seconds after nucleotide addition on a Microcentrifuge, and removal of label from the supernatant was determined (12). For aggregation-competence, as indicated below the abscissa, the ability of cells to form terminal contacts was taken as a criterion.

TWO INDEPENDENT CONTACT-SITE SYSTEMS AT THE SURFACE OF AGGREGATING CELLS

Univalent antibody fragments (Fab) were used as blocking agents in order to demonstrate that cell-to-cell adhesion depends on the activity of specific membrane sites not identical with the bulk of cell surface structures present in aggregating cells. Use of Fab fragments is necessary because IgG causes agglutination of D.discoideum cells and a redistribution of surface antigens (13, 14) similar to cap formation in lymphocytes (15, 16, 17). In what follows, the operational term "contact sites" will be applied to the target sites of Fab species that block cell-to-cell adhesion.

Aggregating cells form preferentially terminal, but also side-by-side contacts, in this way assembling into the stream-like cell associations typical for this developmental stage. Fab directed against growth-phase cells specifically inhibited side-by-side adhesion. In contrast, "aggregation-specific" Fab, i.e. Fab against particle fractions from aggregation-competent cells, absorbed with particles from growth-phase cells, eliminated the preference for terminal adhesion (fig. 2). Thus, two systems for intercellular adhesion functioning independently of each other are active in aggregating cells. Blockage of "contact sites A" eliminates the terminal adhesion, blockage of "contact sites B" the side-by-side adhesion. Both systems of contact sites can be distinguished not only by their immunospecificity, but also by their different EDTA-sensitivity and developmental regulation (tab.1). By use of a quantitative Fab-absorption assay (21) it has been shown that the quantity of contact sites B changes only insignificantly during differentiation to aggregation-competence, whereas contact sites A become first detectable just before the cells begin to form terminal cell adhesions (22).

Binding to the cell surface without any effect on the contact-site-A system was obtained with Fab against heated cells, referred to as "anti-carbohydrate" Fab: its main target antigen is "antigen I", a glycosphingolipid complex containing fucose, N-acetylglucosamine and mannose (18,19). At least part of these sugars are constituents of the immuno-determinants (20).

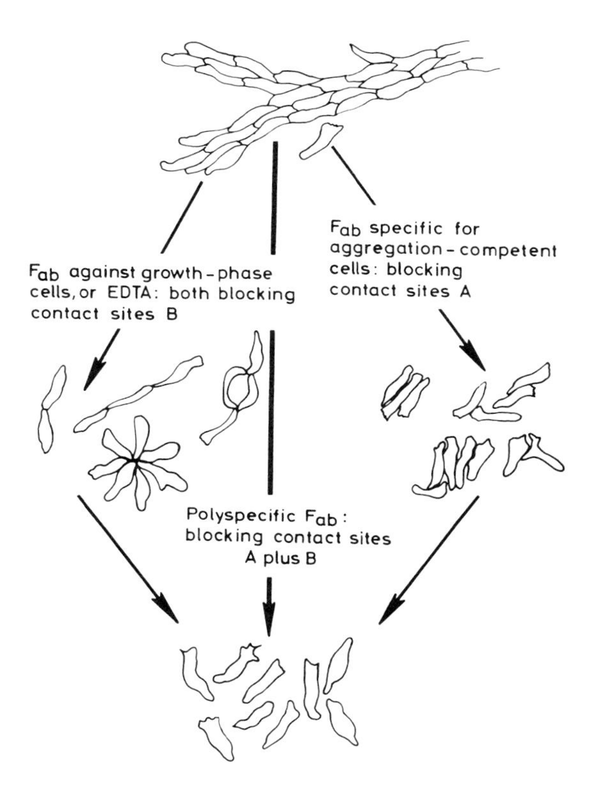

Fig. 2. Diagram showing the effects of Fab fragments of different specificities, and of EDTA, on cell assemblage.

QUANTITATION OF Fab BINDING TO CONTACT SITES A AND OTHER MEMBRANE LOCI

The idea of contact sites implies that inhibition of cell adhesion is not a matter of the total number of Fab molecules bound per cell, but depends on their specificity. This has been tested by use of tritiated Fab with the result that anti-carbohydrate Fab, which did not inhibit contact sites A, was bound up to a level of $2x10^6$ molecules per cell. When, on the other hand, aggregation-specific Fab was used, $3x10^5$ molecules were sufficient to block contact sites A completely. Under these conditions, not more than

2 per cent of the cell surface area was covered by Fab (14).

The finding that anti-carbohydrate Fab of the specificity used did not block contact sites A, does not necessarily mean that carbohydrates are generally not involved in this type of cell interaction. It should be mentioned here that one of our non-aggregating mutants, aggr 20-2, showed increased agglutinability with wheat germ agglutinin, defective fucose incorporation into the glycosphingolipids (19) and concomitant changes in antigen I specificity (23). Since the mutant has multiple membrane defects (24) (fig.4), the influence of the defect in glycolipid synthesis on cell behavior is unclear.

TABLE 1

Properties of the contact site systems A and B

	A	B
Pattern of cell assembly	Cell chains and rosettes formed by terminal contacts	Loose, irregular cell groups, often side-by-side adhesion
Activity in		
growth-phase cells	-	+
aggregation-competent cells	+	+
Fab used for inhibition	"Aggregation-specific"	Anti-growth phase
Resistance to EDTA	+	-

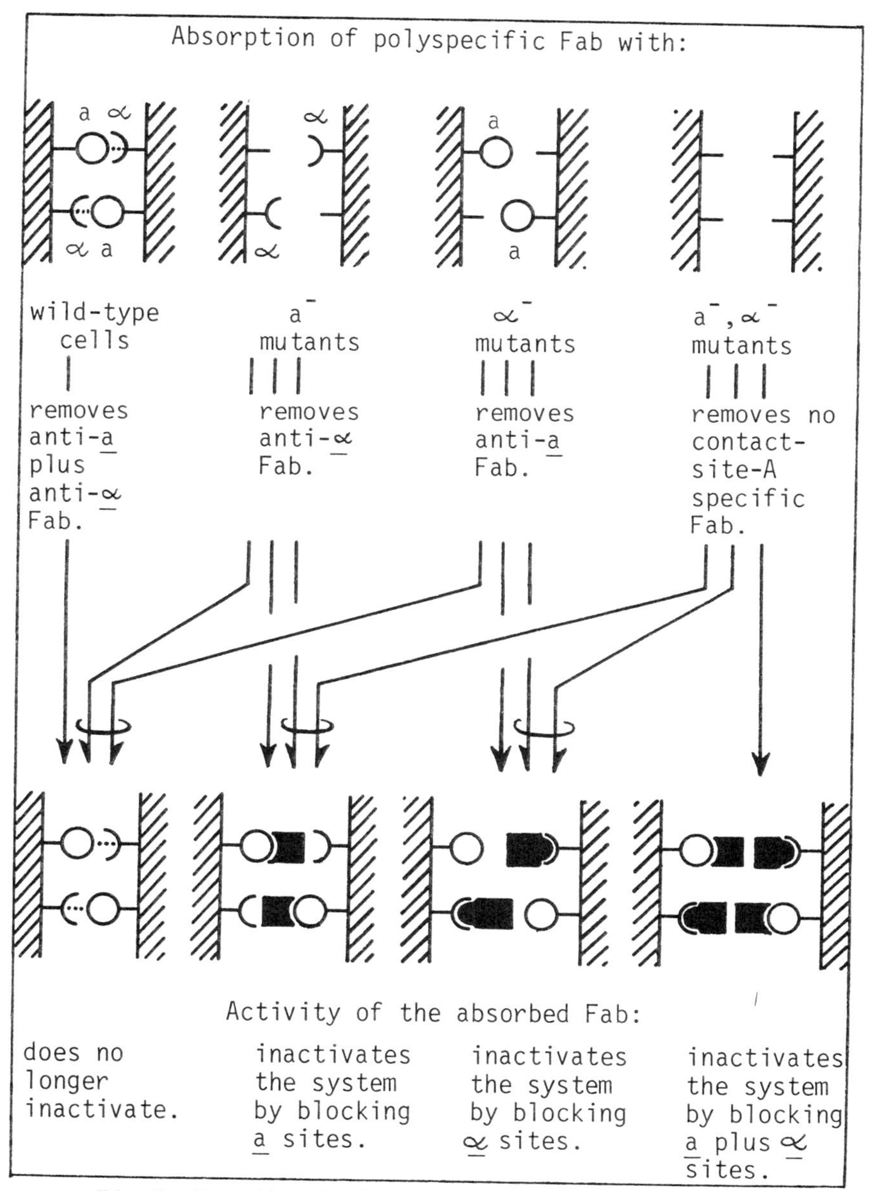

Fig.3. Hypothetical scheme of the contact-site-A system and its analysis by Fab-absorption techniques.

HETEROGENEOUS ANTIBODY TARGET SITES IN THE CONTACT-SITE-A SYSTEM

The classical Tyler-Weiss hypothesis (25, 26) on the molecular basis of cell-to-cell adhesion, recently modified by Roseman (27), proposes two classes of complementary, interacting sites at the surfaces of adjacent cells. The hypothesis predicts that the contact-site-A system can be completely inhibited by blocking either one sub-class of these sites (a in Fig.3) or the other (α), or both. Similarly, non-aggregating mutants could be defective in either a or α, or both. If a Fab preparation contains anti-a and anti-α activity, none of these mutants should completely remove its contact-site-A blocking activity, provided the defective sites have lost their ability to bind antibody. Fab absorbed with such mutants should still be able to fully inactivate the contact-site-A system, because blockage of either a or α would be sufficient.

If two mutants share at least one defect, double absorption with both mutants, again, should not completely remove the blocking activity of the Fab (Fig.3). This conclusion is independent of the number of sub-classes in the system. However, if the same Fab preparation is absorbed with two mutants which have no defect in common, the double absorption would remove the full contact-site-A blocking activity, as absorption with wild-type cells does.

The Fab-absorbing activity of non-aggregating mutants has been tested in order to demonstrate heterogeneous components of the contact-site-A system (28). With respect to the Fab-absorbing activity, three groups of non-aggregating mutants were found: (1) those without any detectable serological defect, (2) those which absorb only part of the contact-site-A blocking Fab species, and (3) those with almost absent absorbing activity for this Fab (Fig.4).

In the combinations hitherto tested (28), double absorption with mutants of group 2 and 3 did not fully remove the contact-site-A blocking activity, indicating that all these mutants shared at least one defect in the contact-site-A system. This is due in part to the fact that in group 3, control mutants are common which do not differentiate from the growth-phase state to aggregation-competence. As shown below, such mutants exhibit pleiotropic defects in developmentally controlled cell surface sites and are possibly devoid of any component of the contact-

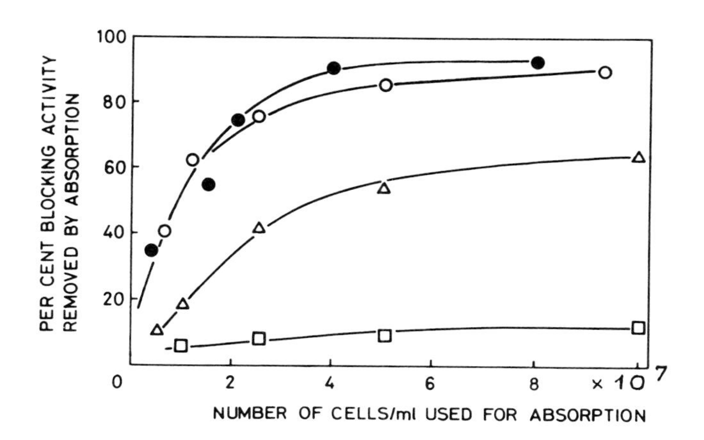

Fig. 4. Absorption of contact-site-A blocking Fab by stationary phase cells of the wild-type M2 (●) and of three non-aggregating mutants: ○, aggr 39; △, ap 66-aggrl; □, aggr 20-2. The solution used for absorption contained 0.8 mg/ml aggregation-specific Fab plus 2 mg/ml non-immune Fab.

site-A system.

Detailed analysis of a mutant belonging to group 2 showed, however, that the contact-site-A system comprises at least two different target sites for antibody, and that blockage of any one of these sites is sufficient for full inactivation of the system. This mutant, ap 66-aggr 1 (Fig.4), reduced the blocking activity of the Fab preparation used up to a plateau where about 70 per cent of the activity was removed. The exhaustively absorbed Fab still fully inactivated the contact-site-A system. This activity could be removed only by absorption with wild-type cells (28). Thus, the intact wild-type system consists of a minimum of two sites, one still serologically active and the other inactive in the mutant. This result is in accord with but does not prove the Tyler-Weiss hypothesis. It has to be shown that the two sites are complementary in the functional sense, and also that they function in the transposition, i.e., that *a* at one cell surface interacts with *α* at the other, and *vice versa*.

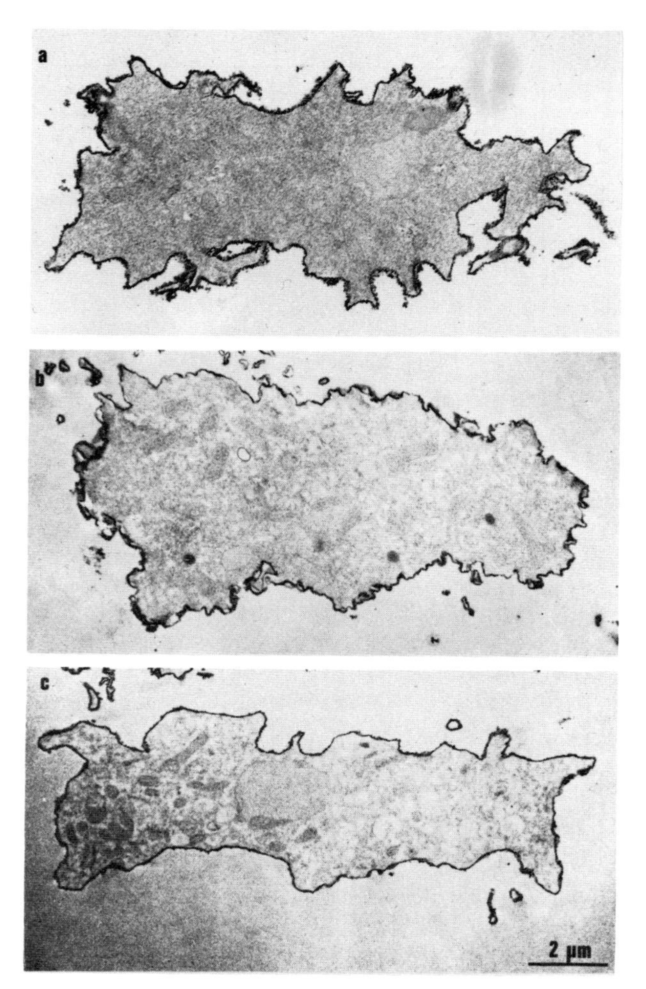

Fig. 5. Peroxidase staining of aggregation-competent cells labelled with either Concanavalin A or IgG. Cells: D.discoideum Ax-2 fixed with 0.25 per cent glutaraldehyde. a, anti-carbohydrate rabbit IgG as first layer, anti-rabbit IgG goat IgG, conjugated with peroxidase (32), as second layer. b, Concanavalin A as first layer, peroxidase as second layer (33). c, aggregation-specific IgG as first layer, second layer as in a.

TWO AND THREE DIMENSIONAL ANTIGEN PATTERN AT THE SURFACE OF AGGREGATING CELLS

A possible reason why anti-carbohydrate Fab does not block terminal cell adhesion could be the absence of its target antigens from the cell ends. The alternative would be that the target antigens are present there and that Fab bound to these sites fits within the space between adjacent cell membranes without blocking adhesion. To distinguish between these possibilities, the distribution of target sites for blocking and non-blocking antibodies was studied by use of IgG labelled with peroxidase, ferritin or fluorescein-isothiocyanate (29). The results proved the second alternative to be correct. Not only were the carbohydrate sites homogeneously distributed over the entire cell surface prior to aggregation (Fig. 5a, b), but brilliant fluorescence of labelled Fab directed against these antigens was retained at the contiguous ends of already aggregated cells (14). This demonstrates that sufficient space is left at the areas of actual adhesion for Fab molecules of a size of 35 x 35 x 60 Å (30) to fit in, indicating that the presence of Fab in these areas does not interfere with adhesion, except if the Fab is bound to the relevant sites.

Aggregation-specific Fab is the most specific Fab preparation yet available for blocking contact sites A. IgG of the same specificity was used for immuno-labelling of cells. This label shows the overall distribution of cell surface antigens characteristic for aggregation-competent cells. If contact sites A represent a major fraction of these sites, their distribution should detectably affect the overall pattern. At the cell ends where contact sites A are active, no denser labelling was observed than at other parts of the cell surface (Fig. 5c). If any inhomogeneity of labelling existed, it was only a very fine clustering of the ferritin molecules (29). These results together with the homogeneous distribution of carbohydrate-antigens renders it improbable that the contact sites are associated into larger patches of specialized surface areas; and they suggest that contact sites A are not specifically localized at the cell ends but rather specifically activated there.

In order to bridge the approx. 150 Å gap between the membranes of aggregated cells (31), contact sites should extend vertically from the membranes into the intercellular space beyond the layer of other cell surface constituents.

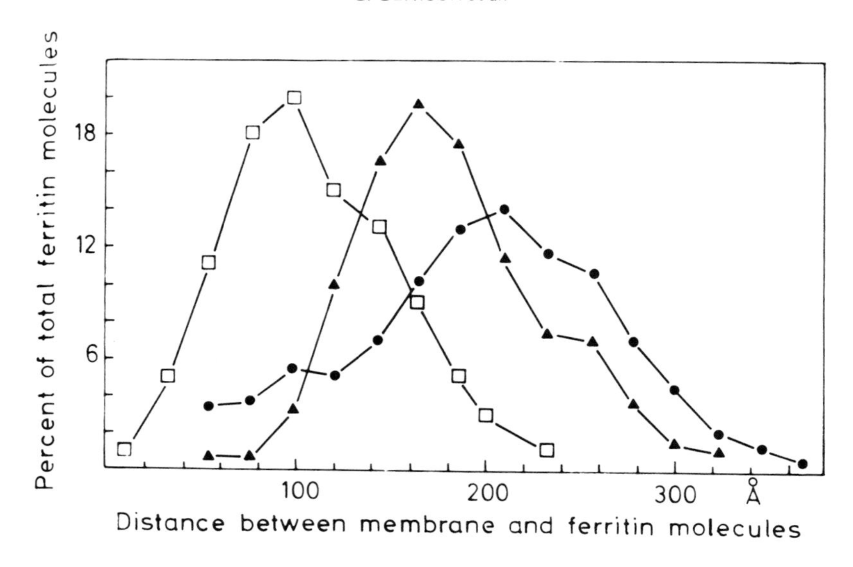

Fig. 6. Distribution curves for the distances between the outer electron dense layer of the plasma membrane and the protein shell of ferritin molecules. □ , Concanavalin A-ferritin conjugate; ▲, anti-carbohydrate rabbit IgG as first layer, ferritin-conjugated anti-rabbit IgG goat IgG as second layer; ●, aggregation-specific IgG as first layer, the same second layer as before. The aggregation-competent D.discoideum Ax-2 cells were fixed in 0.25 per cent glutaraldehyde before labelling.

Under identical conditions, glutaraldehyde-fixed cells were labelled with either anti-carbohydrate IgG or aggregation-specific IgG, and with ferritin labelled anti-rabbit-IgG goat IgG as second layer. The distribution of distances of the ferritin from the membrane (Fig.6) was relatively narrow in the case of anti-carbohydrate IgG, indicating that the glycosphingolipids had short oligosaccharide chains of similar lengths, a conclusion which can be extended to the Concanavalin A receptors. In the case of the aggregation-specific sites, the distribution was broader and shifted to greater distances, in accord with the assumption that part of these sites link the cells by interacting pair-wise within the intercellular space.

MEMBRANE MARKERS OF THE DIFFERENTIATED STATE

The activity of three membrane constituents, cyclic-AMP binding sites, cell-bound phosphodiesterase and contact sites A, increases drastically during cell differentiation from the growth-phase stage to aggregation-competence. These structures, therefore, represent markers of the differentiated state. Additionally to these loci, an aggregation-competent cell contains plasma membrane constituents already present in growth-phase cells, as contact sites B and antigen I, the glycosphingolipid complex. In fact, those membrane antigens which are shared by both developmental stages are, quantitatively, the predominant ones (14). As shown by analysis of an UV-induced mutant and its spontaneous revertant (Tab.2), the manifestation of the differentiated state depends on a genetic system that, without affecting the other membrane constituents, specifically controls the appearance of the membrane markers for aggregation-competence.

Table 2

Cell surface constituents in a non-aggregating mutant and in a revertant showing normal development.

Cell surface sites	Activity in stationary phase cells of	
	mutant aggr 50-2	revertant aggr 50-3 (rev)
Cyclic AMP binding sites	low	high
Cell-bound phosphodiesterase	low	high
Contact sites A	-	+
Contact sites B	+	not tested
Antigen I (Glycosphingolipid)	+	not tested

REFERENCES

(1) D.J.Watts and J.M.Ashworth, Biochem. J. 119 (1970) 171.

(2) G.Gerisch, Roux'Archiv Entwicklungsmechanik 153 (1962) 603.

(3) D.Malchow, B.Nägele, H.Schwarz and G.Gerisch, Eur.J. Biochem. 28 (1972) 136.

(4) J.T.Bonner, D.S.Barkley, E.M.Hall, T.M.Konijn, J.W. Mason, G.O'Keefe and P.B.Wolfe, Developm. Biol. 20 (1969) 72.

(5) G.Gerisch, in: Current Topics in Developmental Biology, Vol.3, eds. A.A.Moscona and A.Monroy (Academic Press, New York-London, 1968) p. 157.

(6) M.H.Cohen and A.Robertson, J. Theoret. Biol. 31 (1971) 119.

(7) G.Gerisch, Naturwissenschaften 58 (1971) 430.

(8) A.Robertson, D.J.Drage and M.H.Cohen, Science 175 (1972) 333.

(9) R.G.Pannbacker and L.J.Bravard, Science 175 (1972) 1014.

(10) D.Malchow, J.Fuchila and B.Jastorff, FEBS Letters 34 (1973) 5.

(11) Konijn,T.M., Advances in Cyclic Nucleotide Research 1 (1972) 17.

(12) D.Malchow and G.Gerisch, Biochem. Biophys. Res. Comm. 55 (1973) 200.

(13) G.Gerisch, in: "Non-specific" factors influencing host resistance, eds. W.Braun and J.Ungar (S.Karger, Basel, 1973) p. 236.

(14) H.Beug, F.E.Katz, A.Stein and G.Gerisch, Proc.Natl. Ac. Sci. USA 70 (1973) 3150.

(15) R.B.Taylor, W.P.H.Duffus, M.C.Raff and S.dePetris, Nature New Biol. 233 (1971) 225.

(16) M.C.Raff and S.dePetris, Federation Proc. 32 (1973) 48.

(17) M.J.Karnovsky and E.R.Unanue, Federation Proc. 32 (1973) 55.

(18) D.Malchow, O.Lüderitz, O.Westphal, G.Gerisch and V. Riedel, Eur. J. Biochem. 2 (1967) 469.

(19) H.Wilhelms, Thesis, Universität Freiburg/Br. (1972).

(20) G.Gerisch, D.Malchow, H.Wilhelms and O.Lüderitz, Eur. J. Biochem. 9 (1969) 229.

(21) H.Beug and G.Gerisch, J. Immunol. Methods 2 (1972) 49.

(22) H.Beug, F.E.Katz and G.Gerisch, J. Cell Biol. 56 (1973) 647.

(23) S.Kempff, Diplomarbeit, Universität Freiburg/Br. (1969).

(24) H.Beug, G.Gerisch, S.Kempff, V.Riedel and G.Cremer, Exptl. Cell Res. 63 (1970) 147.

(25) A.Tyler, West. J. Surg. Obstet. Gynecol. 50 (1942) 126.

(26) P.Weiss, Yale J.Biol.Med. 19 (1947) 235.

(27) S.Roseman, Chem. Phys. Lipids 5 (1970) 270.

(28) A.von Stein, Diplomarbeit, Universität Tübingen (1973).

(29) H.Schwarz, Thesis, Universität Tübingen (1973).

(30) R.C.Valentine and N.M.Green, J. Mol. Biol. 27 (1967) 615.

(31) E.H.Mercer and B.M.Shaffer, J. Biophys. Biochem. Cytol. 7 (1960) 353.

(32) S.Avrameas, Immunochemistry 6 (1969) 43.

(33) W.Bernhard and S.Avrameas, Exptl. Cell Res. 64 (1970) 232.

The work was supported by the Deutsche Forschungsgemeinschaft and Stiftung Volkswagenwerk.

DISCUSSION

I. SCHENKEIN: Could you comment as to how EDTA mimicks the F_{ab} fragment activity in the contact site B inhibition? Is it by an entirely different mechanism?

G. GERISCH: It is right that either EDTA or F_{ab} can be used to block contact sites B. The EDTA-effect can be reversed by Ca^{++}, whereas F_{ab} also blocks with Ca^{++} added. The contact site A system does not seem to depend on divalent cations.

S. ASSAF: Is the spectrophotometer you were using, a double beam spectrophotometer or is it a new instrument for measuring agglutination?

G. GERISCH: It is an Eppendorf filter photometer which has the advantage that the special equipment we use for the agglutination assay can be easily put into the light beam. The instrument and method is described in J. Immunol. Methods 2 (1972) 49.

S. ASSAF: Could you tell us what role cyclic AMP has in slime mold aggregation? You showed that the mutants which have a low level of cAMP do not aggregate, while the others which have high levels of cAMP exhibited considerable cell aggregation. In other cells such as platelets, cyclic AMP plays a role in inhibiting aggregation; perhaps in the slime mold one may conclude that it is promoting aggregation. In other cells such as platelets, cyclic AMP plays a role in inhibiting aggregation; perhaps in the slime mold one may conclude that it is promoting aggregation.

You also showed that phosphodiesterase levels were high in one case and low in the other. Could you elaborate on that?

G. GERISCH: Many of our non-aggregating mutants are defective in the increase of membrane-bound phosphodiesterase (J. Mol. Biol. *74* (1973) 573). However, these mutants may be control mutants like that shown in Table 2, which is defective in a variety of cell surface sites.

There is another mechanism for regulating cyclic AMP which operates through the control of extracellular phosphodiesterase by an inhibitor released into the medium just at the turning point between end of growth and the beginning of differentiation. This inhibitor is trypsin sensitive, has a molecular weight of about 40,000 and is specific for this enzyme. About 50% of our non-aggregating mutants are defective with respect to this inhibitor and so this defect could be the reason for their inability to aggregate. However, most of these mutants are control mutants which are blocked in an early step of differentiation, so that they are missing all the specific factors, including the membrane components, which are formed between end of growth and the aggregation-competent state. Since the inhibitor is only one of these factors, the correlation between its absence and the aggregation-defect does not necessarily mean that the inhibitor itself is really important for aggregation. Nevertheless, under our conditions the extracellular cyclic-AMP phosphodiesterase is drastically regulated at the post-translational level by the macromolecular inhibitor.

E.W. SUTHERLAND: I heard a verbal presentation which suggested that there was a third stage in which cyclic AMP acted on these slime molds - it was the stage at which they learned to recognize their head from their tail, but I have never seen that in print.

G. GERISCH: The cells form ends even when they are not moving along a cyclic AMP gradient. Under submerged culture conditions the cells already have an elongated shape with clear-cut cell ends in the shaken suspension. Of course, cyclic AMP may be present in the shaken suspension, but a concentration gradient can hardly be formed. So the gradient at least does not appear to be necessary.

J. DANIEL: As the F_{ab} antibody titrates the aggregation ability, what happens to the phosphodiesterase activity?

G. GERISCH: F_{ab} of an antiserum from one particular rabbit blocked the membrane-bound phosphodiesterase only partially but cell-to-cell adhesion completely. The same F_{ab} did not block chemotaxis, at least not completely. I think that the phosphodiesterase and the contact sites are not identical.

J. DANIEL: Is this in contrast with the report (Goidl et al. PNAS *69* (1972) 1128) that cyclic AMP phosphodiesterase specific antibodies block aggregation?

G. GERISCH: We did not really test this in detail, but with respect to the question of whether blockage of phosphodiesterase inhibits aggregation, I should say that the phosphodiesterase inhibitor is specific *in vivo* for the extracellular enzyme and does not block the cell-bound phosphodiesterase. When particle fractions from these cells are prepared and when these particles are partially solubilized, the membrane-bound phosphodiesterase becomes increasingly inhibited by the inhibitor, indicating that the cell-bound diesterase possesses the inhibitor binding site, although it is not exposed in the living cells. If the cell-bound enzyme has any function in chemotaxis it would be necessary that it is protected against the inhibitor, because the inhibitor is present just in the aggregation state, where the chemotactic response is necessary.

J. DANIEL: Are the antibodies active against species in which aggregation is not cAMP induced?

G. GERISCH: The contact sites A are species specific. Using *Polysphondylium pallidum*, which according to Bonner does not respond to cAMP, for Iab absorption, we found no material that cross-reacted with contact sites A. Since most of the cellular antigens, soluble as well as membrane-bound ones, are species specific, this result on contact sites A is in accord with a species specific function of these sites in aggregation. However, this is not a strong argument.

THE ADHESION AND AGGREGATION OF BLOOD PLATELETS

G.A. JAMIESON
The American National Red Cross
Blood Research Laboratory
Bethesda, Maryland 20014

A primary step in hemostasis is thought to be the adhesion of circulating blood platelets to collagen, or to basement membrane, which has been exposed by damage to the vascular endothelium. Following this, there is the release of certain specific platelet constituents, notably adenosine diphosphate, which results in the aggregation of further platelets to those already adhering to collagen. Although the biochemical events involved in these two phenomena are not known, recent results suggest that the carbohydrate present on the platelet surface, or on the collagen substrate, may play a major role. This paper will review recent work from this laboratory relating to the role of glycoproteins in platelet function and structure.

The blood platelet is unique insofar as it is formed by the fragmentation of its precursor cell, the megakaryocyte; each megakaryocyte produces about 2,000 individual platelets (1). The platelet is an anucleate cell which contains several types of electron-dense granules as well as a few primitive mitochondria. The platelet membrane is characterized by extensive invaginations which can apparently communicate with different areas of the surface, as shown by freeze-cleave experiments in electron microscopy. This canalicular system gives the platelet a "sponge-like" quality which may account for its ability to transport certain proteins of the intrinsic coagulation system.

The platelet surface is rich in carbohydrate and early ultrastructural data showed that this platelet glycocalyx could be removed by treatment with proteolytic enzymes (2). Subsequently, it was shown that tryptic digestion of intact platelets (3,4) or of isolated membranes (5), resulted in three size classes of glycopeptides. These were designated as glycopeptide I, which was termed a macroglycopeptide because of its molecular weight (120,000), glycopeptide II (M.W. 22,500), and glycopeptide III (M.W. 5,000). Glycopeptide II occurred in only small amounts relative to glycopeptides I and III but, like glycopeptide I, it contained both N- and O-glycosidic linkages, whereas glycopeptide III contained only the N-glycosidic linkages similar to those found in plasma glycoproteins. The macroglycopeptide appeared to be related to the thrombocyte-specific antigen of the platelet surface (6) and to the platelet glycoprotein inhibitor of viral hemagglutination (7). Recently the glycoprotein precursor of the macroglycopeptide (GP-I) has been isolated. It has a molecular weight of about 150,000 and yields the macroglycopeptide as its sole glycopeptide product on tryptic digestion together with a peptide fragment of molecular weight about 30,000. It has an amino acid composition comparable with that of the macroglycopeptide isolated from intact platelets, except for an increased proportion of lipophilic amino acids.

Although there had been previous studies on individual aspects of platelet membranes the first extensive study was by Marcus and associates (8) who utilized conventional techniques of homogenization but obtained a well characterized membrane fraction and studied its composition, particularly with regard to lipid components. Subsequently, a new isolation procedure was developed based on the intracellular loading of platelets with glycerol by sedimentation through a glycerol gradient and the subsequent treatment of the glycerol-loaded platelet pellet with isotonic buffer solution (9). An extensive comparison of various methods for platelet lysis, including the use of various agents for the hardening of the platelet membrane, indicated that this method was most satisfactory in terms of yield and reproducibility (10).

Platelet membranes isolated by this "glycerol-lysis" technique were compared in terms of their procoagulant activity (11), and chemical and enzymatic composition (9). These data showed that the enzymes usually associated with the plasma membrane in other cells; namely, the Na^{+}/K^{+} ATPase, the Ca^{++}/Mg^{++} ATPase, phosphodiesterase and acid phosphatase were obtained in a purification and recovery corresponding to that in other plasma cells. This was of interest in view of a previous suggestion that the platelet plasma membrane might result from certain subcellular structures of the megakaryocyte (12,13) rather than by the formation of demarcation membranes by invagination of its own plasma membranes. These results indicated that, at least in terms of its enzymatic content, as well as chemical composition, the platelet plasma membrane was comparable with that of the plasma membrane fractions obtained from other cells.

Further studies showed that the platelet plasma membrane isolated by this technique was particularly rich in collagen: glucosyltransferase, an enzyme which transfers glucose from UDPG to collagen or to a glucosyl acceptor prepared by the partial acid hydrolysis of collagen, itself(14). This enzyme was obtained in approximately 50% yield and 20-fold purification. In addition, this enzyme appeared to be located on the outer surface of the membrane since it was detected with equal activity in intact platelets and in isolated membranes, while a corresponding galactosyltransferase was detected only in isolated membranes, suggesting that it might be present on the inner aspect of the membrane or might be buried sufficiently deep in the membrane that it was only exposed by the isolation procedure (15). Even higher purifications in the plasma membrane fraction were reported in other studies (16).

A similar glucosyltransferase has been detected in a number of other cells, including kidney cortex (17) where it was shown to have a requirement for free ϵamino groups in the hydroxylysine to which the heterosaccharide chains were joined, and for free sulfhydryl groups in the enzyme, itself. Similar requirements were known to exist for platelet adhesion (18,19).

In addition, there had been previous reports that platelet adhesion was inhibited by aspirin (20,21), by chlorpromazine (22) and by glucosamine (23). Studies with the platelet enzyme showed that it was also inhibited by these three substances. Thus, these experiments indicated that a range of parallels existed between the collagen: glucosyltransferase of human platelet membranes and the phenomenon of platelet: collagen adhesion.

In light of this, an experiment was devised in which collagen was degraded with bacterial collagenase and separated by gel filtration on Sephadex G-25 into two main fractions; one, which emerged close to the column volume, contained virtually all of the collagen carbohydrate (ca. 17% hexose) while the second, which was retarded on gel filtration, had virtually no detectable carbohydrate. Platelet adhesion was measured in the aggregometer by the method of Spaet and Lejnieks (24), in the presence of EDTA in order to differentiate between collagen-induced adhesion and collagen-induced aggregation. Under these circumstances, collagen-induced adhesion was inhibited about 30% by the total collagenase digest of collagen but it was inhibited only about 15% by the carbohydrate free-peptides. On the other hand, adhesion was inhibited about 85% by the collagen glycopeptides at a level of about 4 mg/ml; at levels below 2 mg/ml no inhibitory effect was observed (25).

On the basis of these studies, a hypothesis was proposed (25) suggesting that platelet: collagen adhesion represented a special case of the general hypothesis proposed by Roseman (26) which suggested that intercellular adhesion was mediated by the presence of specific glycosyltransferases on one cell and complementary, incomplete heterosaccharide chains on the apposing cell. Thus, in the case of platelet: collagen adhesion, the enzyme would be present only on the platelet side and would be a specific glucosyltransferase, while a specific identified acceptor (galactosylhydroxylysine) was present on the receptor side.

Further studies in this area have focused on two major problems. First, it was known that there was an endogenous acceptor in the platelet enzyme system. That is, glucose was transferred from UDPG to an endogenous acceptor in the absence of added collagen. Extensive studies were carried out in order to determine whether this might represent incorporation into endogenous glycoprotein of the platelet membrane. However, this endogenous incorporation appears to be entirely dependent upon incorporation into glycogen, due to the presence of small amounts of glycogen synthase which are strongly bound to isolated platelet plasma membranes and to the presence of traces of glycogen in these preparations (27).

The second area of activity has been in attempts to increase the sensitivity of the assay procedure since the earliest studies had required the membranes isolated from the platelets found in approximately 500 ml of blood in order to make a single observation. More recently, a 10-fold increase in the sensitivity of the assay has been achieved by the use of highly purified galactosylhydroxylysine, the characteristic incomplete glycopeptide of collagen (28).

This hypothesis has stimulated additional work to evaluate a possible role for the carbohydrate of collagen in these phenomena. Unfortunately, these studies have, without exception, addressed themselves to the study of platelet aggregation rather than adhesion, so that it is not possible to compare the two types of experiments. Aggregation is a much less specific reaction than adhesion and can be induced, not only by collagen, but also by ADP, thrombin, serotonin, epinephrine, fatty acids, antigen-antibody complexes, bentonite, porous nickel, and a wide variety of other surfaces. However, these studies showed that the ability of collagen to induce aggregation was completely destroyed by modification of the terminal galactose residues of the incomplete heterosaccharide chains of collagen (29) although it has been subsequently found (R. W. Colman, personal communication; H. R. Baumgartner, personal communication) that this is due to a modification of the ability to form microfibrillar collagen which is required for platelet aggregation in the aggregometer. Second, Cunningham and his associates (30) have shown that the ability of collagen to induce platelet aggregation remains intact under conditions whereby all the carbohydrate is destroyed although these authors feel that collagen carbohydrate may function as a "recognition site", rather than an "adhesion site" in platelet aggregation. Katzman and associate (31) separated the cyanogen bromide fragments of collagen and showed that only that fragment which contained the collagen carbohydrate (α1-CB5) was effective in inducing platelet aggregation, although the authors stated that they did not believe the amount of the incomplete heterosaccharide chains (galactosyl residues) in this fraction was insufficient to explain the observed aggregation on the basis of the glycosyltransferase hypothesis. It appears necessary to carry out further studies in this area, and to effect a consistency of experimental design with regard to the measurement of adhesion or aggregation, in order to reconcile these conflicting observations.

Both the collagen: glucosyltransferase of human platelet membranes and the collagen: galactosyltransferase have an optimal metal requirement of 15 mM manganese while the latter shows about one-fifth activity with cadmium ions at the same concentration. During these enzymatic studies it was observed that the agglutination of platelet membranes, or intact platelets, occurred in the presence of either of these divalent cations (32). This raised the possibility that Roseman's hypothesis (26) could be extended to the aggregation, as distinct from the adhesion, of platelets and that this agglutination might be arising from the activation of membrane-bound glycosyltransferases, although there was no experimental evidence to support this in platelets. A platelet mannosyltransferase has been detected which transfers mannose to endogenous acceptors (33) as well as glycoprotein: sialyltransferase which transfers to exogenous acceptors such as desialated fetuin and prothrombin (34).

Studies with manganese and cadmium showed that both induced platelet aggregation which reached a maximum at 15 mM concentration, equal to the optimal concentration for transferase activity. Further addition of manganese was without effect whereas, at cadmium ion concentrations above 25 mM, there was a rapid change in platelet aggregation consistent with the initiation of the platelet release reaction. In addition, even at lower cation concentrations, electron microscopy showed that cadmium ions induced considerable changes in platelet ultrastructure, including extensive vacuolization and pseudopod formation, while manganese had relatively little ultrastructural effect beyond an intensification of the microfibrillar system.

It was also found that there was a critical concentration of cadmium in order to induce these effects; if cadmium was added to concentrations below about 3 mM the subsequent addition of manganese gave aggregation patterns identical with those of manganese alone, whereas at cadmium ion concentrations above about 5 mM the subsequent addition of manganese gave aggregation and release patterns characteristic of cadmium. These results, which were supported by studies on membranes isolated by the glycerol-lysis technique, suggested that there were two types of receptors for divalent cations on the platelet membrane; one available only to cadmium while the other was available to either cadmium or manganese.

The fourth area of study has involved the interaction of lectins with platelets and platelet membranes, the intention here being to test the transferase theory by trying to inhibit the adhesion of platelets to collagen by pre-treating the collagen with lectins specific for terminal galactosyl residues. While this was not achieved, since the patterns of lectin inhibition were relatively non-specific, certain patterns did emerge. First, platelet aggregation was effected only by wheat germ agglutinin (GlcNAc determinant), *Ricinus communis* (galactose determinant), and *Phaseolus coccineus* (determinant unknown); in the case of *Phaseolus coccineus* it may be noted that the erythrocyte-agglutinating activity of the lectin is inhibited by glycopeptide I.3 of the erythrocyte membrane (35) and that similar inhibition is shown for the erythrocyte-agglutinating activity of Phaseolus vulgaris, which is also known to cause platelet aggregation (36). Second, lectins with a specificity directed towards glucose and mannose did not cause platelet aggregation but did result in the release of serotonin. Conconavalin A appeared to be a special case of lectins of this class since it was able to induce the aggregation of isolated membranes and platelets but only after the removal of plasma. Third, a lectin with activity directed towards galNAc (soybean) caused neither aggregation nor the release of serotonin.

In terms of the first group of lectins it appeared that the activity of Ricinus communis was primarily dependent on the release of ADP while that of wheat germ agglutinin and Phaseolus coccineus appeared to depend more on their agglutinating ability. This was supported by the important finding that the agglutinating effect of wheat germ agglutinin was inhibited by the macroglycopeptide of human platelet membranes while the effect of Phaseolus coccineus was completely inhibited by glycopeptide II. The third glycopeptide class isolated from platelet membranes, glycopeptide III, was found to be without effect on either of these lectins. Not only were these three lectins the most effective of the nine tested in causing platelet aggregation but also they were the most effective, at subthreshold levels at which they did not themselves effect aggregation, in inhibiting the aggregation induced by ADP, or serotonin, or the release of these components induced by other lectins.

A major problem in the study of platelet function has been the lack of application of newer methods of instrumentation. Although not directly related to the role of glycoproteins in platelet function, another area of investigation has been the application of microcalorimetry to the study of platelet aggregation. Initial studies (37) have been with washed platelets to eliminate complicating plasma effects. These studies have shown that the reaction of thrombin with washed platelets results in a heat production of approximately 300 microwatts/10^9 platelets, corresponding to a temperature rise of about

0.0125°C in this instrument and totalling 50 millicalories over about 30 min. Studies with metabolic inhibitors (antimycin and 2-deoxyglucose) have shown that the heat production of the initial phases (5 min) of the reaction are mainly dependent on oxidative phosphorylation, while the latter stages are probably due to glycolysis. Since the initial steps in hemostasis are certainly complete within five minutes, these results suggest that oxidative phosphorylation is making the main contribution to platelet metabolism in this period.

The experiments reviewed above on platelet adhesion and on platelet aggregation induced by lectins, divalent cations, and thrombin provide indirect evidence for a role for platelet carbohydrate in platelet function. While extensive further studies are necessary the platelet appears to be a useful model system, not only in terms of its importance in hemostatic phenomena, but it can be used to differentiate betweeen cell-substrate adhesion and cell-cell aggregation and because its release function is a model for the stimulation of other secretory cells.

(1) L. A. Harker, in: The Platelet, eds. K. M. Brinkhous, R. W. Shermer and F. K. Mostofi (The Williams and Wilkins Company, 1971) p. 13.

(2) T. Hovig. Ser. Haemat. 1 (1968) 3.

(3) D. S. Pepper and G. A. Jamieson. Biochemistry, 8 (1969) 3362.

(4) D. S. Pepper and G. A. Jamieson. Biochemistry, 9 (1970) 3706.

(5) A. J. Barber and G. A. Jamieson. Biochemistry, 10 (1971) 4711.

(6) N. Hanna and D. Nelken. Immunol. 16 (1969) 601.

(7) D. S. Pepper and G. A. Jamieson. Nature, 219 (1968) 1252.

(8) A. J. Marcus, D. Zucker-Franklin, L. B. Safier, and H. L. Ullman. J. Clin. Invest. 45 (1966) 14.

(9) A. J. Barber and G. A. Jamieson. J. Biol. Chem. 245 (1970) 6357.

(10) A. J. Barber, D. S. Pepper and G. A. Jamieson. Thromb. Diath. Haemorrh. 26 (1971) 38.

(11) A. J. Barber, D. C. Triantaphyllopoulos and G. A. Jamieson. Thromb. Diath. Haemorrh. 28 (1972) 206.

(12) H. Schulz, Verh. Deut. Ges. Pathol. 50 (1966) 239.

(13) O. Behnke. J. Ultrastruct. Res. 24 (1968) 412.

(14) A. J. Barber and G. A. Jamieson. Biochim. Biophys. Acta, 252 (1971) 533.

(15) A. J. Barber and G. A. Jamieson. Biochim. Biophys. Acta, 252 (1971) 546.

(16) H. B. Bosmann. Biochem. Biophys. Res. Commun. 43 (1971) 1118.

(17) R. G. Spiro and M. J. Spiro. J. Biol. Chem. 264 (1971) 4899.

(18) G. D. Wilner, H. L. Nossel and E. C. LeRoy. J. Clin. Invest. 47 (1968) 2616.

(19) H. Al-Mondhiry and T. H. Spaet. Proc. Soc. Exp. Biol. Med. 135 (1970) 878.

(20) E. J. W. Bowie and C. A. Owen. Circulation, 40 (1969) 757.

(21) G. Evans, M. A. Packham, E. E. Nishizawa, J. F. Mustard and E. A. Murphy. J. Exp. Med. 128 (1968) 877.

(22) J. F. Mustard and M. A. Packham. Pharmacol. Rev. 22 (1970) 97.

(23) Y. Legrand, J. P. Caen and L. Robert. Proc. Soc. Exp. Biol. Med. 127 (1968) 941.

(24) T. H. Spaet and I. Lejnieks. Proc. Soc. Exp. Biol. Med. 132 (1969) 1038.

(25) G. A. Jamieson, C. L. Urban and A. J. Barber. Nature, New Biol. 234 (1971) 5.

(26) S. Roseman. J. Chem. Phys. Lipids, 5 (1970) 270.

(27) J. H. Greenberg, A. P. Fletcher and G. A. Jamieson. Thromb. Diath. Haemorrh. 30 (1973) 307.

(28) D. Smith and G. A. Jamieson. Unpublished observations.

(29) C. Chesney, E. Harper and R. W. Colman. J. Clin. Invest. 51 (1972) 2693.

(30) D. Puett, B. K. Wasserman, J. D. Ford and L. W. Cunningham. J. Clin. Invest. 52 (1973) 2495.

(31) R. L. Katzman, A. H. Kang and E. H. Beachey. Science, 181 (1973) 670.

(32) C. L. Urban, A. P. Fletcher and G. A. Jamieson. Unpublished observations.

(33) L. DeLuca, S. DeLuca, A. J. Barber and G. A. Jamieson. Unpublished observations.

(34) H. B. Bosmann. Biochim. Biophys. Acta, 279 (1972) 456.

(35) J. Kubanek, G. Entlicher and J. Kocourek. Biochim. Biophys. Acta, 304 (1973) 93.

(36) P. W. Majerus and G. N. Brodie. J. Biol. Chem. 247 (1972) 4253.

(37) P. D. Ross, A. P. Fletcher and G. A. Jamieson. Biochim. Biophys. Acta, 313 (1973) 106.

DISCUSSION

S.A. ASSAF: Since you believe that glycosyltransferase is responsible for controlling platelet adhesiveness, how is this enzyme regulated? Why is not this platelet enzyme always bound by collagen if it is really at the surface of the platelet? Do you have any hypothesis or speculations about the regulation of glucosyltransferase activity?

G.A. JAMIESON: Platelets do not adhere to endothelial cells or to undamaged vascular endothelium. When the vascular endothelium is damaged, then collagen or basement membrane is exposed and we believe that the glucosyltransferase on the platelet surface is then able to react with the carbohydrate of the collagen. It is not a question of enzyme regulation, as much, but the fact that the substrate is not exposed to the flowing blood under normal conditions of the vessel wall. The slide I showed was of a microthrombus which had formed by adhesion of a single platelet to collagen through a greatly enlarged endothelial gap.

S.A. ASSAF: Could you then comment on the work of David Puett and his colleagues from Vanderbilt on the work that just appeared in JCI (i.e. Cunningham's group) October issue, vol. 52, 2495 (1973).

G.A. JAMIESON: I think that Drs. Cunningham and Puett's work is related to whether the carbohydrate of collagen, even when exposed, can be a substrate. He disagrees with me on whether the collagen carbohydrate is the substrate, but I do not think he disagrees on the fact that collagen itself is a substrate and that it is exposed only by damage to the endothelium.

A.H. REDDI: Is there any specificity exhibited by collagen from different sources? Is skin collagen the same as subendothelial collagen?

G.A. JAMIESON: That is an interesting question, but there is no quantitative evidence regarding adhesion to collagens from different sources.

R. BERNACKI: You mentioned manganese and cadmium effects on your enzymes. Have you looked at the effects of calcium on these transferases?

G.A. JAMIESON: No, we haven't looked at that.

BIOCHEMICAL PROPERTIES AND IMMUNOCHEMICAL-GENETIC RELATIONSHIPS OF ALLOANTIGENS OF THE MOUSE H-2 MAJOR HISTOCOMPATIBILITY COMPLEX

S. G. NATHENSON, S. E. CULLEN and J. L. BROWN
Departments of Microbiology and Immunology and Cell Biology
Albert Einstein College of Medicine

Abstract: H-2K and H-2D alloantigens are primary products of genes of the K and D regions which map at opposite ends of the mouse H-2 MHC (major histocompatibility complex). These membrane located antigens are glycoproteins (protein, 90%; carbohydrate, 10%) of approximately 45,000 in molecular weight after reduction in SDS. They can also be isolated as dimers of 90,000 probably composed of disulfide linked identical monomers. In non-ionic detergents, larger aggregates may be present. Peptide comparisons of products of the alleles of either the H-2K or H-2D genes showed considerable uniqueness existed for these products. For example, in comparison of products of alleles of D vs K, or alleles of D vs D or K vs K from about 30 to 60% of peptides were found to be identical. Such diversity of primary protein structure is consistent with the extreme polymorphism of the H-2K and H-2D gene system, and suggests special genetic mechanisms may be involved.

Ia alloantigens are products of genes mapping in the I region of the H-2 MHC. They are also membrane located, but probably occur only on subpopulations of lymphocytes. These antigens were found to be glycoproteins of about 30,000 M.W. after reduction in SDS, but also were found as dimers.

INTRODUCTION

The major histocompatibility gene complex of mice (H-2) (1,2,3) is comprised of 4 regions defined by recombination which map in the following order from the centromere: K, I, S and D (4). The K (H-2K) and D (H-2D) regions are separated by an 0.5% recombination frequency and contain the structural genes controlling serologically detected cell membrane antigens located on many tissues, but present in high density on lymphoid tissues. The S (SS-Slp) region (5) contains a gene controlling the quantity of a serum protein thought to be involved in the complement system (6), and the I region contains one or more genes involved in the immune responses to certain protein and polypeptide antigens (7), to lymphocytic proliferative responses (cf. Ref. 8 for review), and susceptibility to certain types of mouse leukemias (9).

The H-2K and H-2D genes are centrally involved in tissue graft rejection. However, the true physiological function of these membrane integrated macromolecules has not yet been established. One property certainly related to their antigenicity and possibly related to their function is their extraordinary polymorphism. The extent of this polymorphism is attested to by the fact that 41 distinct H-2 haplotypes have been described in laboratory and wild mice (10), and this number clearly is a minimum estimate.

It is the purpose of this communication to review some results of studies from our laboratory on the biochemical parameters of the products of the H-2K and H-2D genes and the results of preliminary experiments on the Ia antigens — a newly described series of antigens detected by serological techniques which are the putative products of genes residing in the I region.

These studies have been carried out because of the assumption that the products of the H-2 system comprise a group of immunologically distinct membrane glycoproteins whose biological activity, membrane affinity, reactivity with alloantibodies and genetic variability must all rest on structural features which can be chemically defined.

GENETIC AND SEROLOGICAL PROPERTIES OF H-2 ANTIGENS

Table 1 describes some of the salient features of a selected list of H-2 antigens. Each mouse strain of a particular haplotype (e.g. $H\text{-}2^b$) has an H-2K and H-2D gene. Each H-2 product has an unique or private specificity and one or more public or shared specificity.

TABLE 1

Simplified version of H-2 chart (selected strains)*

Chromosome symbol (haplotype)	Genotype description	Specificity		
		H-2K private**	H-2D private**	Public
$H\text{-}2^b$	$H\text{-}2K^b, H\text{-}2D^b$	33	2	5,6,27,28,29 35,36
$H\text{-}2^d$	$H\text{-}2K^d, H\text{-}2D^d$	31	4	3,6,13,27,28 29,35,36,41 42,43
$H\text{-}2^k$	$H\text{-}2K^k, H\text{-}2D^k$	23	32	1,3,5,8

*Data taken from Ref. 8.

**A specificity is defined as that portion of an antigen molecule which reacts with the combining site of the antibody molecules used to define that specificity. Each H-2 gene product carries one unique or private specificity, and also may carry one or more public specificities which are not uniquely found on one gene product.

BIOCHEMICAL PARAMETERS OF H-2 ANTIGENS

The studies which we are describing in this paper have utilized two complementary approaches. In our earlier work (11) H-2 antigen activity was released by limited proteolysis of membranes or whole cells and the antigens which had been converted to a water soluble form were purified by physical and chemical separation methods. Using a more recently developed approach the membrane was solubilized by the non-ionic detergent NP-40 (Nonidet P-40) and the detergent soluble but water insoluble antigen was isolated by a serological technique (12,13).

I. Chemical and Physical Properties of the H-2 Alloantigens.

A. Chemical properties, molecular size, and subunit structure.

The glycoprotein nature of the molecules bearing the H-2 alloantigenic sites was first established using the H-2 alloantigen fragments released from spleen cell membranes by papain digestion. In these studies (11) two different sized fragments, called Class I and Class II, were recovered after the enzymatic solubilization step. The Class I fragments had a molecular weight of approximately 37,000 and were about 90% protein and 10% carbohydrate. Class II fragments were about 28,000 in molecular weight and had a similar composition.

Since the water-soluble antigen fragments were glycoprotein in nature, it was expected that the detergent solubilized, water-insoluble antigen molecules would also be glycoproteins. This was confirmed by a double labeling experiment. Antigen from cells radiolabeled with both ^{14}C-leucine and ^{3}H-fucose was solubilized and reacted in an indirect immunoprecipitation procedure using anti-H-2 alloantiserum and anti-mouse IgG serum. The precipitate was dissolved in SDS (sodium dodecyl sulfate) with mercaptoethanol, and electrophoresed on an SDS-polyacrylamide gel. The amino acid and monosaccharide labels co-electrophoresed in a single band (13). The protein and sugar label could be separated only if the antigen were subjected to exhaustive pronase digestion to reduce the protein to amino acids and peptides, and it was concluded that the NP-40 solubilized material was in fact glycoprotein.

An estimate of the molecular size of the intact H-2 alloantigen solubilized by the detergent NP-40 was obtained from measuring its elution position during molecular seive chromatography on agarose 5M columns in 0.5% NP-40 buffer (13). By comparison to the elution volumes of reference standards, an approximate molecular size of 380,000 daltons was calculated. This figure must be considered only an estimate since it is not known how the interaction of hydrophobic molecules with detergents may affect their elution behavior and hence the molecular weight determination.

The larger apparent molecular weight could be reduced by using the stronger detergent SDS. When SDS alone was used as the dissociating agent, two forms of the antigen were found, with molecular weights of approximately 90,000 and 45,000 daltons. These molecular weights were determined by comparing the migration of radioactively labeled antigen to that of reference standards during SDS-polyacrylamide gel electrophoresis. The larger form, tentatively called a "dimer", was converted into the smaller "monomer" form when reducing agents were present. Of interest is the further finding that the molecular weights for the $\underline{H\text{-}2D^d}$ and the $\underline{H\text{-}2K^d}$ gene products were found to be slightly different, i.e., $\underline{H\text{-}2D^d}$ was about 43,000 molecular weight and $\underline{H\text{-}2K^d}$ was about 47,000 daltons.

The molecular weight estimates obtained by the SDS method can be considered firm since the use of SDS prevents anomalous migration rates by obliterating configurational peculiarities. A possible interaction of carbohydrate with the gel as a cause of anomalous migration rate is unlikely since determinations were carried out in both 5 and 10% gels with similar results.

The finding of a relatively large molecular weight for the antigen in the NP-40 solubilized state is of interest since it suggests that the antigen molecules of 45,000 may be associated into larger complexes *in situ*. These large aggregates may be related to possible functional properties.

The relationship between the molecular weight of the H-2 antigen fragments solubilized by papain, and the antigen molecule solubilized by NP-40 was determined by direct comparison on SDS gels. The papain solubilized fragment migrated more rapidly than the intact molecule, and the difference in molecular weight was calculated to be about 3000-6000 daltons. It can be thus concluded that the minimally cleaved Class I papain fragment represents the major part of the native glycoprotein. Thus the loss of a relatively small portion of the intact antigen confers aqueous solubility on the papain fragment. It seems reasonable to assume that this small region which appears to be antigenically silent is necessary for membrane integration of the molecule.

An additional factor regarding the subunit structure of the H-2 glycoproteins is suggested from recent findings from the HL-A antigen system (14,15,16). An 11,000 M.W. protein, β-2 microglobulin (β-2m) has been shown to be non-covalently associated with the purified HL-A glycoprotein fragment when solubilized by papain digestion. The β-2M is thought to be invariant in primary sequence and bears a small but significant homology to IgG heavy chain domains (17). As its association with the HL-A glycoprotein is non-covalent, it was postulated that the β-2m which carries no alloantigen activity provided a structural function related to binding the antigenically active HL-A glycoprotein in the membrane matrix (15).

In the early studies (11) we noted that upon SDS polyacrylamide electrophoresis of purified H-2 alloantigen fragments solubilized by papain, there was a major protein staining band of approximately 35,000 M.W., and a minor staining band with a M.W. of the order of 10,000 to 15,000. The smaller protein bore no radioactive carbohydrate label and had no antigenic activity if dissociated by mild alkali or acid from the carbohydrate containing larger glycoprotein fragment reported upon earlier in this review (Yamane, K., Shimada, A. and Nathenson, S.G., unpublished observation).

This observed protein may be similar to the β-2m protein found to be associated with the HL-A glycoprotein. While proof of this possibility is at most only suggestive it is tempting to speculate that under certain conditions H-2 may exist as a heterogeneous complex of multiples of two non-covalently bound chains, each of which is genetically determined by a separate gene. The larger glycoprotein chain would represent the polymorphic portion and carry the antigenic sites, while the β-2m molecule would provide an invariant part of the structure. The role of the associated β-2m protein might be to stabilize the H-2 product or in some unknown manner regulate its true physiological function.

B. Carbohydrate composition and structure.

The overall composition of the carbohydrate of proteolytically cleaved H-2 glycoprotein fragments was determined by paper chromatography and by gas/liquid chromatography (11,18). Galactose, glucosamine, mannose, fucose, and sialic acid (mainly N-acetyl neuraminic acid) were present. Galactosamine and glucose were absent.

For preliminary structural studies (19) H-2 glycoproteins were isolated from tumor or spleen cells previously radiolabeled with radioactive monosaccharides. The alloantigens were recovered by specific immune complex formation and the radioactive sugars present in the antigen molecule were analyzed by paper chromatography of a hydrolyzed sample. In studies with different radiolabeled monosaccharides it was confirmed that galactose, glucosamine, fucose and mannose were incorporated into the H-2 glycoproteins of cells from two different strains ($\underline{H\text{-}2^b}$ and $\underline{H\text{-}2^d}$), but that glucose, rhamnose, arabinose and xylose, while incorporated by the cells, were absent from the specific H-2 glycoprotein.

Glycopeptides were generated from intact, detergent solubilized or papain solubilized serologically purified H-2 antigen by exhaustive pronase digestion of the protein backbone. These radioactive glycopeptides chromatographed on Sephadex G-50 columns in a single peak, with an elution position corresponding to about 3300 $\pm$ 500 daltons as estimated by comparison with the behavior of standard glycopeptides. The H-2 glycopeptide, which was estimated to carry at most one or two residues of amino acid and 12-15 monosaccharide residues, was of a unique size in contrast to the heterogeneous array of sizes observed for glycopeptides extracted from crude membrane fractions.

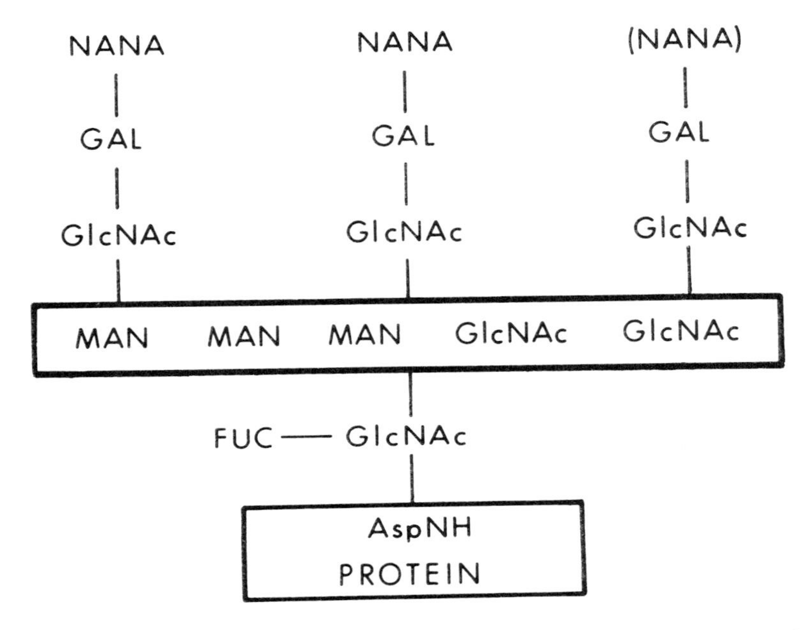

Fig. 1. Hypothetical model of the carbohydrate chain of the H-2 glycoproteins. The evidence for this model is presented in reference 18. Abbreviations are: NANA = sialic acid; GAL = galactose; GlcNAc = N-acetylglucosamine; MAN = mannose; FUC = fucose.

The number of carbohydrate chains per molecule has not yet been established. However, since approximately 10-12% of the weight of the papain solubilized glycoprotein is carbohydrate, one can estimate that at least one, but not more than two, carbohydrate chains are present on each H-2 glycopeptide.

Glycopeptides from different H-2 haplotypes were found to be similar by the criteria of gel filtration and DEAE-Sephadex ion exchange chromatography. This similarity was also observed when the glycopeptides from normal spleen cells or from tumor cells were compared. Thus, easily detectable differences in overall size or charge dependent on the H-2 haplotype or on cell source (normal vs tumor) were not observed.

These results, taken together with other studies reported below, support the contention that the carbohydrate structure does not influence the antigenic properties of the alloantigens.

Enzymatic digestion and limited acid and alkali treatment allowed the construction of a model shown in Fig. 1 for the structure of the carbohydrate chains. This model applies to the structure of both H-2b and H-2d glycopeptides, which have shown almost identical properties in all the studies which we have carried out.

II. The nature of the antigenically active sites of the H-2 glycoproteins.

A. Does the carbohydrate moiety determine antigenic specificity?

If the carbohydrate chains had antigenic activity, one would expect that: 1) glycopeptides isolated from H-2 glycoproteins might have antibody binding ability; 2) protein moieties of the antigen (after removal of some or most of the carbohydrate) might show lack of antibody binding capacity; 3) comparative sequencing of the carbohydrate chains from nonidentical haplotypes might show differences in carbohydrate structure to account for the antigenic differences.

These three approaches have been carried out and the results are consistent with the hypothesis that the carbohydrate itself does not directly express the antigenic specificity. Thus, the isolated glycopeptides of H-2 alloantigens have always been inactive when tested for antigenic activity (18). Enzymatic removal of the carbohydrate, furthermore, was ineffective in destroying antigenic activity. For example, removal of 100% of the sialic acid, 70% of the galactose, and 25% of the glucosamine (or a loss of about 35% of the total carbohydrate) did not produce any change in the H-2 specificities measured. The outer sugars, which were the ones principally affected, are the more likely immunodominant residues (18).

In addition, comparative structural studies on the carbohydrate chains from different H-2 glycopeptides using a relatively crude enzymic digestion method showed identical patterns for release of sugars (18). Such studies, of course, are not quantitative, and more precise studies of sequence and linkage may show differences.

B. Does the protein structure express antigenic specificity?

If the protein portion of the H-2 molecule expresses the antigen activity, one would expect that: 1) protein denaturants might cause loss of antigen activity due to a change in conformation around the antigen sites; 2) protein modification by reagents affecting specific amino acid residues, might cause loss of antigenic activity either of all or of particular antigenic sites; 3) comparative peptide mapping of molecules from nonidentical haplotypes might show unique peptides, since a model in which antigen activity resides in the protein predicts differences in amino acid sequence which might be revealed in a peptide map.

The extreme lability of antigen activity to reagents which alter protein conformation has been demonstrated in early studies on the H-2 alloantigens (20). For example, nearly all antigenic activity is lost after treatment with protein denaturants such as 6M urea, or when the antigen is subjected to extremes of pH (below pH 3 and above pH 10), or to proteolytic digestion. Thus protein conformation must be intact in order for the appropriate antigenic activity to be expressed.

When the H-2 glycoproteins were chemically modified (21) by selective reagents which alter only certain amino acid residues, loss of antigenic specificities was found. When N-acetylimidazole and tetranitromethane were used as modifying reagents, both of which affect tyrosine residues under the conditions used, the altered H-2 alloantigen preparations showed a loss of all tested antigenic specificities.

The use of modification procedures such as reductive methylation which affect amino groups rather than tyrosine residues also produced a loss of antigenic activity. However, only certain antigenic specificities were destroyed. With the H-2d glycoprotein, specificity H-2.4 was almost totally lost, while H-2.31 was nearly completely retained. Such effects were noted under conditions where the 90-95% of the lysine residues were altered to dimethyllysine.

Therefore, the loss of capacity to bind antibody to only some antigenic sites upon alteration of lysine amino groups suggests that for these determinants a lysine with a free epsilon amino group is necessary for antigenic activity. Other antigenic sites, however, are not inactivated by reductive methylation, and apparently do not contain lysine residues in a critical position for their antigenic activity. However, the altered amino group does not necessarily have to be in the antigenic site, but could be located at some distance from the site if it has an effect on the three dimensional conformation of the polypeptide region bearing the determinant.

These results support the basic hypothesis that protein determines antigenic specificity. In addition, the differential susceptibility of some antigenic sites to lysine modification supports the contention that these sites are different from unaffected sites.

Peptide mapping studies of the protein portion of antigenically distinct H-2 glycopeptides showed distinct and reproducible differences between products of different H-2 genes — differences which would be expected to underlie serological differences. Such findings also support the hypothesis that antigenic sites are determined by protein structure. The details of these experiments are discussed in the next section where their immunochemical and genetic implications are explored.

III. Studies in the peptide composition of selected H-2D and H-2K gene products.

An understanding of the primary structure of the H-2 polypeptides will provide concrete information on the fundamental properties of these molecules, including the basis for their membrane affinity, the basis for their expression of antigenic sites, and properties of their genetic determinants. It may be possible to show, for example, the extent of homology between the K and D genes, and to determine whether or not the extensive polymorphism is due to genetic hypervariable regions.

Analysis of the peptides produced by trypsin digestion can provide a qualitative approach to investigation of primary structure. In earlier studies we compared the peptide composition of two papain solubilized purified alloantigen fragments using two dimensional thin layer cellulose chromatography (22). Using cyanogen bromide and trypsin digested materials it was found that approximately 80-85% of the peptides of $H\text{-}2^b$ and $H\text{-}2^d$ materials had similar chromatographic behavior. Clearly 15-20% of the peptides from each strain were found to be unique, and this was a minimum estimate due to the insensitivity of this method.

In our more recent studies (23) we attempted to obtain information on two different questions. First, what are the similarities and differences in peptide composition among molecules determined by alleles of the same gene (e.g. $H\text{-}2K^b$ vs $H\text{-}2K^d$)? Second, what are the similarities and differences in peptide composition between products of the alleles of the H-2K and H-2D genes of a single haplotype (e.g. $H\text{-}2K^d$ vs $H\text{-}2D^d$)? For these studies we utilized ^3H or ^{14}C-arginine or lysine labeled, intact NP-40 solubilized H-2 glycoproteins isolated by indirect immunoprecipitation and purified by SDS Biogel chromatography. Tryptic peptides of products were compared in a double label technique by ion exchange chromatography.

As an example of such a comparison, an ion exchange chromatograph of a mixture of ^{14}C-arginine labeled $H\text{-}2K^b$ peptides (private specificity H-2.33) and ^3H-arginine labeled $H\text{-}2K^d$ (private specificity H-2.31) is shown in

Fig. 2. Approximately 11 peaks from H-2Kb and 15 peaks from H-2Kd antigen are visualized, each peak presumably containing one peptide, or at most, two. Only four of the peptides can be said to coincide on this map, or about 25-35%. Thus approximately 65-75% of the peptides are unique.

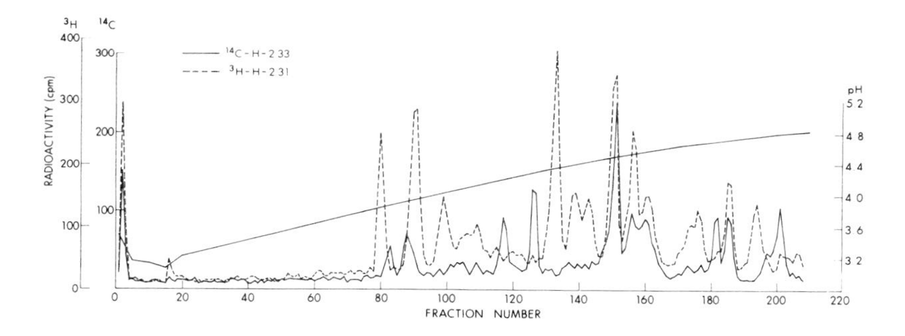

Fig. 2. Ion exchange chromatographic double-label comparison of arginine tryptic peptides of H-$2K^b$ (H-2.33) and H-$2K^d$ (^{14}C-arginine) and H-$2K^d$ (^3H-arginine) were prepared, and fractionated on Bio-Gel 0.5 M columns in 0.5% SDS. H-2 alloantigen peaks were treated with trypsin, and applied to the Spherex Resin XX8-60-0 column (Phoenix Co.) and eluted with a pyridine-acetic buffer with the pH gradient indicated. Full details are presented in reference 23.

Studies of products of alleles as shown in this example (Fig. 2) have been carried out on all H-2D and H-2K products from two strains, H-2^d and H-2^b, for both lysine and arginine labeled peptides. A general summary of the results is as follows: Peptide comparisons of products of alleles of the same gene (e.g. K^b vs K^d, D^b vs D^d) showed about 35% similarities. Peptide comparisons of the products of alleles of H-2K vs H-2D showed somewhat divergent results since K^b vs D^b showed 60% similarity, while K^d vs D^d showed about 40% similarities (23).

These results are most striking since they show an extreme degree of diversity between the products of alleles of the same gene. In fact, differences between products of alleles of D and K genes were somewhat less than the differences of alleles of the same genes. While these are still preliminary studies, and pertain to only two haplotypes of the known 41 haplotypes, the results, if applicable to other strains point out the extraordinary uniqueness of these products.

Of course, the techniques of peptide mapping tend to over-estimate differences in protein structure, since a single amino acid exchange could alter the chromatographic behavior of a peptide consisting of many conserved amino acids. For example, peptide maps of the kappa and lambda light chains can show almost complete nonidentity while sequence studies show 30-40% homology (24).

The finding of considerable uniqueness for the products of alleles of the H-2 genes is possibly not unexpected, since the polymorphism of these genes has been their hallmark. In fact, the complex serological profiles associated with products of different alleles suggests that there must be a complex and variable structural basis. For example, there are five defined public antigenic differences between H-$2K^b$ and H-$2K^d$ including the private specificities H-2.33 and H-2.31 respectively.

The more refined studies, therefore, substantiate our previous findings of peptide differences (22), and hence amino acid sequence differences, between the products of nonidentical H-2 genes. This variation is consistent with the multiple antigenic differences found by H-2 serological testing. It may be that a special genetic mechanism may be involved in generating and/or maintaining such genetic and structural diversity. Further investigations, including amino acid sequence data, are necessary to answer the many questions about the precise chemical features responsible for the biological, immunological, and genetic properties of the H-2 glycoproteins.

MEMBRANE MOLECULES DETERMINED BY THE H-2 ASSOCIATED I REGION: ISOLATION AND SOME PROPERTIES.

The Ia (I associated) antigens (25) have recently been defined (26,27,28,29,30) as a system of cell surface antigens determined by one or more genes of the I region of the H-2 MHC. So far these antigens have been found only on lymphoid cells. As mentioned in the Introduction, the I region is of particular interest because it has genes controlling several immune functions. These include the Ir-1A and Ir-1B genes which control the capacity to respond to certain antigens (7), the gene or genes which control the proliferative response of allogeneic lymphocytes (MLR or mixed lymphocyte response (8)), and the gene which is important in leukemogenesis (9).

In our studies (31) of the Ia antigens, anti-Ia antisera (26) have been used in an indirect immunoprecipitation method described previously for the isolation of detergent solubilized H-2 molecules (12). ^{3}H-leucine labeled spleen cells from an appropriate strain (here B10.S, $H\text{-}2^{s}$) were solubilized with the non-ionic detergent NP-40 and this preparation was reacted with the alloantiserum (A.TL anti-A.TH) and precipitated. Precipitates solubilized in SDS and electrophoresed in SDS on polyacrylamide gels showed two specific radioactive peaks of materials with which the alloantiserum had reacted (Fig. 3c). The controls containing either normal serum or antisera with inappropriate specificity showed no peaks (Figs. 3a,b,d). The migration rates for the two specific peaks with respect to reduced and alkylated H and L chain markers indicated molecular weights of approximately 60,000 and 30,000.

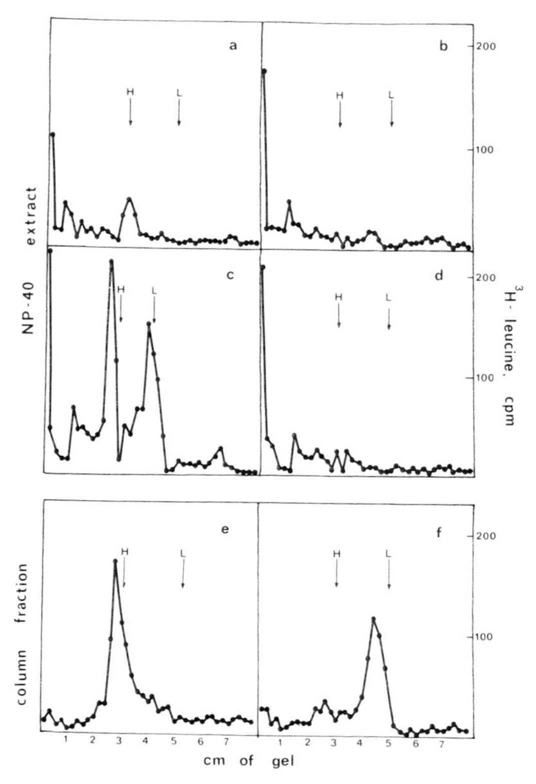

Fig. 3. Ia molecules demonstrated by gel electrophoresis

(a-d) Electrophoresis patterns of indirect precipitates of ^{3}H-leucine labeled B10.S antigen with four mouse sera: a) normal serum; b) A.TH anti-A.TL; c) A.TL anti-A.TH; d) A.TL anti-A.AL. Precipitates were dissolved in 2% SDS and were not reduced. The positions of marker H and L chains indicate that only the serum A.TL anti-A.TH (c) precipitates with molecules showing a molecular weight of 61,000 and 30,000 daltons under these electrophoresis conditions.

(e-f) Results of electrophoresis of unreduced (e) and reduced (f) aliquots of the 61,000 dalton peaks as isolated by column chromatography on an SDS-Bio-Gel 1.5 M column.

When precipitates similar to that shown in Fig. 3c were reduced with 2-mercaptoethanol, only a single peak of molecular weight 30,000 was observed. This suggested a dimer/monomer relationship. Therefore, the 60,000 M.W. species was isolated from an immune precipitate subjected to fractionation on an 0.5% SDS, Biogel 1.5M column. Upon electrophoresis in SDS, the isolated 60,000 molecular weight ran true (Fig. 3e), but addition of 2-mercaptoethanol converted the putative dimer to the 30,000 molecular weight form (Fig. 3f). Subsequently most precipitates were reduced to simplify the electrophoresis patterns.

The Ia antigens were shown to be membrane located by demonstrating that spleen cell membrane preparations contained the 30,000 M.W. radiolabeled molecules reacted with anti-Ia alloantisera, and that these molecules were absent from the cell sap.

Evidence that Ia antigens were glycoproteins came from experiments in which spleen cell extracts were labeled with radioactive fucose, galactose, or mannose. All three sugars were incorporated into materials which precipitated with specific antiserum and electrophoresed in the 30,000 M.W. region.

The anti-Ia antisera were often found to precipitate products which upon SDS polyacrylamide electrophoresis showed relatively broad peaks with occasional prominent shoulders. In addition, the anti-Ia antisera showed cross-reactivity with mouse strains differing in genotype. Hence, it was suspected that multiple antigenic specificities were involved.

Since the Shreffler group (26) demonstrated by absorption techniques that a single antiserum could contain several different antibody populations, we examined the possibility that more than a single antigen molecule could be present in a homozygous animal using the strain B10.HTT. Strain B10.HTT radiolabeled antigen preparations were reacted with the two reciprocal anti-Ia sera, A.TL anti-A.TH and A.TH anti-A.TL. Three different indirect precipitations were carried out using these two antisera and a normal mouse serum control with B10.HTT antigen. After removal of the precipitates, the supernatant materials were tested by indirect precipitation for the capacity to react with the same two antisera and normal mouse serum (cf. Fig. 2, reference 31). The supernatant from the A.TH anti-A.TL reaction showed no further reactivity with the A.TH anti-A.TL antiserum, showing complete clearing of these Ia molecules. However, when the same supernatant was tested with A.TL anti-A.TH, a positive reaction was found. In the converse experiment, an A.TH anti-A.TL serum was positive for molecules not removed by previous precipitation with A.TL anti-A.TH antiserum. These experiments thus show that these antisera reacted with separate molecules.

Thus, the Ia molecules which were isolated by the use of the indirect immunoprecipitation method were found to be carbohydrate-containing proteins of approximately 30,000 M.W. A larger size (60,000) was found when precipitates were not reduced. The Ia molecules were differentiable from H-2 molecules in terms of size, and were separable from them by serological methods. Since molecules bearing different Ia specificities were found to be separable in at least one strain, it is possible that the _I_ region contains multiple genes controlling these antigens.

REFERENCES

(1) G.D. Snell and J.H. Stimpfling, in: Biology of the Laboratory Mouse, ed. E. Green (McGraw-Hill, Inc., New York, 1966), pp. 457-491.

(2) P. Ivanyi. Curr. Top. Microbiol. Immunol. 53 (1970) 1-90.

(3) J. Klein and D.C. Shreffler. Transp. Rev. 6 (1971) 3-29.

(4) J. Klein, P. Demant, H. Festenstein, H.O. McDevitt, D.C. Shreffler, G.D. Snell and J.H. Stimpfling. Immunogenetics (1974). In press.

(5) H.C. Passmore and D.C. Shreffler. Genetics 60 (1968) 210-211.

(6) E. Hinzova, P. Demant and P. Ivanyi. Folia Biol. (Praha) 18 (1972) 237-243.

(7) B. Benaceraff and H.O. McDevitt. Science 175 (1972) 273-279.

(8) P. Demant. Transp. Rev. 15 (1973) 164-200.

(9) F. Lilly. J. Natl. Can. Inst. 49 (1973) 927-934.

(10) J. Klein. Transplantation 13 (1972) 291-299.

(11) A. Shimada and S.G. Nathenson. Biochemistry 8 (1969) 4048-4062.

(12) B.D. Schwartz and S.G. Nathenson. J. Immunol. 107 (1971) 1363-1367.

(13) B.D. Schwartz, K. Kato, S.E. Cullen and S.G. Nathenson. Biochemistry 12 (1973) 2157-2164.

(14) P.A. Peterson, L. Rask and J.B. Lindholm. Proc. Natl. Acad. Sci. (1974), in press.

(15) H.M. Grey, R.T. Kubo, S.M. Colon, M.D. Poulik, P. Cresswell, T. Springer, M. Turner and J.L. Strominger. J. Exp. Med. (1974), in press.

(16) K. Nakamuro, N. Tanigaki and D. Pressman. Proc. Natl. Acad. Sci. 70 (1973) 2863-2865.

(17) P.A. Peterson, B.A. Cunningham, I. Berggard and G.M. Edelman. Proc. Natl. Acad. Sci. 69 (1972) 1697-1701.

(18) S. G. Nathenson and T. Muramatsu, in: Glycoproteins of Blood Cells, eds. G.A. Jamieson and T.J. Greenwalt (Lippincott Co., Philadelphia, 1971).

(19) T. Muramatsu and S.G. Nathenson. Biochemistry 9 (1970) 4875-4883.

(20) A.A. Kandutsch and V. Reinert-Wenck. J. Exp. Med. 105 (1959) 125-139.

(21) S. Pancake and S.G. Nathenson. J. Immunol. 111 (1973) 1086-1092.

(22) A. Shimada, K. Yamane and S.G. Nathenson. Proc. Natl. Acad. Sci. 65 (1970) 691-696.

(23) L.J. Brown, K. Kato, J. Silver and S.G. Nathenson. (1974) submitted for publication.

(24) W.C. Barker, P.J. McLoughlin and M.O. Dayhoff, in: Atlas of Protein Sequence and Structure, ed. M.O. Dayhoff (Academic Press, New York, 1972).

(25) D.C. Shreffler, C. David, D. Gotze, J. Klein, H.O. McDevitt and D. Sachs. Immunogenetics (1974), in press.

(26) C.S. David, D.C. Shreffler and J.A. Frelinger. Proc. Natl. Acad. Sci. 70 (1973) 2509-2514.

(27) V. Hauptfeld, D. Klein and J. Klein. Science 181 (1973) 167-168.

(28) D. Gotze, R.A. Reisfeld and J. Klein. J. Exp. Med. 138 (1973) 1003-1008.

(29) D.H. Sachs and J.L. Cone. J. Exp. Med. (1974), in press.

(30) G.J. Hammerling, B.D. Deak, G. Mauve, V. Hammerling and H.O. McDevitt. Immunogenetics (1974), in press.

(31) S.E. Cullen, C.S. David, D.C. Shreffler and S.G. Nathenson. Proc. Natl. Acad. Sci. (1974), in press.

DISCUSSION

S. ROSENBERG: In view of Shreffler's demonstration of the existence of gene products of the IR gene on the cell surface, have you tested by immunoprecipitation, or any other methods, whether the H-2K gene products and the IR-gene products are associated in one molecule?

S. NATHENSON: We have not tested this and I don't know if anyone else has; it's an interesting thought and certainly quite relevant.

S. ROSENBERG: Would you care to comment on whether it is the IR gene product or the H2 gene product which is important in transplantation in the mouse?

S. NATHENSON: It's very complicated, but it is thought that the H-2K and H-2D gene products should be present in order to get transplantation rejection. There may be another gene in the I region which itself is a transplantation antigen. On the other hand, there are many genes in this region which are not involved in transplantation rejection; they are involved in other responses. That's all I can say without going into all the biology of it.

M. HOROWITZ: Is transplantation rejection in the mouse system based largely on humoral antibodies or cell-bound antibodies? If it is based on cell-bound antibodies, then one could not eliminate the carbohydrates as possible antigenic determinants involved in transplantation rejection because only the soluble antibody system was used to test the modified glycoproteins.

S. NATHENSON: Well, again, the biology of the situation is very complicated; it is certainly clear that cell-mediated immunity is involved in transplantation rejection, and it is clear that we are not directly measuring transplant rejection when we are doing these studies using the humoral antibody. The humoral antibody just goes along with the cell-mediated immunity, so I really can't answer the question beyond that; carbohydrate may be involved in some way.

W.W. FULLERTON: You said that the two monomers were similar or identical; have you taken either of the monomers and tested them for antigenicity?

S. NATHENSON: We haven't done that directly because to make the monomer you have to reduce, alkylate and run on SDS gels, and the H-2 glycoprotein loses almost all its activity in the process. M. Hess has reduced and alkylated papain-solubilized fragments and obtained an antigenic monomer of the papain fragment.

CELL-CONTACT AND TRANSFORMATION-INDUCED CHANGES IN THE DYNAMIC ORGANIZATION OF NORMAL AND NEOPLASTIC CELL PLASMA MEMBRANES AND THEIR ROLE IN LECTIN-MEDIATED TOXICITY TOWARD TUMOR CELLS

GARTH L. NICOLSON
Cancer Council and Electron Microscopy Laboratories
The Salk Institute for Biological Studies
San Diego, California 92112

Abstract: Differences in lectin-mediated agglutination and direct cell cytotoxicity were related to the relative mobility of lectin receptors in the fluid membrane environment as indicated by lectin-induced redistribution. Normal 3T3 fibroblasts show a cell contact-dependent 2.5-3 fold increase in *Ricinus communis* agglutinin (RCA) receptors, while RCA-mediated 3T3 cell agglutination decreases at contact. Highly agglutinable SV3T3 or 3T12 transformed fibroblasts show decreases in RCA-mediated agglutinability at cell-contact, but not quantitative changes in RCA receptors. The number of RCA receptors on transformed cells remains equivalent to normal confluent 3T3 cells. Using fluorescent-RCA the relative mobility of RCA receptors on 3T3 cells indicated by lectin-induced receptor clustering into "patches" was low compared to the mobility on transformed SV3T3 or 3T12 cells. Cytoplasmic membrane-associated components (microfilaments/microtubules) may be involved in restricting the movement of lectin receptors on 3T3 cells after cell-contact. Confluent 3T3 cells labeled with ferritin-Con A at 0°C and incubated further at 37°C do not show lectin-induced redistribution in regions where membrane-associated structures are tightly apposed to the cell membrane. SV3T3 cells are more sensitive than 3T3 cells to the toxic action of RCA_{II}, one of the two castor bean lectins.

Cell toxicity is probably due to inhibition of protein synthesis which occurs rapidly in cell free systems, but takes 30-60 min in intact cells. The time lag for inhibition of protein synthesis and the differential sensitivity of SV3T3 cells is probably due to a requirement for clustering and subsequent endocytosis of the RCA_{II} molecules. Experiments with ferritin-RCA_{II} indicated that it binds very rapidly to SV3T3 cells, is clustered on the cell surface, and then endocytosed. Once inside the SV3T3 cells in endocytotic vesicles, it gradually appears in the cytoplasm, probably via a breakdown of the endocytotic vesicles.

INTRODUCTION

Mammalian cells are surrounded by a dynamic fluid plasma membrane that forms the permeability barrier of the cell and is also responsible for a variety of cellular functions such as transport, recognition, movement, etc. (reviews: 1,2). Although the plasma membrane is thought to be basically constructed of an interrupted lipid bilayer (3-12) with certain proteins and glycoproteins tightly interacting with, and intercalated to various degrees into, the bilayer (1,2,13-17) (the integral membrane zone [2]), surfaces of mammalian cells are more complexly organized and even controlled by structures within the cell cytoplasm (2). This control is mediated, in part, through peripheral membrane components that are attached to the inner surface of the integral membrane zone of the plasma membrane, probably by non-hydrophobic forces. The peripheral components (mainly proteins) form the peripheral membrane zone on the cytoplasmic side of the membrane (2) (Fig. 1), although peripheral membrane zones may exist on both sides of the integral membrane zone. In the cell cytoplasm there are, additionally, components that interact transiently with the cytoplasmic side of the plasma membrane at either the integral or peripheral organizational level. These are the membrane-associated components (18) that may be microfilaments, microtubules or other contractile proteins. These components form the membrane-associated zone (2) of the plasma membrane, and their association with the membrane depends on cell-density or contact, cell cycle, cell energy levels, etc. This

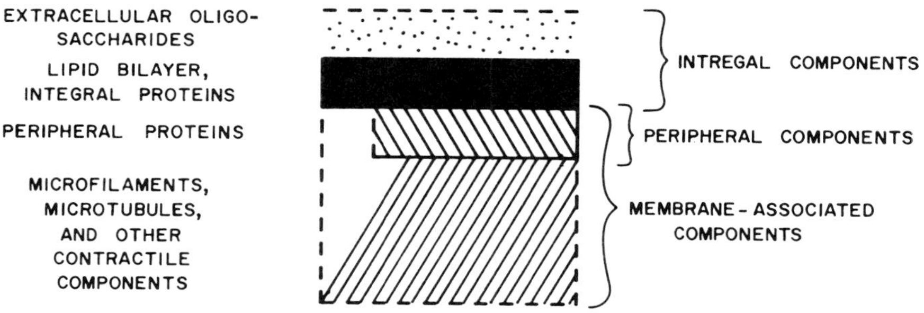

Fig. 1. Levels of plasma membrane organization. See text for details.

zone is probably involved in controlling cell shape and movement, cell-contact, adhesion, endocytosis and other phenomena as well.

Mobility of Cell Surface Components in the Membrane Plane. Several laboratories have now documented that certain membrane components are capable of rapid lateral diffusion in the membrane plane, while other components are not. Membrane phospholipids belong to the highly mobile class of components in mammalian plasma membranes and model membrane systems under physiological conditions. Estimates of phospholipid planar diffusion using nuclear magnetic and electron paramagnetic resonance spectroscopy (9-11,19-23) indicate rapid planar motion ($\underline{D} \cong 10^{-8}$ cm sec^{-1}), but little or no "flip-flop" of phospholipids from one side of the membrane to the other (12).

Other components such as membrane glycoproteins are also capable of lateral diffusion, but their rates of movement appear to be much lower than phospholipids. Frye and Edidin (24) found that it takes 30-40 min to completely intermix human and mouse antigens after Sendai virus-induced fusion to form cell heterokaryons. Several investigators have found that cell surface antigens can be aggregated by antibodies into "patches" and "caps" (25-29).

Using micropipette techniques for applying small aliquots of fluorescent-labeled antibody to localized regions of cultured muscle fiber surfaces, Edidin and Fambrough (30) were able to estimate a lateral diffusion constant of $\underline{D} \cong 10^{-9}$ cm sec^{-1} for an uncharacterized muscle glycoprotein antigen. Glycoprotein oligosaccharides can be quickly aggregated by low concentrations of plant lectins on certain cells (31-38), but under similar conditions the same lectins do not aggregate their receptors on other cells (32-36). Thus, cell membranes contain a variety of different components, each with different lateral rates of motion from very fast to very slow (c.f. 19-24 and 25-38). The mobility of each component is related to its molecular structure, and also to the existence of specialized structural restraints that impede lateral motion (2,18,29,34-36). These restraints can be applied from outside the cell (e.g., by cell-to-cell junctions [39-41]) or inside the cell (e.g., by peripheral or membrane-associated proteins [2,42-45]).

Possible Relationship of Receptor Mobility to Lectin Toxicity. Several lectins are notably toxic to mammalian cells, and some of these (among them concanavalin A [Con A] [46,47], P. vulgaris [PHA] [48,49], Robinia pseudoacacia agglutinin [50], one of the Ricinus communis agglutinins [RCA_{II}] [51-53], etc.) differentially kill tumor cells. Con A and RCA_{II} show differential killing properties under conditions where there is little or no difference in the number of lectin receptors on confluently grown transformed cells compared to their normal counterparts (35,54-58; one exception: 59). Differences in the mobility of Con A (33,34,36) and RCA (35) receptors on virus-transformed fibroblasts shown by surface aggregation experiments with fluorescent- and hemocyanin-labeled lectins suggests thtat the subsequent movement of the lectin receptors after lectin binding may be involved in lectin-mediated toxicity (53).

Dynamic Structure of Cell Membrane Is under Cytoplasmic Control. In a modified version (Fig. 2, from ref. 2) of the Fluid Mosaic Membrane Model (1), the bulk lipid is proposed to form the matrix of the integral membrane zone and is in rapid lateral motion. Certain integral proteins and glycoproteins are moving slower than

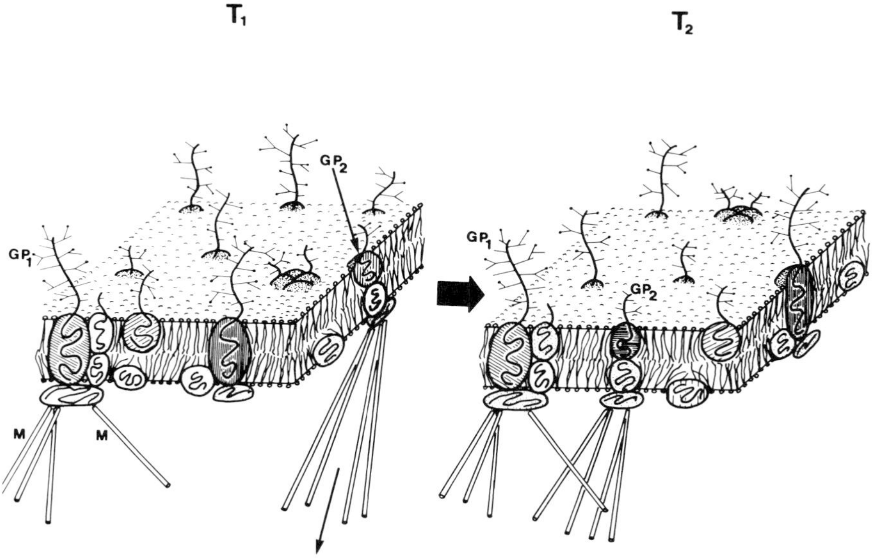

Fig. 2. Modified version of the Fluid Mosaic Model of cell membrane structure (ref. 2). T_1 and T_2 represent different points in time. Certain hypothetical integral membrane glycoprotein components are free to laterally diffuse in the membrane plane formed by a lipid bilayer while others such as the integral glycoprotein-peripheral protein macromolecular complex (GP_1) are impeded by membrane-associated components (M). Under certain conditions some membrane macromolecular complexes (GP_2) can be laterally displaced by membrane-associated contractile components in an energy-dependent process.

the bulk lipids, but considerably faster than other glycoproteins (such as the macromolecular complex GP_1 in Fig. 2) that are relatively "frozen" by cytoplasmic restraints. In the latter case mobility is representationally restricted by the attachment of a cell membrane-associated, microfilament/microtubule system. Under certain conditions membrane-associated microfilament/microtubule contractile systems may laterally displace attached membrane components (such as the macromolecular complex GP_2 in Fig. 2) in an energy-dependent process (2).

The existence of a membrane-associated zone in most mammalian cells is well established (60,61). Membrane-associated contractile components have been implicated in controlling capping of receptors on lymphocytes (25,26,28, 29), cell motility (25,26,60,61), endocytosis (25,26,28, 62) and transport processes (63). Berlin and coworkers (44,45) have shown that the microtubule-disrupting drugs colchicine, colcemid and vinblastine affect the agglutinability of polymorphonuclear lymphocytes and fibroblasts, and it has been shown recently that dibutyryl-cyclic AMP modifies fibroblast agglutinability with wheat germ agglutinin, a process that can be reversed by microtubule-disrupting drugs (64,65).

EXPERIMENTAL

Murine Balb/c 3T3 (A·31) fibroblasts and their SV40 virus-transformed derivative, SV3T3, were obtained from Dr. S. Aaronson (N.I.H.) and spontaneously-transformed 3T12 fibroblasts from Dr. S. Roth (Johns Hopkins University). Con A, RCA_I and RCA_{II} were purified according to Agrawal and Goldstein (66) and Nicolson and Blaustein (67), respectively. Fluorescent- and ferritin-labeled lectins were synthesized and purified as described (34,35,68,69). Lectins were radioisotope-labeled with ^{125}I as in previous studies (58,70). Cell agglutination was performed in small wells on a rotary table and scored visually after 15 or 30 min (35,58,70).

Labeling procedures using short incubations at 4°C with ^{125}I-Con A, ^{125}I-RCA_I and ^{125}I-RCA_{II} are described elsewhere (35,53,58,70). Fluorescent-RCA_I labeling was performed according to ref. 34-35. Ferritin-Con A and ferritin-RCA_{II} labeling were performed as follows: Cells were labeled _in situ_ or in suspension with the ferritin conjugates (1-4 mg/ml protein) at 0°C for ∿15 min. The samples were washed once, and then incubated further at 0°C or at 37°C for 15-20 min prior to fixation with 1% buffered glutaraldehyde. The labeled, fixed samples were post-fixed in 1% buffered osmium tetroxide and embedded in 'Epon 812'. Controls for Con A labeling experiments contained α-methyl-D-mannopyranoside (αMM) (66); those for RCA contained β-lactose (67).

RESULTS AND DISCUSSION

Cell Contact-Dependent Changes in Lectin Receptors. Lectin agglutinability changes during cell growth in vitro. Spontaneously-transformed 3T6 cells which have lost density-dependent inhibition of growth are less agglutinable by Con A after cell confluency has been reached (71). Decreases in Con A-mediated agglutination of 3T3 and SV3T3 cells also occur in culture at cell confluency, and it has been noted that the decrease occurs after cell-contact (35). However, a change does not occur in the number of Con A receptors after cell-contact (35). Thus, the difference in Con A agglutinability must be due to a contact-dependent change in the cell surface that reduces the tendency for cell aggregation. As discussed elsewhere (2,35), differences in cell agglutinability can be explained by a number of different surface changes involving receptor distribution, mobility and accessibility, cell deformability and surface structures (microvilli), cell zeta potential, etc.

Similar to Con A, the agglutinability of fibroblasts by RCA_I decreased at cell contact (70). In contrast with Con A receptors, the number of RCA_I receptors increases 2.5-3 times after cell-contact on normal 3T3 cells, but not on SV3T3 or 3T12 cells (70). The increase in RCA_I-binding sites on 3T3 cells at contact occurs concurrently with the decrease in RCA_I agglutinability (Fig. 3) demonstrating a lack of correlation between lectin-mediated agglutination and the total number of cell surface lectin-binding sites. These quantitative changes found in 3T3 cells were independent of plating conditions and serum concentrations suggesting that a cell-cell glycosidase system might be involved (72), but alternative possibilities such as enhanced plasma membrane turnover after cell-contact (73) are just as likely.

Difference in Relative Mobility of Lectin Receptors on Normal and Transformed Cells. Using ferritin-Con A staining of mounted unfixed 3T3 and SV3T3 plasma membranes at 22°C, Con A receptors were found to be more clustered

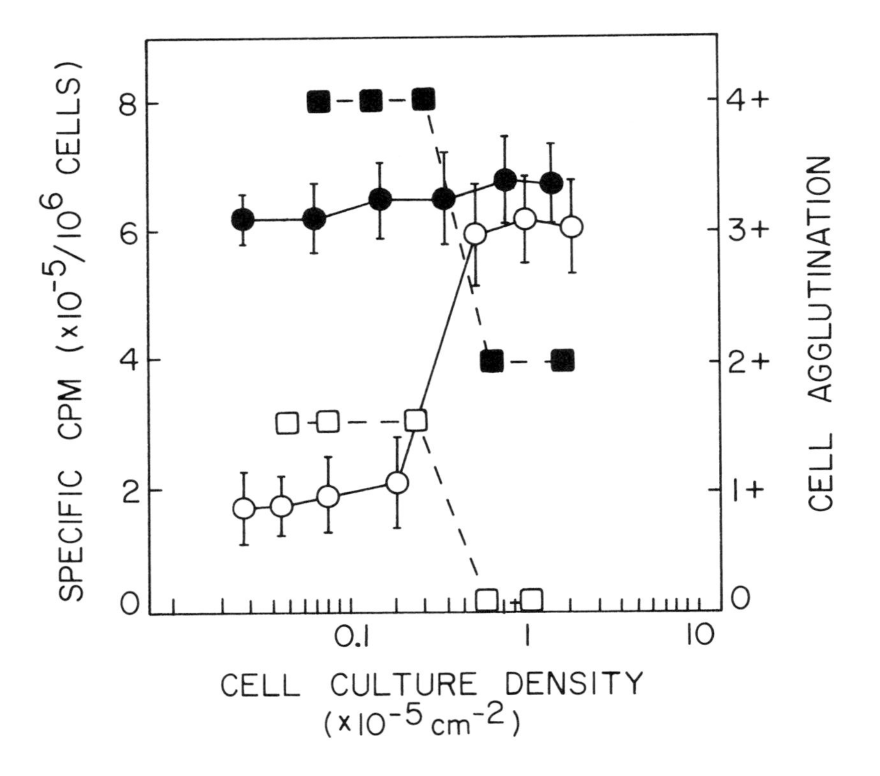

Fig. 3. Relationship of RCA_I agglutinability and number of RCA_I receptors to 3T3 cell density in culture. o——o, quantitative binding of ^{125}I-RCA_I (100 μg/ml) to 3T3 cells; •——•, quantitative binding of ^{125}I-RCA_I (100 μg/ml) to SV3T3 cells; □- - -□, agglutination of 3T3 cells by 3 μg/ml RCA_I; ■- - -■, agglutination of SV3T3 cells by 3 μg/ml RCA_I (data adapted from refs. 35, 70).

on transformed cells correlating with their increased lectin agglutinability (74). This was also found using Con A-peroxide labeling techniques *in situ* (75,75) and for trypsinized normal cells as well. Briefly trypsinized normal cells that are highly lectin agglutinable were labeled with ferritin-Con A, and they showed a more clustered topographic distribution of Con A receptors similar to untrypsinized transformed cells. After 6 hr recovery in fresh culture media, the trypsinized cells regained their normal low agglutinable state and showed dispersed distributions of Con A receptors (32). Direct evidence for the involvement of the clustered lectin receptors in cell agglutination was demonstrated by agglutinating unfixed, intact cells at 22°C with ferritin-Con A and carefully removing the small cell-aggregates by micromanipulation as they formed. The clustered ferritin-Con A was found at the sites of cell-contact in the small cell aggregates (33,35). The ferritin-conjugates were not non-specifically trapped between the cells and could be removed by their specific saccharide inhibitors. This was also shown by agglutinating cells with RCA_I in the presence of ferritin-Con A and its inhibitor, αMM. In either experiment almost all ferritin-lectin molecules were removed from the cells by inhibitory saccharides (33,35). However, these experiments did not properly control for the lectin-induced lateral mobility of the receptors after lectin binding.

Using aldehyde-fixed cells or labeling at low temperature (usually 0°C), it has been shown that the inherent distribution of lectin receptors on normal and transformed cells is essentially dispersed, but the receptors can quickly aggregate by lectin-induced clustering on the transformed cells at or above room temperature (31,33, 35-37, 77). For example, fluorescent-labeled RCA_I molecules bind to SV3T3 cells at 0°C in a dispersed distribution identifiable by a characteristic "ring" fluorescence of the cell (Fig. 4a). However, the uniform fluorescence labeling pattern quickly changes to a "patchy" distribution after raising the temperature to 37°C for 15-20 min (Fig. 4b). The labeling is specific as β-lactose abolishes fluorescent-RCA_I binding to SV3T3 cells (Fig. 4c) and fixation prevents redistribution to a "patchy" fluorescence pattern (35). In our hands normal cells do not show "patchy" fluorescence distributions after the cells have been labeled at 0°C, washed and then further incubated for 15-20 min at 37°C (34,35). This indicates that the <u>relative mobility</u> of these lectin receptors is higher on SV3T3 cells than on normal 3T3 cells. These findings

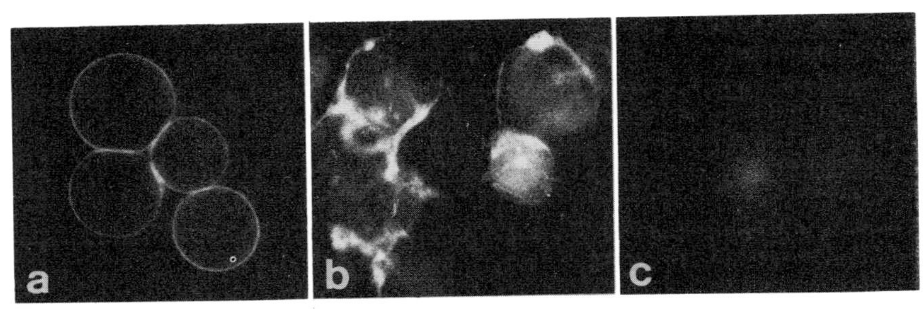

Fig. 4. (a) SV3T3 cells were treated with 1 μg/ml R. communis agglutinin at 0°C for 15 min, washed and then fixed in 2% formaldehyde (first incubation). The cells were incubated for 10 min at 37°C (second incubation) and then labeled with fluorescent anti-RCA for 60 min at 0°C (third incubation) and washed and examined with a light microscope under UV illumination. The cells show uniform or ring fluorescence. (b) Same as in (a) except that the cells were not fixed in 2% formaldehyde until after the second incubation. The cells show patchy or discontinuous fluorescence. (c) Same as in (a) except that concanavalin A (20 μg/ml) was substituted for RCA. Arrow indicates some cell autofluorescence. X 2340. (from ref. 35)

are similar to those for Con A (33,34,36,77), although one report claims that there is essentially no difference in the clustering of Con A receptors on normal and transformed cells (38).

Lectin Receptor Mobility and the Role of Cytoplasmic Membrane-Associated Components. The submembrane surface structure of cells such as normal and transformed fibroblasts changes during growth in culture. Microfilament and microtubule systems are present in sparse fibroblast cultures, but their presence and membrane-association is dramatically increased after cell contact (60,61). The role of microtubules in controlling the expression of Con A receptors and others as well has been demonstrated with drugs such as colchicine and vinblastine (63,63,78). Microfilaments in some systems are sensitive to the drug cytochalasin B (79), and this drug has been reported to affect cell agglutination (80); however, reports to the contrary have appeared (38).

To examine the possibility that the interactions of submembrane microfilaments or microtubules affect the distribution and/or mobility of lectin receptors, we labeled sparse and confluent 3T3 fibroblasts with ferritin-Con A and allowed lectin-induced clustering to occur before fixation. On sparse cells without extensive microfilament/microtubule systems, the relative mobility of lectin receptors indicated by lectin-induced ferritin-Con A clustering was greater than on confluent cells with extensive membrane-associated microfilament/microtubule systems. This was clearly apparent by the lack of ferritin-Con A clustering on the upper surfaces of cells where the microfilament systems were the most extensive (Fig. 5). In other regions of the same cell, ferritin-Con A clustering was noticeable, indicating the mobility of lectin receptors may be different in different regions of the cell surface. This may be another example of transmembrane control of the cell surface by peripheral and membrane-associated components (42,43).

Mechanism of Ricinus communis Agglutinin Toxicity. The toxicity of Ricinus communis proteins to tumor cells is well known (51,52), but its mechanism of action had remained elusive until Lin and his collaborators found that Ricin D, the toxin of R. communis beans, selectively inhibited protein synthesis (52). Later it was found that Ricin D, and a subunit isolated from the molecule, inhibits protein synthesis by blocking peptide elongation on polyribosomes (81-83).

Working with the lectins of R. communis, we have found that RCA_{II}, but not RCA_{I}, inhibits cell protein synthesis in situ, or in cell-free systems (53). Although RCA_{II}, isolated by affinity chromatography (67), was thought to be equivalent to Ricin D isolatable by conventional means (84), this is now under question by Lugnier and Dirheimer (85). Tumor cells such as SV3T3 are more sensitive to the toxic effects of RCA_{II} than are normal cells, at least in the concentration range from 1-100 ng/ml (53). RCA_{II} can be used as a selective agent to obtain cell variants (86), and we have been investigating the properties of such variants.

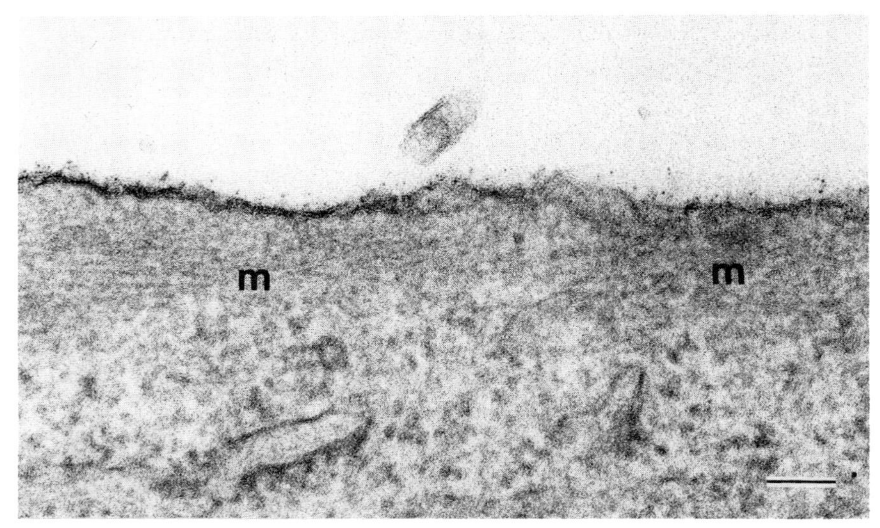

Fig. 5. Confluent 3T3 cells were labeled *in situ* with ferritin-Con A for 15 min at 0°C, washed once, and incubated at 37°C for 15-20 min. Cells were fixed and processed for 'Epon' embedding after staining in block with uranyl acetate. Lectin-induced clustering is not apparent in surface regions with extensive membrane-associated (m) components. Bar equals 0.1 μm.

One important question which we have asked is why are transformed cells such as SV3T3 more sensitive to the toxic effects of RCA_{II}? The difference in sensitivity is not due to a difference in the number of binding sites at saturating concentrations of lectin (58) or at low subsaturating RCA_{II} concentrations (53). There is an apparent difference, however, in the relative mobility of RCA_{I} receptors (see above and ref. 35). Assuming RCA_{II} cell receptors to be similar to RCA_{I} receptors (a very good assumption for cells such as certain murine lymphomas [87]), a difference in toxicity could be related to clustering and subsequent endocytosis of RCA_{II} receptors. This is currently under investigation. Another question that remains is: Does the lectin molecule enter the cell to act directly on the polysomes or does the toxicity occur by some transmembrane phenomenon? By using ferritin-RCA_{II} we have found that the former is more likely as the ferritin-RCA_{II} molecules are fairly quickly aggregated and

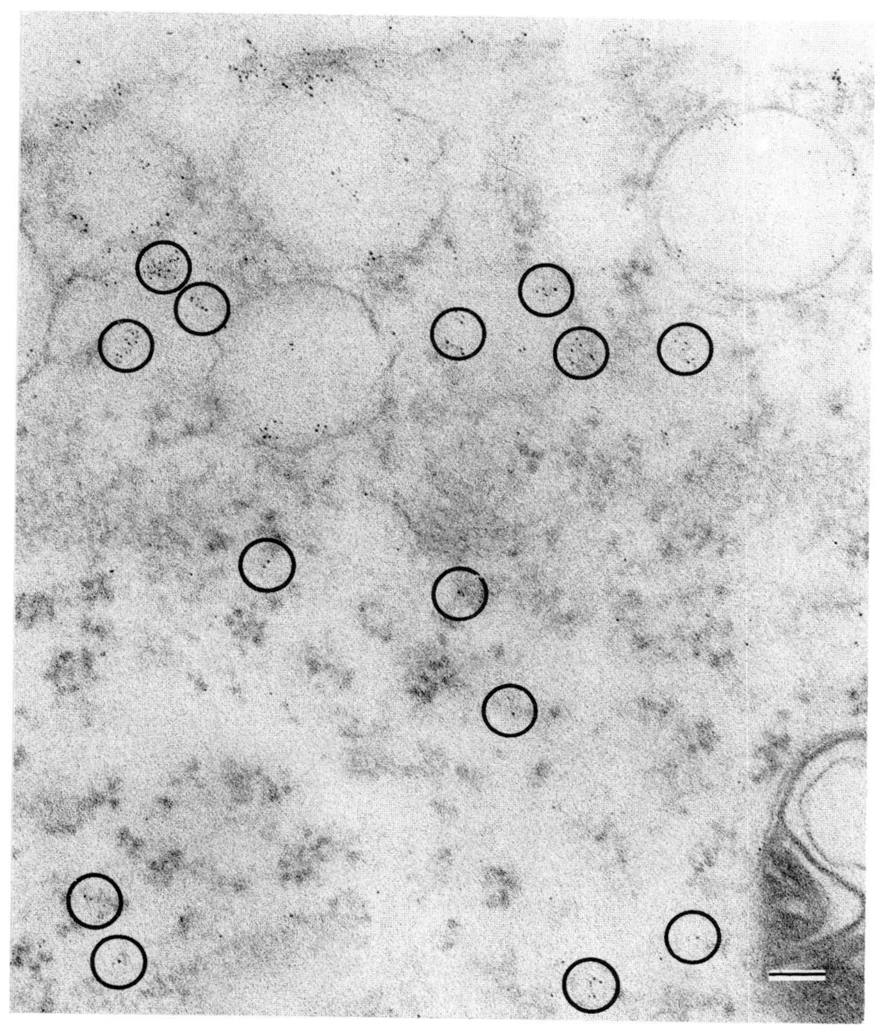

Fig. 6. Endocytosis of ferritin-RCA_{II}. SV40-transformed 3T3 cells were labeled with ferritin-conjugated RCA_{II} for 10 min at 5°C. The cells were washed and incubated for 60 min at 37°C and then fixed and processed for 'Epon' embedding. Ferritin-RCA_{II} molecules have entered the cell by an endocytosis mechanism (10-30 min), but by 60 min some ferritin-lectin molecules are present in the cell cytoplasm (circles). Bar equals 0.1 μm.

endocytosed from SV3T3 cell surfaces (53). The endocytosis occurs within 10-30 min at 37°C, and at that time the ferritin is entirely inside endocytotic vesicles. After 30-60 min, however, some of the endocytotic vesicles appear to rupture and ferritin can be located in the cell cytoplasm (Fig. 6). The time for binding, surface aggregation, endocytosis and release into the cell cytoplasm (30-60 min) correlates with the time required to inactivate protein synthesis in intact cells (∿60 min), but not in cell-free systems (<1-2 min) (53). Cell surface binding occurs very rapidly (<10 min)(53,87), and the toxicity can be completely reversed at this point by saccharide inhibitors suggesting that transport of the toxic molecules inside the sensitive cells is a necessary step in lectin-mediated killing by inhibition of protein synthesis (53).

ACKNOWLEDGMENTS

Supported by a contract from the Tumor Immunology Program of the National Cancer Institute (CB-33879) and grants from the Human Cell Biology Program of the National Science Foundation (GB-34178) and the New York Cancer Research Institute.

[Figure 4 is reproduced by permission of the Cold Spring Harbour Laboratory]

REFERENCES

(1) S.J. Singer and G.L. Nicolson, Science, 1975 (1972) 720.

(2) G.L. Nicolson, Intern. Rev. Cytol. (1974) in press.

(3) J.K. Blasie and C.R. Worthington, J. Mol. Biol., 39 (1969) 417.

(4) D.M. Engelman, J. Mol. Biol., 47 (1970) 115.

(5) A.E. Blaurock, J. Mol. Biol., 56 (1971) 35.

(6) M.H.F. Wilkins, A.E. Blaurock and D.M. Engelman, Nature New Biol., 230 (1971) 72.

(7) J.M. Stein, M.E. Tourtellotte, J.C. Reinert, R.N. McElhaney and R.L. Rader, Proc. Nat. Acad. Sci. USA, 63 (1969) 104.

(8) W.L. Hubbell and H.M. McConnell, Proc. Nat. Acad. Sci. USA, 64 (1969) 20.

(9) A.D. Keith, A.S. Waggoner and O.H. Griffith, Proc. Nat. Acad. Sci. USA, 61 (1970) 819.

(10) M.E. Tourtellotte, D. Branton and A. Keith, Proc. Nat. Acad. Sci. USA, 66 (1970) 909.

(11) R.D. Kornberg and H.M. McConnell, Proc. Nat. Acad. Sci. USA, 68 (1971) 2564.

(12) R.D. Kornberg and H.M. McConnell, Biochemistry, 10 (1971) 1111.

(13) A. Morawiecki, Biochim. Biophys. Acta, 83 (1964) 339.

(14) R.F. Winzler, in: The Red Cell Membrane: Structure and Function, eds. G.A. Jamieson and T.J. Greenwalt (Lippincott Company, Philadelphia, 1969) p. 157.

(15) V.T. Marchesi, T.W. Tillack, R.L. Jackson, J.P. Segrest and R.E. Scott, Proc. Nat. Acad. Sci. USA, 69 (1972) 1445.

(16) J.P. Segrest, I. Kahne, R.L. Jackson and V.T. Marchesi, Arch. Biochem. Biophys., 155 (1973) 167.

(17) A. Ito and R. Sato, J. Biol. Chem., 243 (1968) 4922.

(18) S. Fleischer, W.L. Zahler and H. Ozawa, in: Biomembranes, Vol. 2, ed. L.A. Manson (Plenum Press, New York, 1971) p. 105.

(19) A.G. Lee, N.J.M. Birdsall and J.C. Metcalfe, Biochemistry, 12 (1973) 1650.

(20) C.J. Scandella, P. Devaux and H.M. McConnell, Proc. Nat. Acad. Sci. USA, 69 (1972) 2056.

(21) P.C. Jost, O.H. Griffith, R.A. Capaldi and G. Vanderkooi, Proc. Nat. Acad. Sci. USA, 70 (1973) 480.

(22) P. Devaux and H.M. McConnell, J. Am. Chem. Soc., 94 (1972) 4475.

(23) P. Jost, L.J. Libertini, V.C. Hebert and D.H. Griffith, J. Mol. Biol., 59 (1971) 77.

(24) L.D. Frye and M. Edidin, J. Cell Sci., 7 (1970) 319.

(25) R. Taylor, P. Duffus, M. Raff and S. de Petris, Nature New Biol., 233 (1971) 225.

(26) S. de Petris and M. Raff, Eur. J. Immunol., 2 (1972) 524.

(27) W.C. Davis, Science, 175 (1972) 1006.

(28) M.J. Karnovsky, E.R. Unanue and M.J. Leventhal, J. Exp. Med., 136 (1972) 907.

(29) M. Edidin and A. Weiss, Proc. Nat. Acad. Sci. USA, 69 (1972) 2456.

(30) M. Edidin and D. Fambrough, J. Cell Biol., 47 (1973) 27.

(31) P.M. Comoglio and R. Guglielmone, FEBS Letters, 27 (1972) 256.

(32) G.L. Nicolson, Nature New Biol., 239 (1972) 193.

(33) M. Inbar and L. Sachs, FEBS Letters, 32 (1973) 124.

(34) G.L. Nicolson, Nature New Biol., 243 (1973) 218.

(35) G.L. Nicolson, in: Control of Proliferation in Animal Cells, eds. B. Clarkson and R. Baserga (Cold Spring Harbor Laboratory, New York, 1974) in press.

(36) J.Z. Rosenblith, T.E. Ukena, H.H. Yin, R.D. Berlin and M.J. Karnovsky, Proc. Nat. Acad. Sci. USA, 70 (1973) 1625.

(37) P.M. Comoglio and G. Filogamo, J. Cell Sci., 13 (1973) 415.

(38) S. de Petris, M.C. Raff and L. Mallucci, Nature New Biol., 241 (1973) 257.

(39) J.P. Revel, A.G. Yee and A.J. Hudspeth, Proc. Nat. Acad. Sci. USA, 68 (1971) 2924.

(40) D.A. Goodenough and W. Stoeckenius, J. Cell Biol., 54 (1972) 641.

(41) P. Pinto da Silva and N.B. Gilula, Exptl. Cell Res., 71 (1972) 393.

(42) G.L. Nicolson and R.G. Painter, J. Cell Biol., 59 (1973) 395.

(43) G.L. Nicolson, J. Supramol. Struct., 1 (1973) 410.

(44) R.D. Berlin and T.E. Ukena, Nature New Biol., 238 (1972) 120.

(45) H.H. Yin, T.E. Ukena and R.D. Berlin, Science, 178 (1972) 867.

(46) J. Shoham, M. Inbar and L Sachs, Nature, 227 (1970) 1244.

(47) M. Inbar, H. Ben-Bassat and L. Sachs, Int. J. Cancer, 9 (1972) 143.

(48) M. Gail and C. Boone, Exptl. Cell Res., 70 (1972) 33.

(49) P. Dent, J. Nat. Cancer Inst., 46 (1971) 763.

(50) M. Aubery, J. Font and R. Bourrillon, Exptl. Cell Res., 71 (1971) 59.

(51) J-Y. Lin, K-Y. Tserng, C. Chi-Ching, L-T. Lin and T-C. Tung, Nature, 227 (1970) 292.

(52) J-Y. Lin, K. Lin, C-C. Chen and T-C. Tung, Cancer Res., 31 (1971) 921.

(53) G.L. Nicolson, M. Lacorbiere and T. Hunter, in preparation.

(54) M.J. Cline and D.C. Livingston, Nature New Biol., 232 (1971) 155.

(55) B. Ozanne and J. Sambrook, Nature New Biol., 232 (1971) 156.

(56) D.J. Arndt-Jovin and P. Berg, Virology, 8 (1971) 716.

(57) M. Inbar, H. Ben-Bassat and L. Sachs, Proc. Nat. Acad. Sci. USA, 68 (1971) 2748.

(58) G.L. Nicolson, J. Nat. Cancer Inst., 50 (1973) 1451.

(59) K.D. Noonan and M.M. Burger, J. Biol. Chem., 248 (1973) 4286.

(60) N.S. McNutt, L.A. Culp and P.H. Black, J. Cell Biol., 50 (1971) 691.

(61) J.F. Perdue, J. Cell Biol., 58 (1973) 265.

(62) R.D. Berlin, Nature New Biol., 235 (1972) 44.

(63) T.E. Ukena and R.D. Berlin, J. Exp. Med., 136 (1972) 1.

(64) A.W. Hsie and T.T. Puck, Proc. Nat. Acad. Sci. USA, 68 (1971) 358.

(65) A.W. Hsie, C. Jones and T.T. Puck, Proc. Nat. Acad. Sci. USA, 68 (1971) 1648.

(66) B.B.L. Agrawal and I.J. Goldstein, Biochim. Biophys. Acta, 147 (1967) 267.

(67) G.L. Nicolson and J. Blaustein, Biochim. Biophys. Acta, 266 (1972) 543.

(68) G.L. Nicolson and S.J. Singer, Proc. Nat. Acad. Sci. USA, 68 (1971) 942.

(69) G.L. Nicolson and S.J. Singer, J. Cell Biol., 60 (1974) 236.

(70) G.L. Nicolson and M. Lacorbiere, Proc. Nat. Acad. Sci. USA, 70 (1973) 1672.

(71) M. Goto, K. Katooka, K. Goto and H. Sato, Gann, 63 (1972) 505.

(72) S. Roth and D. White, Proc. Nat. Acad. Sci. USA, 69 (1972) 485.

(73) J.B. Baker and T. Humphreys, Science, 175 (1972) 905.

(74) G.L. Nicolson, Nature New Biol., 233 (1971) 244.

(75) A. Martinez-Palomo, R. Wicker and W. Bernhard, Int. J. Cancer, 9 (1972) 676.

(76) R. Bretton, R. Wicker and W. Bernhard, Int. J. Cancer, 10 (1972) 397.

(77) M. Inbar, H. Ben-Bassat, C. Huet and L. Sachs, Biochim. Biophys. Acta, 311 (1973) 594.

(78) G.M. Edelman, I. Yahara and J.L. Wang, Proc. Nat. Acad. Sci. USA, 70 (1973) 1442.

(79) N.K. Wessells, B.S. Spooner, J.F. Ash, M.O. Bradly, M.A. Lunduena, E.L. Taylor, J.T. Wrenn and K.M. Yamada, Science, 171 (1971) 135.

(80) I. Kaneko, H. Satoh and T. Ukita, Biochem. Biophys. Res. Commun., 50 (1973) 1087.

(81) S. Olsnes and A. Pihl, FEBS Letters, 20 (1972) 327.

(82) S. Olsnes and A. Pihl, FEBS Letters, 28 (1972) 48.

(83) S. Olsnes, R. Heiberg and A. Pihl, Mol. Biol. Reports, 1 (1973) 15.

(84) M. Ishiguro, T. Takahashi, G. Funatsu, K. Hayashi and M. Funatsu, J. Biochem. (Tokyo), 55 (1964) 587.

(85) A. Lugnier and G. Dirheimer, FEBS Letters, 35 (1973) 117.

(86) R. Hyman, M. Lacorbiere, S. Stavarek and G.L. Nicolson, J. Nat. Cancer Inst. (1974) in press.

(87) G.L. Nicolson, J. Blaustein and M.E. Etzler, Biochemistry (1974) in press.

DISCUSSION

S. ROSENBERG: Do microtubule or microfilament blocking agents such as colchicine or cytochalasin inhibit clustering or aggregation of sites in your system?

G.L. NICOLSON: We are doing those experiments right now.

G. KOCH: When you measured inhibition of protein synthesis by incorporation of labelled amino acids, did you examine whether your conditions affect uptake of amino acids by the cell?

G.L. NICOLSON: We checked only leucine, which is the label that we use, and we did not see any differences in initial transport.

M. LUBIN: Did you measure the effect of cell density on agglutination and on binding of the lectin at the same temperature? I know binding was done at 4°.

G.L. NICOLSON: Agglutination was performed at room temperature.

M. LUBIN: Burger and Noonan have reported that binding varies greatly with temperature in certain cells. Thus, it is a point worth checking because you might find quite a different binding characteristic at the agglutination temperature.

G.L. NICOLSON: We haven't done binding at temperatures higher than 4° C, because we checked our experiments with ferritin-conjugated lectins in parallel with I^{125} labelled lectins. If we increased the temperature to 20° or 37° C, endocytosis occurred and it is very difficult to determine what fraction of the extra binding that occurs at room temperature or above is due to this endocytosis. Although Dr. Burger has reported that endocytosis of lectins is completely blocked with azide, we preferred to use labelling procedures that do not permit endocytosis and which show very high labelling specificity. When lectins or antibodies are bound at room temperature or above I am reasonably sure that endocytosis will be a problem.

J. ROTH: Have you done any of the binding experiments with a ferritin labelled-"univalent" con A or ricin?

G.L. NICOLSON: We have tried to make the so called "univalent" con A and we found that it wasn't very univalent. We found that protease treatment of con A produced a population of molecules some of which appear to remain tetravalent and some of which appear to be trivalent, divalent, etc. We also obtained all sorts of fragments which could be seen on SDS-polyacrylamide gels.

J. ROTH: You didn't use the "univalent" lectin with ferritin label to test for cluster formation or endocytosis?

G.L. NICOLSON: No, we have not tried to use the "univalent" lectins in labelling experiments.

M. CZECH: In order to help clarify the role of microfilament structures in agglutinability in the plane of the membrane, have you looked at this phenomena in isolated membranes which may not have these structures?

G.L. NICOLSON: Yes, we saw aggregation in the plane of the membrane in some isolated membranes, but it is very difficult when using isolated membranes to know whether some membrane-associated components are present or not.

A COMPARATIVE STUDY OF SIALIC ACID INCORPORATION INTO ENDOGENOUS ACCEPTORS BY NORMAL AND POLYOMA VIRUS TRANSFORMED HAMSTER CELLS

T. SASAKI* and P.W. ROBBINS
Department of Biology
Massachusetts Institute of Technology

*Present address: Cancer Research Institute, Sapporo Medical College

Abstract: Cultured hamster cells harvested by the use of chelating agents incorporated sialic acid into endogenous acceptors when cells were incubated with CMP-sialic acid in buffered saline. CMP-sialic acid directly participated in the reaction as sialic acid donor. The pretreatment of cells with neuraminidase resulted in a three fold stimulation of the incorporation. One major protein component was observed in the product upon polyacrylamide gel electrophoresis. Ten to twenty percent of the incorporated radioactivity was found in the lipid fraction and tentatively identified as hematoside.

Pronase digestion of the cells yielded sialic acid containing glycopeptides. The product formed by control cells and by polyoma virus transformed cells were compared by Sephadex G-50 chromatography of glycopeptides. When neuraminidase-pretreated cells were used for the incubation, glycopeptides from transformed cells were relatively enriched in high molecular weight components as compared to those from control cells. When cells were incubated without the neuraminidase-pretreatment, a pattern relatively enriched in low molecular weight components was obtained on the G-50 chromatography.

Experiments were performed to define the role of carbohydrate moieties on the cell surface in the density dependent inhibition of cell division. A glycopeptide fraction prepared from the cell surface was found to stimulate the growth of a confluent culture of normal cells.

INTRODUCTION

One of the important properties of transformed cells is an alteration in intercellular interactions (1-3). Carbohydrate containing molecules on the cell surface have been presumed to be involved in intercellular adhesion (4-6), and many comparative studies have been performed to define differences in glycolipids and glycoproteins accompanying transformation (7-13). However, functional change of carbohydrate containing molecules in the membrane will not always require any quantitative or qualitative change of these molecules. Changes in the interaction of these molecules with other membrane components may also result in functional change (14-16).

The incorporation of labeled sugars from corresponding nucleotide sugars into whole animal cells has been studied by Roth et al. (17),

Roth and White (18), and Bosmann (19). Evidence has been presented which indicates that glycosyltransferase reactions can take place on the outer surface of cells. Roth and White (18) found a difference in cell surface glycosyltransferase reactions between untransformed and transformed Balb/c mouse cell lines. Their results indicate that product formation in these reactions will reflect not only the nature of endogenous acceptors and glycosyltransferases but also the mutual interaction of these components within the membrane.

In the present study, the [^{3}H] sialic acid incorporation from CMP-[^{3}H]sialic acid into whole culture hamster cells was studied and the products were partly characterized. The [^{3}H] sialic acid-labeled glycopeptides obtained on exhaustive pronase digestion of the products were examined by chromatography on a Sephadex G-50 column. Comparisons were made between those products formed by control cells and those formed by polyoma virus transformed cells.

EXPERIMENTAL PROCEDURE

Materials CMP-[^{3}H]sialic acid (CMP-[^{3}H]NAN) (25 μCi per μmole) labeled with tritium in the acetyl group was prepared according to the procedure described by Grimes (20). CMP-[^{3}H]NAN was isolated from the reaction mixture first by Dowex-1 column chromatography in the triethylamine bicarbonate system (21), then by the paper chromatography in Solvent B described below. The paper chromatographic purification was repeated. Before the elution of CMP-[^{3}H]NAN with 2 mM Tris-HCl (pH 8.0), the second paper chromatogram was developed in 95% ethanol to remove ammonium acetate from the paper. UDP-D-[^{14}C]glucose (52.5 μCi per μmole) was prepared by the method of Wright and Robbins (22). UDP-N-acetyl-D-[^{14}C] galactosamine (3.3 μCi per μmole) was prepared according to the method of O'Brien (23), and purified by paper chromatography as described by Carminatti and Passeron (24). Contaminating salts in UDP-D-[^{14}C]glucose and UDP-N-acetyl-D-[^{14}C] galactosamine preparations were removed by paper chromatography first in Solvent B, then in 95% ethanol. UDP-N-acetyl-D-1-[^{14}C]glucosamine (42 μCi per μmole) was purchased from New England Nuclear.

Neuraminidase from _Vibrio cholerae_ (500 units per ml) was purchased from Calbiochem. The preparation was essentially free from protease activity when assayed as described below (less than one ng equivalence of pronase per 0.2 ml of the preparation). Pronase (B grade) and venom phosphodiesterase from _Crotalus adamanteus_ were purchased from Calbiochem and Worthington Biochemical Corporation, respectively.

Cell culture The cells used in this experiment were a hamster fibroblast clone, Nil 1Cl, and Nil 1Cl cells transformed with polyoma virus, Nil py1Cl (25). Experiments were performed on cells passaged less than 20 times (maximum of 100 generations) after the establishment of clones. The cells were cultured in a medium composed of Eagle's minimal essential medium with four times the usual concentration of vitamins and amino acids, 10% fetal bovine serum, and penicillin and streptomycin at concentrations of 75 units and 50 μg per ml, respectively. Cells were grown at 37° in Bellco roller bottles with 1410 cm^2 or 840 cm^2 of cell-growing area. Experiments were performed on cells grown for more than three days after transfer.

Buffered salines The following buffered salines were used. Buffered saline A; 0.137 M NaCl, 9 mM potassium phosphate buffer pH 6.5, 0.5 mM $MgCl_2$, 0.5 mM $CaCl_2$, and 10 mM NaN_3. Buffered saline B; 0.137 M NaCl, 5.1 mM KCl, 24.75 mM Trizma base 10 mM $MnCl_2$, 0.5 mM $MgCl_2$, 0.5 mM $CaCl_2$, 10 mM NaN_3, and maleic acid to bring the pH to 6.5. Phosphate buffered saline (pH 7.4): 0.137 M NaCl, 2.68 mM KCl, 8.1 mM Na_2 HPO_4, and 1.47 mM KH_2PO_4. EGTA-solution; above described phosphate buffered saline containing 0.5 mM (ethylene bis (oxyethylenenitrilo)) tetraacetic acid.

Whole cell preparation Whole cells used for the incubation were prepared as follows. After removal of the medium the cell sheet in a large roller bottle was washed twice with 50 ml of phosphate buffered saline. Then 20 ml of EGTA-solution was added and the bottle was rotated at 37° until cells came off the glass. Cells were sedimented by centrifugation at 600xg for 7 min, and washed once with 40 ml of the same buffered saline that was to be used for the incubation procedure. Where indicated cells were pretreated with neuraminidase as follows: cells (6 ~ 20 mg protein per ml) were incubated in buffered saline A containing neuraminidase (40 units per ml of reaction mixture) at 37° for 15 min. After incubation the reaction mixture was centrifuged at 600xg for 7 min. Sedimented cells were washed twice with buffered saline A containing 2 mM EDTA and finally washed once with the buffered saline which was to be used in the incubation for the $[^3H]$ NAN incorporation from CMP-$[^3H]$ NAN. The pretreatment with neuraminidase released 5.2 µg and 2.8 µg of sialic acid per mg of total cellular protein from Nil 1Cl and Nil py1Cl cells, respectively.

Sialyltransferase reaction Standard incubation mixture A consisted of buffered solution A containing 2 mM EDTA, 0.41 mM CMP-$[^3H]$ NAN and whole

Nil 1Cl or Nil py1Cl cells prepared as described above. Standard incubation mixture B was the same as the standard incubation mixture A except that buffered saline B was used instead of buffered saline A and EDTA was omitted. Incubation was carried out at 37° for the indicated period.

In order to measure the [^{3}H] NAN incorporation, 30 volumes of cold 5% TCA were added to the reaction mixture. An appropriate amount of Nil cells was added as carrier. Precipitated material was washed four times with cold 5% TCA, then dissolved in 1 ml of 2% SDS. The suspension was immersed in a boiling water bath for 3 min. and counted in a liquid scintillation spectrometer after the addition of 10 ml of Triton X100-toluene scintillation solution (26).

Preparation of pronase glycopeptides from the [^{3}H] NAN incorporation products In order to obtain glycopeptides of the [^{3}H] NAN incorporation products, the incubation mixture was centrifuged at 1000xg for 10 min. The cell pellet was washed once with buffered saline B without NaN_3, and suspended in 1~2 ml of 0.2 M Tris-HCl (pH 7.8) containing 1.5 mM $CaCl_2$. The suspension was immersed in a boiling water bath for 100 seconds. Then the [^{3}H] NAN incorporation product was digested with pronase for five days at 37° under a layer of toluene. Pronase was added every day for four days at the concentration of 1 mg per ml. On the third day sodium dodecyl sulfate (SDS) was added at the concentration of 0.1%. After the pronase digestion, toluene was removed from the surface and the digested material was centrifuged at 13,000 xg for 40 min. to sediment any insoluble material. The clear supernatant fluid was concentrated to about 0.5 ml for chromatographic analysis on a Sephadex G-50 column.

Sephadex G-50 gel filtration A column of Sephadex G-50 fine (0.8 cm x 96 cm) was equilibrated and eluted with 0.1 M Tris-acetate buffer (pH 9.0), 0.1% SDS, 0.01% EDTA, and 0.1% mercaptoethanol as described by Buck et al. (9). Blue Dextran 2000 and phenol red were included in each sample as excluded and included markers, respectively. Fractions of 0.7 ml were collected at a rate of 3 ml per hour. The whole volume of each fraction was counted in a liquid scintillation spectrometer after the addition of 7 ml of Triton X100-toluene scintillation solution.

Hydrolysis of CMP-[^{3}H] NAN with venom phosphodiesterase To 20 µl of 5 mM CMP-[^{3}H] NAN in 30 mM Tris-HCl (pH 8.0), 1 µl of 0.3 M $MgCl_2$ and one flake of venom phosphodiesterase were added. The mixture was incubated at 37° for two hours in a sealed capillary. Incubation was terminated by immersing the capillary in a boiling water bath for 3 min. Denatured protein was removed by centrifugation. The supernatant was used as the hydrolyzed CMP-[^{3}H] NAN solution after the addition of 3 µl of 0.1 M EDTA. On paper chromatography of the product in Solvent A and B only one radioactive peak was found at the position of authentic N-acetylneuraminic acid.

Assay of protease activity The reaction mixture contained 1.3 ml of phosphate buffered saline (pH 7.4), 5 mg of Azocoll (general proteolytic substrate, Calbiochem), 15 to 100 ng of pronase or sample in a volume of 1.5 ml. A few drops of toluene were added to prevent bacterial growth. After 12 to 48 hours of incubation at 37°, the mixture was filtered and the optical density of the filtrate was determined.

Preparation of glycopeptides from the surface of Nil 1Cl cells Nil 1Cl cells (about 10^9) growing in seven large roller bottles were washed with Tris-buffered saline and treated for one hour with 20 ml per bottle of Tris-buffered saline (pH 7.3), containing 0.5 mM each of $MgCl_2$ and $CaCl_2$, and 0.1% pronase at 37°. Cells were sedimented by centrifugation at 600xg for 10 min. The supernatant fluid was dialyzed against water at 4°C, then freeze-dried. The freeze-dried material was dissolved in 10 ml of 0.2 M Tris-HCl (pH 7.8) containing 1.5 mM $CaCl_2$, and exhaustively digested with pronase at 37° over four days under a layer of toluene. Five mg of pronase was added on the first day, and 3 mg each was added on the second and the third days. The pronase digested material was treated with an equal volume of 50% phenol for 15 min. at 4°C. After centrifugation, the water phase was removed and lyophilized after removal of phenol by ether-extraction. Freeze-dried material was dissolved in 3 ml of water. Thirty ml of absolute ethanol was added to the solution. Precipitated material was collected by centrifugation, dried under the stream of nitrogen and dissolved in 2 ml of the medium without serum. No protease activity was detectable in the preparation when assayed as described above (less than 0.17 μg pronase equivalence per ml). The preparation was sterilized by placing it under a germicidal lamp for 15 min.

Paper chromatography Descending paper chromatography was performed on Schleicher and Schuell orange ribbon 589C paper at room temperature. Solvents used were: Solvent A, butanol-pyridine-water (6:4:3); Solvent B, 1 M ammonium acetate (pH 7.5)-95% ethanol (3:7). N-Acetylneuraminic acid was visualized by Ehrlich's reagent (27). Reducing sugars were located by the silver nitrate-sodium hydroxide method.

<u>Analytical methods</u> Protein was determined according to the method of Lowry et al. (28), with bovine serum albumin as standard. Carbohydrate concentration was determined by the phenol-sulfuric acid reaction (29).

RESULTS

<u>Nan incorporation into whole cells from CMP-NAN</u>

When Nil 1Cl cells, a cloned hamster fibroblast line, were harvested in isotonic EGTA-solution and incubated with CMP-[^{3}H]NAN in isotonic buffered saline containing sodium azide, [^{3}H]NAN was incorporated into endogenous acceptor present in cells (Fig. 1). The incorporation continued for at least four hours at 37°. After five hours of incubation, paper chromatographic analysis of the reaction mixture revealed 79% of the total radioactivity as CMP-[^{3}H]NAN, 19% as [^{3}H]NAN and about 0.5% as product. Incubation with [^{3}H]NAN prepared by the phosphodiesterase treatment of CMP-[^{3}H]NAN resulted in essentially no incorporation (Fig. 1). When unlabeled NAN was added to the reaction mixture in an 8.8 fold excess, no decrease in the [^{3}H]NAN incorporation was observed. The pretreatment of EGTA-harvested cells with neuraminidase from <u>Vibrio cholerae</u> caused an increase in the rate of the NAN incorporation of about three fold (Fig. 1). This stimulation of the incorporation by pretreatment with neuraminidase was specific for the NAN incorporation from CMP-NAN and was not found for N-acetylglucosamine, glucose, galactose, or N-acetylgalactosamine incorporations from their respective nucleotide sugar derivatives (Table 1). As can be seen in Table 1, the presence of Mn^{2+} in the reaction mixture resulted in a higher level of NAN incorporation. When EDTA was included in the reaction mixture containing Mn^{2+} to chelate divalent cations the NAN incorporation

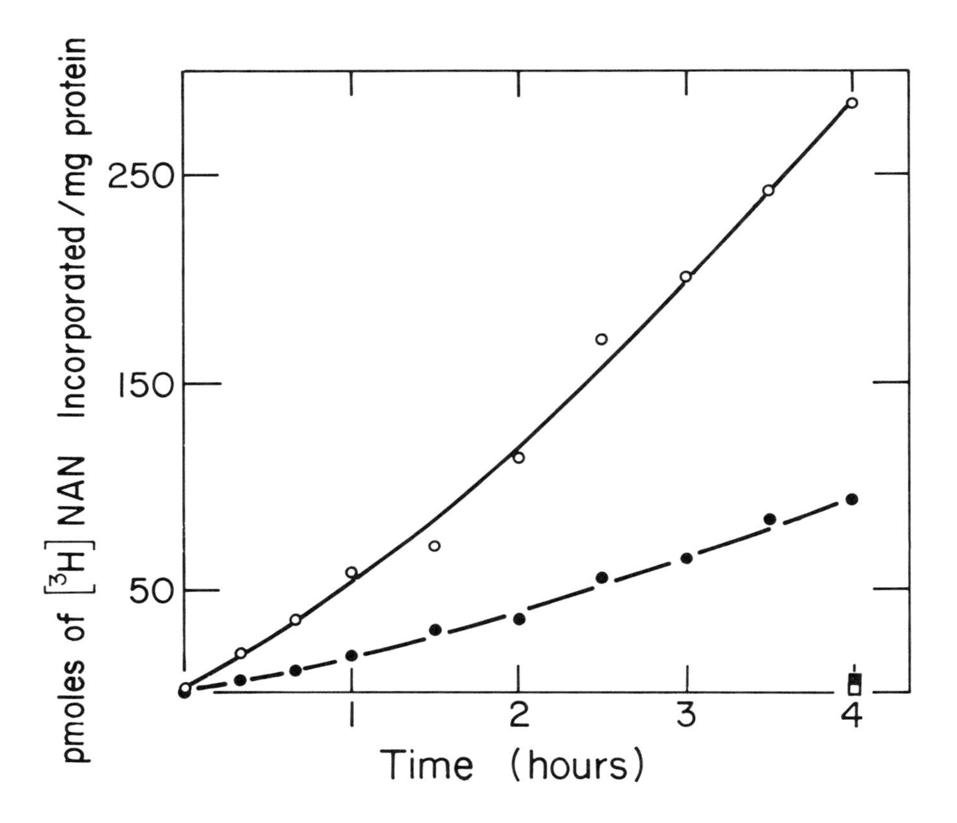

Fig. 1. Time course of NAN incorporation from CMP- ^{3}H NAN into whole Nil 1Cl cells. Incubation was carried out in the standard incubation mixture A with EGTA-harvested Nil 1Cl cells, either with (o———o, and □) or without (●———●, and ■) pretreatment with neuraminidase in a total volume of 60 μl for the time indicated. Whole cells pretreated with neuraminidase (0.577 mg protein) or whole cells without pretreatment (0.923 mg protein) were added to each reaction mixture. The neuraminidase pretreatment and subsequent washing removed less than 5% of the cellular protein. For

the incubations shown by □ and ■, CMP-[^{3}H] NAN pretreated with phosphodiesterase was used. Total incorporation into cold 5% TCA insoluble material was measured as described in experimental procedure.

Table 1

Specificity of Neuraminidase Pretreatment Procedure

Substrate	Pre-incubation	Reaction Mixture	Labeled sugar incorporated pmoles/mg protein
NiL 1Cl cells			
CMP[^{3}H]NAN	none	A	84
	+	A	259
	-	A	104
CMP[^{3}H]NAN	none	B	162
	+	B	417
	-	B	164
UDP[^{14}C] Glc-NAc	none	B	28
	+	B	101
	-	B	71
UDP[^{14}C] Glc + UDP[^{14}C] Gal	none	B	166
	+	B	166
	-	B	154
Nil PylCl cells			
CMP[^{3}H]NAN	none	A	40
	+	A	155
	-	A	52
UDP[^{14}C] Glc-Nac	none	B	55
	+	B	84
	-	B	73

cont. next page

(Cont'd...)

Nil PylCl cells			
UDP[^{14}C]Glc UDP[^{14}C]Gal	none	B	50
	+	B	252
	-	B	223
UDP[^{14}C] Gal-Nac	none	B	77
	+	B	107
	-	B	117

Cells in subconfluent culture were harvested by EGTA-solution as described in experimental procedure. Before washing, cells were divided into three parts. One part was used immediately after washing (preincubation: none). The other portion of cells was preincubated with (+) or without (-) neuraminidase for 15 min. at 37° and washed as described in experimental procedure. Reaction mixture A indicates incubation in the standard incubation mixture A, while reaction mixture B indicates incubation in the standard incubation mixture B. The reaction mixtures for N-acetylglucosamine, glucose-galactose, or N-acetylgalactosamine incorporation contained 17 μM UDP[^{14}C]GlcNAc, 9.1 μM UDP-[^{14}C]Glc and 0.2 unit of UDP-glucose-4-epimerase, or 65 μM UDP-[^{14}C]GalNAc in place of CMP-[^{3}H]NAN. Incubation was carried out at 37° for four hours in a total volume of 60 μl with 160 ~ 300 μg of whole cell protein. Incorporation into the cold 5% TCA insoluble material was measured as described.

was lowered to about half of the level without EDTA (Figure 2). The addition of EDTA to the reaction mixture without Mn^{2+} did not cause any difference in the level of incorporation. As can be seen in Figure 2, the specific activity of NAN incorporation was about two fold higher in control cells than in the transformed counterpart.

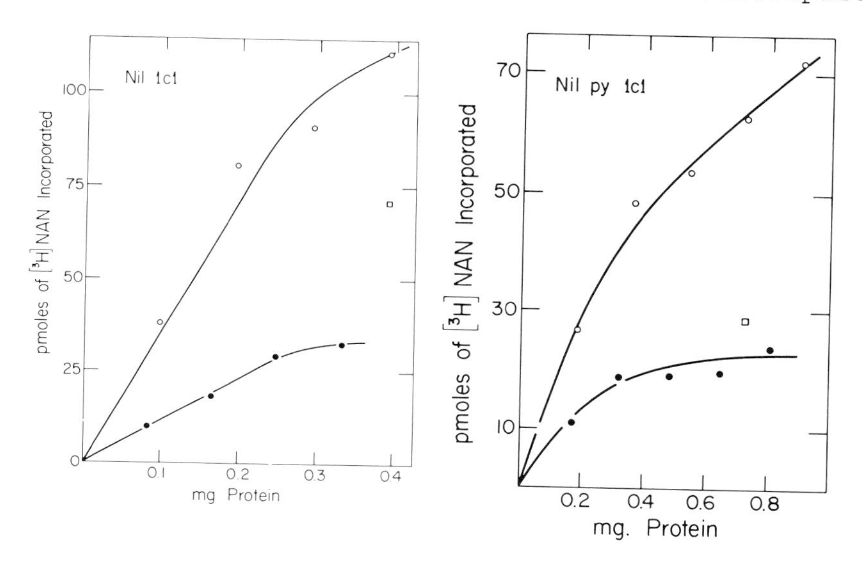

Fig. 2. The dependence of NAN incorporation on the amount of whole Nil 1Cl or Nil py1Cl cells. EGTA-harvested whole cells were prepared from a confluent Nil 1Cl or Nil py1Cl cell culture. Washed cells were divided into two parts. One part was used directly for NAN incorporation (●——●). The other part of the cells was pretreated with neuraminidase (○——○). Reactions were performed at 37° for 3 hours with the indicated amount of whole cells in 60 μl of the standard incubation mixture B except that the CMP-[^{3}H]NAN concentration was 0.27 mM. In the reaction marked □, 12.4 mM EDTA was included with neuraminidase-treated cells. Incorporation into cold 5% TCA insoluble material was measured.

The NAN incorporation was only partly sensitive to pretreatment of cells with trypsin. A pair of Nil-py1Cl cell cultures were prepared. Cells in one culture were harvested with EGTA-solution. Cells in the other culture were harvested with 0.1% trypsin solution. Trypsin inhibitor was added to both harvests. The trypsin harvested cells were preincubated with EGTA solution before use. The trypsin-harvested cells had 37% less activity for NAN incorporation than the EGTA-harvested cells. Trypsinization for 10 or 45 min. gave the same result.

When the NAN incorporation into whole cells and cell homogenates was compared, almost the same initial rate of the activity was found (Table 2). Incubation for longer than one

Table 2

Comparison of Whole Cell and Homogenate Endogenous Incorporation of NAN

Enzyme sources	Addition of Triton X100	NAN incorporation at 30 min.	at 60 min.
		pmoles/mg protein	
whole cells	-	17	29
	+	22	36
cell homogenate	-	20	24
	+	7	12

Whole Nil 1Cl cells were prepared from a confluent culture. Cells were used without the neuraminidase-pretreatment. A portion of cells was homogenized by equilibration for 20 min. at 800 psi in a nitrogen pressure homogenizer followed by rapid return of the cell suspension to atmospheric pressure. Each reaction was performed in 70 μl of the standard incubation mixture B except that the CMP [^{3}H] NAN concentration was 0.61 mM and Triton X100 was included where indicated at the concentration of 0.1%. Whole cells (0.25 mg protein) or the cell homogenate were added to each reaction. The incorporation into cold 5% TCA insoluble material was measured.

hour resulted in a much higher level of the incorporation by whole cells. The addition of Triton X100 at the concentration of 0.1% caused some stimulation in the whole cell reaction and some inhibition in the cell homogenate reaction.

Identification of the incorporated radioactive compound as NAN

The incorporated radioactivity was shown to be completely in NAN residues as follows. After the incubation of whole cells with CMP-[^{3}H] NAN, cells were exhaustively digested with pronase and [^{3}H]-labeled glycopeptides were obtained by chromatography of the digested material on a Bio-gel P-6 column. When a portion of the combined glycopeptide fraction was treated with neuraminidase and the treated sample was chromatographed on paper in Solvent A, all the radioactivity was found at the position of carrier N-acetylneuraminic acid (Fig. 3). When cold 5% trichloroacetic acid insoluble incorporation product from CMP-[^{3}H] NAN was hydrolyzed in 0.1 N sulfuric acid at 80° for one hour, 85% of the radioactivity was found at the position of carrier N-acetylneuraminic acid by paper chromatography and paper electrophoresis (pyridinium acetate-acetic acid, pH6.5, 66 V per cm for 1.6 hours). The 15% of the radioactivity found on paper electrophoresis in faster and slower moving components represents degradation products formed by acid hydrolysis. When cold ethanol-insoluble in a boiling water bath for 3 min. to destroy pronase activity and incubated at 37° for six hours with 13 units of neuraminidase, 17 mM Tris-maleate (pH 6.6), and 1 mM $CaCl_2$ in a total volume of 230 µl. After incubation, the whole reaction mixture was spotted on paper as a 6 cm band with carrier NAN. On a separate paper NAN, glucose, and galactose were spotted. These papers were chromatographed twice with an intermediate drying in Solvent A until the Solvent front reached the tip of the paper. The central 5 mm of the chromatogram was cut out and used to locate carrier NAN. The rest of the chromatogram was cut into 5 mm segments and the radioactivity on these segments was counted in a liquid scintillation spectrometer.

incorporation product from CMP [^{3}H] NAN was treated with neuraminidase, all the radioactivity was found at the position of carrier N-acetylneuraminic acid by paper electrophoresis.

Characterization of the NAN incorporation product

EGTA-harvested, neuraminidase-pretreated whole Nil 1Cl cells (8.8 mg protein) were incubated at 37° for four hours with 0.25 mM CMP- [^{3}H] NAN in buffered saline B in a total volume of 1.2 ml. The whole incubation mixture was homogenized as described in Table 2 and the homogenate was centrifuged at 10^4xg for one hour. The supernatant fluid was concentrated and chromatographed on a Sephadex G-50 column as described in experimental procedure. The [^{3}H] labeled material equivalent to 201 pmoles of [^{3}H] NAN was eluted in the excluded fraction. None of the [^{3}H]-labeled product was found in the included fraction. When the pellet obtained on centrifugation at 10^4xg was exhaustively digested

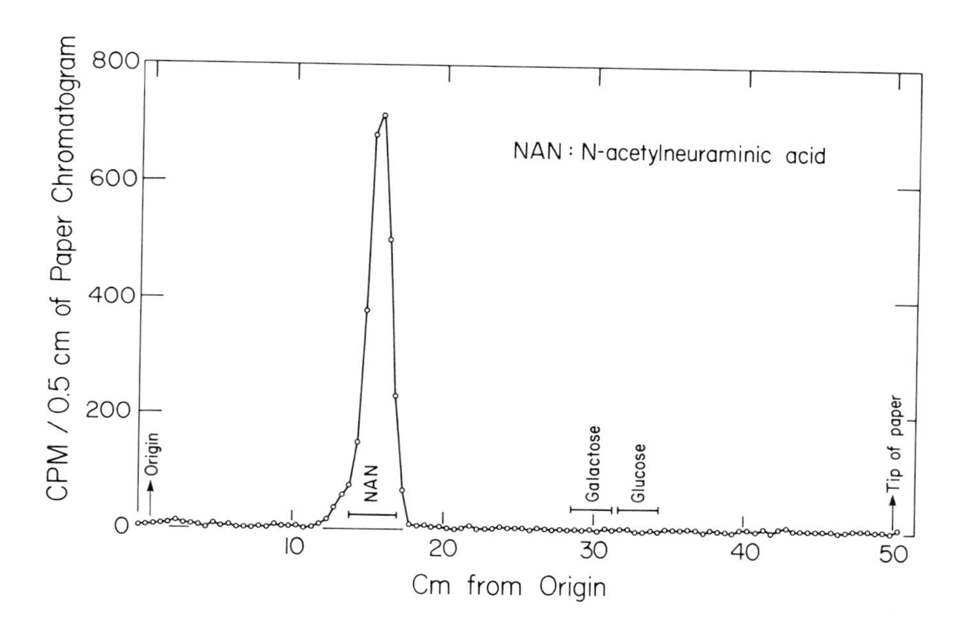

Fig. 3. Release of the radioactivity by neuraminidase from glycopeptides of the incorporation product. Tritium-labeled glycopeptides from the [^{3}H]-NAN incorporation product were obtained as described in Fig.6 by incubation with neuraminidase-treated whole Nil py1C1 cells and pronase digestion. The glycopeptides were separated from residual CMP-[^{3}H] NAN and its degradation products by chromatography on a Biogel P-6 column (0.8 cm x 111 cm, 100 ~ 200 mesh) in water. A portion of the combined [^{3}H]-labeled glycopeptide fraction was immersed

with pronase and the digested material was chromatographed on a Sephadex G-50 column, [^{3}H] labeled glycopeptides equivalent to 425 pmoles of [^{3}H] NAN were found in the included fraction. Thus, on homogenization of the reaction mixture, 68% of the NAN incorporation product was found in the particulate membrane fraction, while 32% of the product was found in the soluble phase.

When [^{3}H] NAN incorporation product from whole cells was solubilized in the sample buffer containing 1% SDS as described by Fairbanks et al. (30) and the solubilized material was analyzed by polyacrylamide gel electrophoresis in SDS, one major peak of [^{3}H]-labeled material and smaller fast moving peaks were found (Fig. 4). The major peak formed by Nil 1Cl cells and Nil pylCl cells had a similar mobility. The product formed by Nil 1Cl cells contained a larger amount of the fast moving components (39% of the total incorporated radioactivity) than the product by Nil pylCl cells. When the solubilized material was chromatographed on a Sephadex G-50 column in the presence of 0.1% SDS as described in experimental procedure, all the product was eluted in the excluded fraction.

When the cold 5% trichloroacetic acid insoluble product of NAN incorporation was extracted with chloroform-methanol (2:1 v/v) 25% of the total radioactivity incorporated by Nil 1Cl cells and 12% of that from Nil pylCl cells was found in the extractable fraction. Eighty percent of the extractable radioactivity was found at the position of hematoside when analyzed by thin layer chromatography on sialic gel plate in chloroform-methanol 2.5 N ammonia (60:35:8, v/v).

Glycopeptides obtained by exhaustive pronase digestion of the NAN incorporation product were

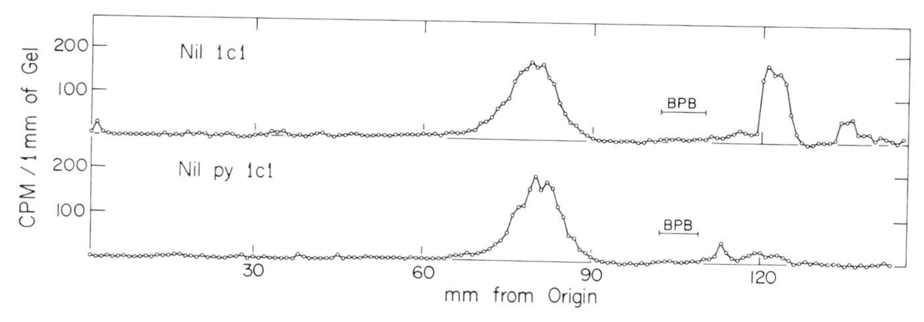

Fig. 4. Polyacrylamide gel electrophoresis in SDS of the [^{3}H] NAN incorporation product. EGTA-harvested whole Nil 1Cl and Nil py1Cl cells were prepared from confluent cultures without pretreatment with neuraminidase. Nil 1Cl cells (2.7 mg protein) or Nil py1Cl cells (10.0 mg protein) were incubated in 0.3 ml of the standard incubation mixture B at 37° for four hours. At the end of the incubation, the reaction mixtures were centrifuged at 700xg for 7

min. and sedimented cells were washed once with buffered saline B containing 3 mM α-toluenesulfonyl fluoride. The cell pellet was dissolved in 0.3-0.4 ml of the sample buffer and electrophoresis was performed in 7.5% polyacrylamide gels in SDS according to the method of Fairbanks et al. (30). Sample (30 ∼ 50 μl) was applied to a 6 mm gel. Bromphenol blue (BPB) was used as a tracking dye.

chromatographed on a Sephadex G-50 column. The pattern of elution of the glycopeptides from the column was compared using glycopeptides from control and from polyoma virus transformed Nil cells. As can be seen in Fig. 5, glycopeptides labeled with [^{3}H] NAN from Nil py1Cl cells were found to be relatively enriched in high molecular weight components when compared to those from Nil 1Cl cells. In these G-50 gel filtrations, CMP-[^{3}H] NAN and its degradation products in the digested material were eluted as a large peak before the phenol red marker, and the position of these materials is indicated by an arrow in each chromatogram. The extraction of lipids from the digested material by chloroform-methanol (2:1, v/v) did not cause any change in the pattern of elution from the G-50 column.

The pattern of elution from a Sephadex G-50 column of [^{3}H] NAN-containing glycopeptides formed in the presence of Mn^{2+} is qualitatively similar to that of glycopeptides formed in the absence of Mn^{2+} (Fig. 6). When sparsely growing Nil 1Cl cells were used for the reaction, [^{3}H] NAN-containing glycopeptides were relatively enriched in the high molecular weight components compared to the glycopeptides from confluent cultures (Fig. 7). The product formed by whole cells

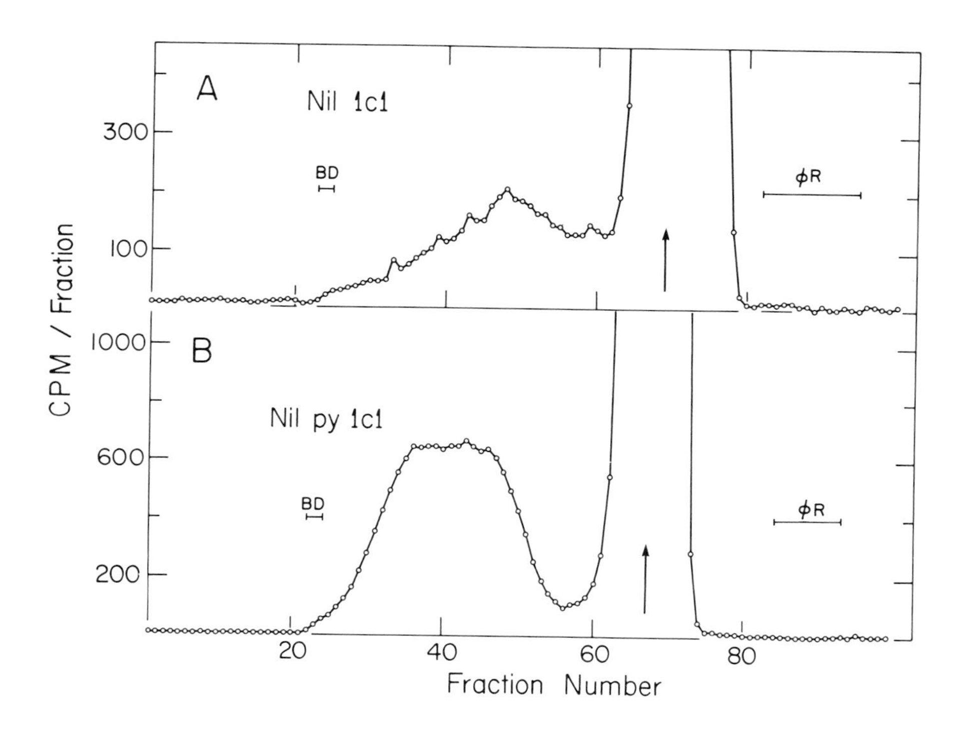

Fig. 5. Sephadex G-50 chromatography of glycopeptides from the [^{3}H] NAN incorporation product formed in the absence of Mn^{2+} by neuraminidase-treated whole Nil 1Cl or Nil py1Cl cells. EGTA-harvested whole cells were prepared from one large roller bottle of a confluent Nil 1Cl cell culture and one small roller bottle of a confluent Nil py1Cl cell culture, and were pretreated with neuraminidase as described in experimental procedure. Nil 1Cl cells (5.66 mg protein)

or Nil py1Cl cells (10.8 mg protein) were incubated in 0.85 ml or 1.0 ml, respectively, of the standard incubation mixture A at 37° for four hours. Pronase glycopeptides of the NAN incorporation product were prepared and chromatographed on a Sephadex G-50 column as described in experimental procedure. ϕR and BD indicate included (phenol red) and excluded (blue dextran 2000) markers, respectively.

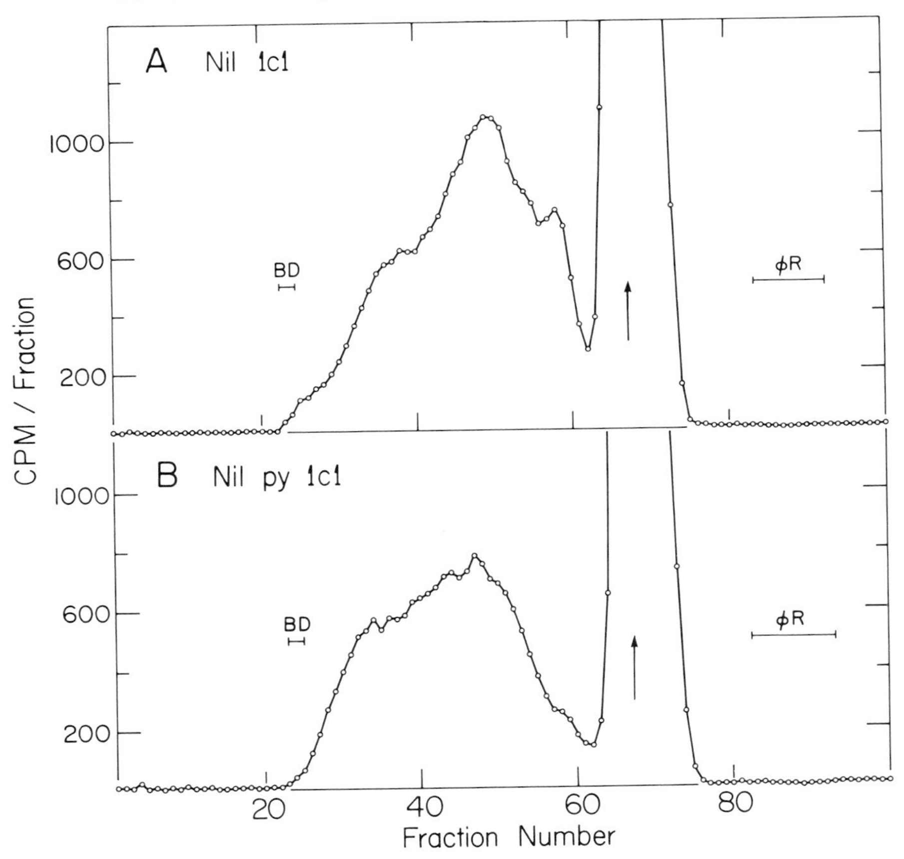

Fig. 6. Sephadex G-50 chromatography of

glycopeptides from the [^{3}H] NAN incorporation product formed in the presence of Mn^{2+} by neuraminidase-treated whole Nil 1Cl or Nil py1Cl cells. EGTA-harvested whole cells were prepared from one large roller bottle of confluent Nil 1Cl cell culture and one small roller bottle of confluent Nil py1Cl cell culture, and were pretreated with neuraminidase. Nil 1Cl cells (3.29 mg protein) or Nil py1Cl cells (4.25 mg protein) were incubated in 0.85 ml of the standard incubation mixture B at 37° for four hours. Pronase glycopeptides of the NAN incorporation product were prepared and chromatographed on a Sephadex G-50 column.

without neuraminidase-pretreatment was different from that obtained from cells pretreated with neuraminidase (Fig. 8). The [^{3}H] NAN-containing glycopeptides from cells without neuraminidase-pretreatment gave a pattern of elution relatively enriched in low molecular weight components.

Release of growth inhibition by cell surface glycopeptides

While we were interested in the possibility that carbohydrate containing molecules on the cell surface might play a role in specific intercellular adhesion (4-6), Baker and Humphreys reported that the immediate effect of the addition of serum to a confluent culture of chick embryo fibroblasts was the disruption of adhesive cell contacts and an increase in the rate of cell movement (31). Studies on the stimulation of cell division by wounding a confluent culture also indicate that the release from cell-to-cell contacts controls the initiation of cell division (3, 32). We wished to test the hypothesis that the "unlocking" of intercellular adhesion between cells by the addition of glycopeptides from the

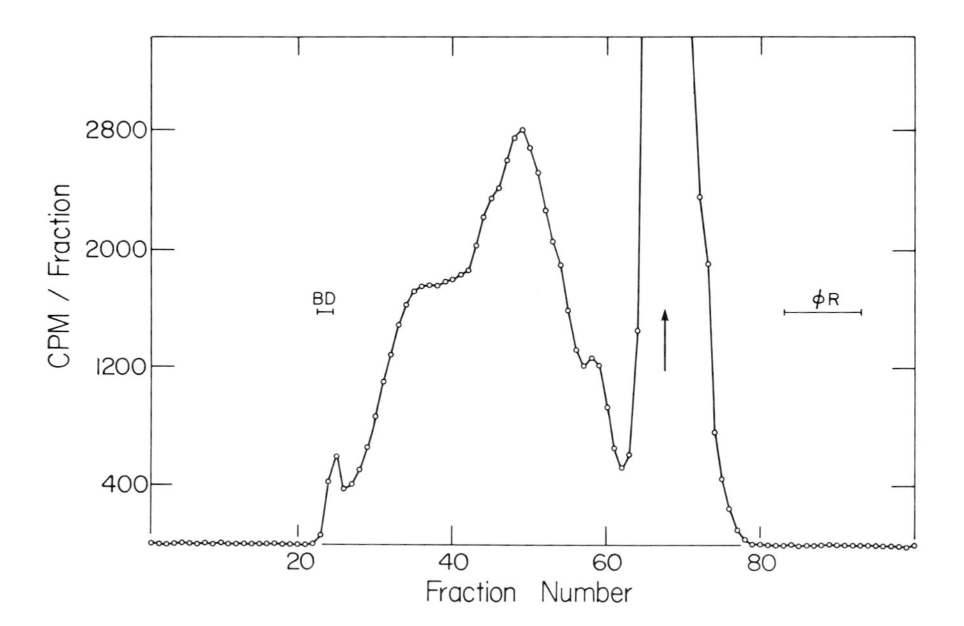

Fig. 7. Sephadex G-50 chromatography of glycopeptides from the [^{3}H] NAN incorporation product formed in the presence of Mn^{2+} by neuraminidase-treated whole Nil 1Cl cells from sparsely growing cultures. EGTA-harvested whole Nil 1Cl cells were prepared from 3 large roller bottles of sparsely growing culture and were pretreated with neuraminidase. Nil 1Cl cells (18 mg protein) were incubated in 0.96 ml of the standard

incubation mixture B at 37° for four hours. Pronase glycopeptides of the NAN incorporation product were prepared and chromatographed on a Sephadex G-50 column.

cell surface might stimulate cell movement and cell division. It seemed possible that glycopeptides from the cell surface might "unlock" intercellular adhesion by a mechanism analogous to the haptenic inhibition of antigen-antibody binding if carbohydrate residues are directly involved in the specific intercellular adhesion. As can be seen in Fig. 9, experimental support was obtained for this hypothesis. When carbohydrate-rich glycoprotein from Nil 1Cl cells prepared according to the method of Marchesi and Andrews (33) was used for the overgrowth experiment instead of glycopeptides, stimulation of cell division was not observed.

DISCUSSION

A difference in the level of the sialyltransferase activity between transformed and untransformed cells was found by Grimes (20, 34). In these studies exogenous acceptors prepared from mucin and serum glycoprotein were used to measure the sialyltransferase activity. Since cultured fibroblasts probably contain several sialyltransferases with different acceptor specificities, the use of these soluble glycoprotein acceptors allows one to study only a part of the cellular sialyltransferases. Warren et al. (35, 36) found a specific sialyltransferase by the use of desialylated glycopeptides prepared from the surface of transformed cells. This sialyltransferase is apparently different from the one which uses desialylated fetuin and bovine submaxillary gland mucin as acceptor.

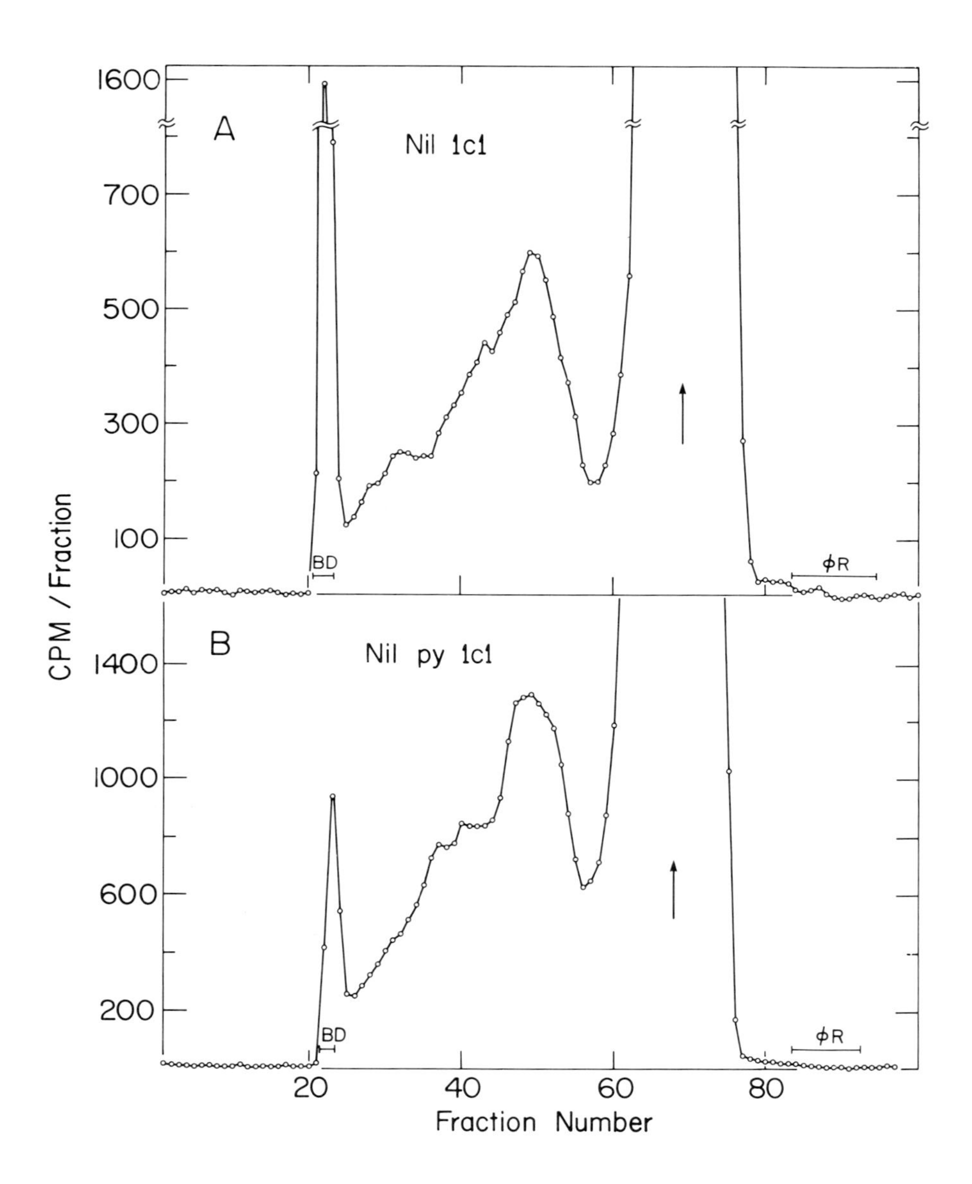

Fig. 8. Sephadex G-50 chromatography of glycopeptides from the [^{3}H] NAN incorporation

product formed in the presence of Mn^{2+} by whole Nil 1Cl or Nil py1Cl cells without neuraminidase-pretreatment. EGTA-harvested whole cells were prepared from two large roller bottles of confluent Nil 1Cl cell culture and one large roller bottle of confluent Nil py1Cl cell culture. Cells were used without pretreatment with neuraminidase. Nil 1Cl cells (33 mg protein) or Nil py1Cl cells (76 mg protein) were incubated in 1.2 ml of the standard incubation mixture B at 37° for four hours. Pronase glycopeptides of the NAN incorporation product were prepared and chromatographed on a Sephadex G-50 column.

In the present study we compared the NAN incorporation from CMP-[^{3}H] NAN into endogenous acceptors by control and polyoma virus transformed cells. When the incubation was performed for more than one hour, the incubation with whole cells resulted in the incorporation of NAN from CMP-NAN into endogenous acceptors with considerably higher efficiency than the incubation with the cell homogenate. Furthermore, the pretreatment of whole cells with neuraminidase stimulated the rate of the NAN incorporation, presumably as a result of increase in acceptor sites for NAN. The pretreatment with neuraminidase not only stimulated the rate of the incorporation but also caused a change in the product formed as shown by the change of the elution pattern of [^{3}H]NAN-containing glycopeptides from a Sephadex G-50 column. When neuraminidase-pretreated cells were used, [^{3}H] NAN-containing glycopeptides of higher molecular weight were formed compared to glycopeptides obtained from cells without the pretreatment.

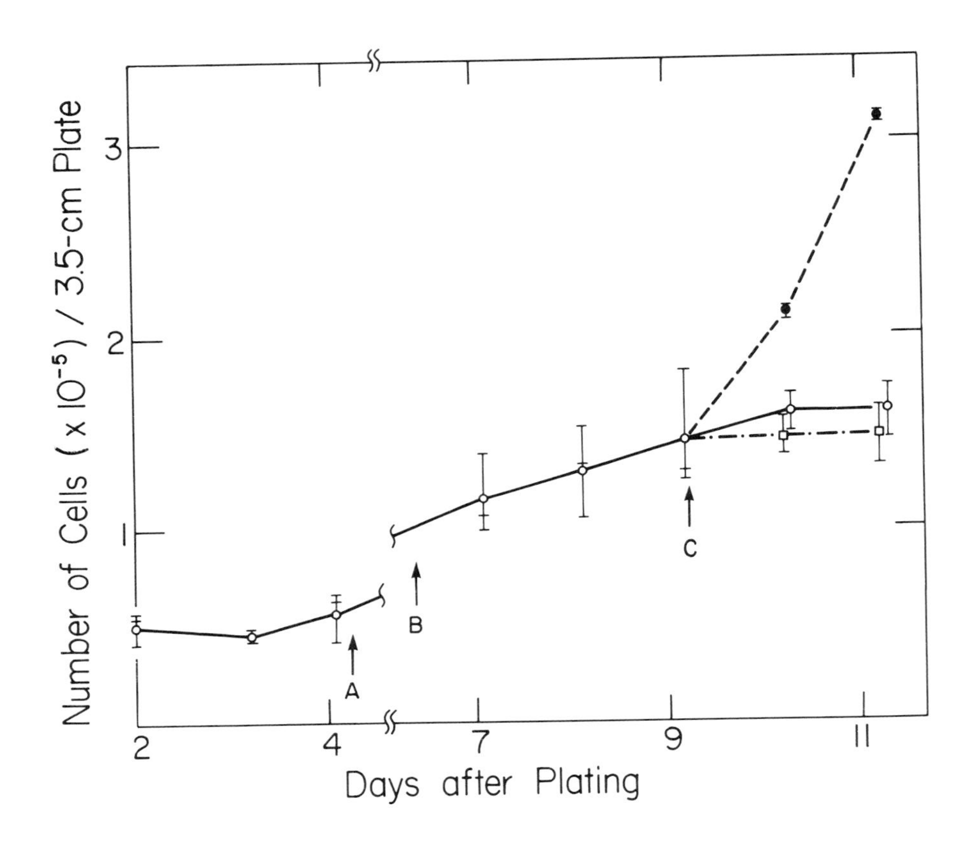

Fig. 9. Release of growth inhibition by cell surface glycopeptides. Nil 1C1 cells were plated in 3.5 cm plastic plates in the medium described in experimental procedure except that the concentration of fetal bovine serum was 3%. The increase in cell number was followed by trypsinization and counting in a hemocytometer. Two or three plates were used for determination at each point. Triplicate counts were made for each plate. Spread in the cell number

obtained from two or three plates is shown by a vertical line at each time point. At the time indicated by arrows A and B, the medium was changed to medium containing 3% fetal bovine serum, which was previously conditioned by supporting the growth of confluent Nil 1Cl cells for one day. The culture became confluent on the seventh day after plating the cells. On the ninth day (indicated by the arrow C), 0.4 ml of the medium without serum (□—·—·—□) or test materials in 0.4 ml of medium without serum were added. The test materials were glycopeptides from the surface of Nil 1Cl cells (●---------●) and control material (○————○). These additions contained carbohydrate at the following concentrations (mg glucose equivalence per ml of solution); medium without serum, 1.05 mg; medium containing glycopeptides, 2.67 mg; medium containing control material, 0.99 mg. Glycopeptides from the surface of Nil 1Cl cells were prepared as described in experimental procedure. The control material was prepared exactly as described for the preparation of glycopeptides from Nil 1Cl cells except that empty roller bottles were used for the pronase treatment instead of roller bottles containing cells.

The absence of NAN incorporation from phosphodiesterase-treated CMP-[^{3}H] NAN and the lack of dilution of the incorporation following addition of excess NAN shows that the incorporation occurred directly from the nucleotide sugar. The effect of neuraminidase-pretreatment on the NAN incorporation suggests that the site of the sialyltransferase reaction was the cell surface. The NAN incorporating activity was only partly reduced by pretreatment of cells with trypsin.

Carbohydrate containing molecules on the cell surface are a likely candidate for involvement in the intercellular recognition and the intercellular adhesion reactions (4-6). Several lines of evidence suggest that abnormality in intercellular adhesion is a basic property of transformed cells. Transformed cells are mutually less adhesive (37). Further, release from cell-to-cell contacts and increase in the rate of cell movement are closely related to the initiation of cell division (3, 31).

Although the active material in the glycopeptide preparation used for the overgrowth experiment was not rigorously characterized, from the method used in preparation it seems not unlikely that glycopeptides caused the overgrowth. One possible explanation of the results is that glycopeptides stimulated cell division by a kind of haptenic inhibition of intercellular adhesion and by stimulation of cell movement. Further work is continuing on the isolation and characterization of the active components.

The abbreviations and trivial names used are: NAN, N-acetylneuraminic acid; EGTA, (ethylene bis (oxyethylenenitrilo)) tetraacetic acid; TCA, trichloroacetic acid; EDTA, (ethylenedinitrilo) tetraacetic acid; SDS, sodium dodecyl sulfate; hematoside, N-acetylneuraminylgalactosylglucosyl ceramide.

REFERENCES

(1) M. Abercrombie and J.E. M. Heaysman. Exp. Cell Res. 6 (1954) 293.

(2) R. Dulbecco. Nature 227 (1970) 802.

(3) R. Dulbecco and M.G.P. Stoker. Proc. Nat. Acad. Sci. US 66 (1970) 204.

(4) S.B. Oppenheimer, M. Edidin, C.W. Orr and S. Roseman. Proc. Nat. Acad. Sci. US 63 (1969) 1395.

(5) S. Roth, E.J. McGuire, and S. Roseman. J. Cell Biol. 51 (1971) 525.

(6) S. Roseman. Chem. Phys. Lipids 5 (1971) 270.

(7) H.C. Wu, E. Meezan, P.H. Black, and P.W. Robbins. Biochemistry 8 (1969) 2509.

(8) E. Meezan, H.C. Wu, P.H. Black, and P.W. Robbins. Biochemistry 8 (1969) 2518.

(9) C.A. Buck, M.C. Glick, and L. Warren. Biochemistry 9 (1970) 4567.

(10) K. Onodera and R. Sheinin. J. Cell Sci. 7 (1970) 337.

(11) S. Hakomori and W.T. Murakami. Proc. Nat. Acad. Sci. US 59 (1968) 254.

(12) P.T. Mora, R.O. Brady, R.M. Bradley, and V.W. McFarland. Proc. Nat. Acad. Sci US 63 (1969) 1290.

(13) P.W. Robbins and I. Macpherson. Nature 229 (1971) 569.

(14) G.L. Nicolson. Nature New Biology 233 (1971) 244.

(15) M. Inbar, H. BenBassat, and L. Sachs. J. Membrane Biol. 6 (1971) 195.

(16) K.J. Isselbacher. Proc. Nat. Acad. Sci. US 69 (1973) 585.

(17) S. Roth, E.J. McGuire, and S. Roseman. J. Cell Biol. 51 (1971) 536.

(18) S. Roth and D. White. Proc. Nat. Acad. Sci. US 69 (1972) 485.

(19) H.B. Bosmann. Biochem. Biophys. Res. Commun. 48 (1972) 523.

(20) W.J. Grimes. Biochemistry 9 (1970) 5083.

(21) D.G. Comb, D.R. Watson, and S.J. Roseman. J. Biol. Chem. 241 (1966) 5637.

(22) A. Wright and P.W. Robbins. Biochim. Biophys. Acta 104 (1965) 594.

(23) P.J. O'Brien. Methods in Enzymol. 8 (1966) 147.

(24) H. Carminatti and S. Passeron. Methods in Enzymol. 8 (1966) 108.

(25) H. Sakiyama, S.K. Gross and P.W. Robbins. Proc. Nat. Acad. Sci. US 69 (1972) 872.

(26) M.S. Patterson and R.C. Green. Anal. Chem. 37 (1965) 854.

(27) E. Svennerholm and L. Svennerholm. Nature 181 (1958) 1154.

(28) O.H. Lowry, N.J. Rosebrough, A.L. Farr, and R.J. Randall. J. Biol. Chem. 193 (1951) 265.

(29) F. Smith and R. Montgomery. Methods of Biochem. Anal. 3 (1956) 182.

(30) G. Fairbanks, T.L. Steck, and D.F.H. Wallach. Biochemistry 10 (1971) 2606.

(31) J.B. Baker and T. Humphreys. Proc. Nat. Acad. Sci. US 68 (1971) 2161.

(32) T. Gurney, Jr. Proc. Nat. Acad. Sci. US 62 (1969) 906.

(33) V.T. Marchesi and E.P. Andrews. Science 174 (1971) 1247.

(34) W.J. Grimes. Biochemistry 12 (1973) 990.

(35) L. Warren, J.P. Fuhrer, and C.A. Buck. Proc. Nat. Acad. Sci. US 69 (1972) 1838.

(36) L. Warren, J.P. Fuhrer, and C.A. Buck. Fed. Proc. 32 (1973) 80.

(37) J.G. Edwards, J.A. Campbell, and J.F. Williams. Nature New Biology 231 (1971) 147.

DISCUSSION

A. VAHERI: In a similar system to Dr. Robbins, involving normal chick embryo fibroblasts and cells transformed with Rous sarcoma virus, we see changes on transformation that may be related to Dr. Robbins observations. Using an immunochemical approach we have detected an antigen complex, specific for fibroblasts, which disappears on transformation. It is an integral component of the membrane of fibroblasts and is composed of at least two polypeptides, one of which has a molecular weight of 170,000 and is glycosylated. It may be that our results are expressions of the general phenomenon that exposed glycoproteins disappear when fibroblasts are transformed by any of several viruses.

POLYPRENOL SUGARS AND GLYCOPROTEIN SYNTHESIS

NICOLAS H. BEHRENS
Instituto de Investigaciones Bioquímicas
"Fundación Campomar" and Facultad de
Ciencias Exactas y Naturales, Buenos Aires, Argentina

The biosynthesis of the carbohydrate chain of glycoproteins has been studied intensively in recent years in many laboratories. Incorporation of labelled sugars "in vivo" or in intact cell preparations, followed by subcellular particle isolation, has resulted in the familiar picture of sequential addition of the sugars. Starting just after or during the last steps of aminoacid polymerization on membrane bound ribosomes, addition of the sugars has been shown to occur first in rough microsomes, then in smooth microsomes and finally in the Golgi bodies (1,2). "In vitro" work carried out with sugar nucleotides as donors, endogenous or added acceptors and tissue homogenate enzyme preparations has confirmed this picture. It is generally believed that the monosaccharides are added sequentially from the corresponding sugar nucleotides and that the structure of the oligosaccharides formed would be dictated by the specificity of the transferases.

Three main kinds of sugar aminoacid bonds have been described, that of xylose or N-acetylgalactosamine to serine or threonine, that of galactose to hydroxylysine and that of N-acetylglucosamine to asparagine. The biosynthetic studies made can be classified appropriately considering these three kinds of bonds.

The glycoproteins in which the oligosaccharide is linked to serine or threonine include the chondroitin sulfates and similar substances frequently called proteoglycans. The polypeptide backbone carries many saccharide side chains with the sequence:

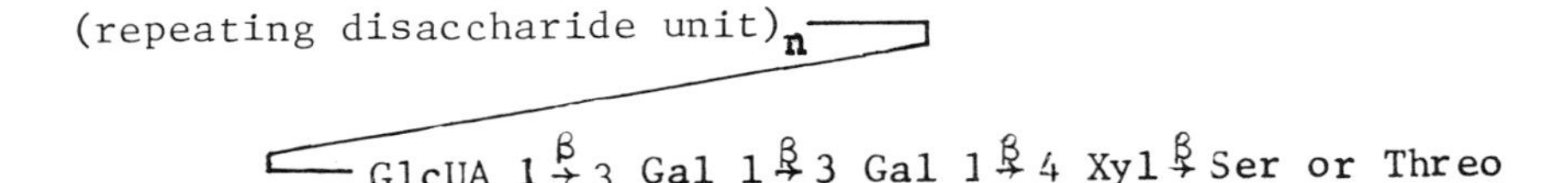

In chondroitin 4-sulfate N-acetylgalactosamine is joined to the glucuronic acid residue and the disaccharide is repeated many times. Transfer of xylose from UDP-xylose to endogenous acceptors has been obtained with hen oviduct, mast cell tumor and chick epiphyseal cartilage (3). Both galactose residues and the first glucuronic acid have been transferred to exogenous acceptors. This study has shown that the reactions are rather specific for the acceptors added (4). Once the linkage region is completed the disaccharide repeating units are formed by alternate addition of N-acetylgalactosamine and glucuronic acid. Sulfate groups are added at a later stage. A similar mechanism is believed to lead to the formation of other proteoglycans.

Another glycoprotein in which the oligosaccharide is joined to serine or threonine is submaxillary mucin. In ovine submaxillary mucin the sugar residue is as follows:

$$\text{NANA } 2 \xrightarrow{\alpha} 6 \text{ GalNAc} \xrightarrow{\alpha} \text{Ser or Threo}$$

The neuraminic acid residue can be removed with neuraminidase and the desialidated compound now serves as acceptor for neuraminic acid from CMP-NANA (5). Similarly the N-acetylgalactosamine residue can also be removed and then the protein serves as acceptor for N-acetylgalactosamine from UDP-N-acetylgalactosamine.

The oligosaccharides of porcine submaxillary mucin have a slightly more complicated structure. The most complex oligosaccharide isolated was the pentasaccharide (5):

$$\begin{array}{lllll} \text{GalNAc } 1 \xrightarrow{\alpha} 3 & \text{Gal } 1 \xrightarrow{\beta} 3 & \text{GalNAc} \rightarrow \text{Ser or Threo} \\ & \uparrow 2 & \uparrow 6 \\ & | 1\ \alpha & | 2\ \alpha \\ & \text{Fuc} & \text{NGNA} \end{array}$$

Another type of glycoprotein which has been

extensively studied from the point of view of structure and biosynthesis is collagen. The sugar residue of collagen and of the kidney glomerular basement membrane is joined to hydroxylysine as follows (6):

$$\mathrm{Glc}\ 1\xrightarrow{\alpha}2\ \ \mathrm{Gal}\xrightarrow{\beta}\mathrm{Hyl}$$

Removal of the glucose residue from collagen by mild acid hydrolysis yields a product which serves as acceptor of glucose from UDP-G. The galactose residue has also been removed, by oxidation with periodate and acid, yielding an acceptor for galactose (3).

There are numerous glycoproteins in which the saccharide is linked to asparagine. They contain mannose, N-acetylglucosamine and may also have galactose, sialic acid and fucose. A typical example is ovalbumin the structure of which is shown in Table 1.

TABLE 1. Structure of some oligosaccharides linked to asparagine in glycoproteins (7).

1. Ovalbumin

$$\begin{array}{l}(\mathrm{GNAc})_{0,1}\ \text{or}\ 2\xrightarrow{\beta}\mathrm{Man}\xrightarrow{\beta}\mathrm{GNAc}\xrightarrow{\beta}\mathrm{GNAc}\xrightarrow{\beta}\mathrm{Asn}\\ \qquad\qquad\qquad\qquad\qquad\qquad\nearrow\alpha\\ (\mathrm{GNAc})_{0}\ \text{or}\ 1\xrightarrow{\beta}\mathrm{Man}\xrightarrow{\alpha}(\mathrm{Man})_{3}\\ \qquad\qquad\qquad\qquad\nearrow\alpha\\ \qquad\qquad\quad(\mathrm{Man})_{0\ \text{or}\ 1}\end{array}$$

2. γG myeloma protein

$$\begin{array}{l}\mathrm{GNAc}\ 1\xrightarrow{\beta}2\ \mathrm{Man}\ 1\xrightarrow{\alpha}6\ \mathrm{Man}\rightarrow\mathrm{GNAc}\xrightarrow{\beta}\mathrm{GNAc}\xrightarrow{\beta}\mathrm{Asn}\\ \qquad\qquad\qquad\qquad\qquad\quad\uparrow 3\qquad\qquad\ \ \uparrow 3\ \text{or}\ 4\\ \qquad\qquad\qquad\qquad\qquad\ \ /1\alpha\qquad\qquad\ \ |\alpha\\ \mathrm{Gal}\ 1\xrightarrow{\beta}6\ \mathrm{GNAc}\ 1\xrightarrow{\beta}2\ \mathrm{Man}\qquad\qquad |1\\ \qquad\qquad\qquad\qquad\qquad\qquad\qquad\quad\ \mathrm{Fuc}\end{array}$$

3. Bromelain

$$\begin{array}{l}\mathrm{Man}\ 1\rightarrow2\ \mathrm{Man}\ 1\rightarrow2\ \text{or}\ 6\ \mathrm{Man}\ 1\xrightarrow{\beta}3?\mathrm{GNAc}\ 1\xrightarrow{\beta}4\ \mathrm{GNAc}\xrightarrow{\beta}\mathrm{Asn}\\ \qquad\qquad\qquad\qquad\nearrow 6\ \text{or}\ 2\qquad\quad\uparrow\\ \qquad\qquad\qquad\mathrm{Fuc}\ 1\qquad\qquad\quad\ \mathrm{Xyl}\end{array}$$

4. α-Amylase

$$\underbrace{\mathrm{Man}\ 1\rightarrow6\ \mathrm{Man}}_{(\mathrm{Man})_4}\ 1\xrightarrow{\beta}4\ \mathrm{GNAc}\ 1\xrightarrow{\beta}4\ \mathrm{GNAc}\xrightarrow{\beta}\mathrm{Asn}$$

Another representative is thyroglobulin which has three kinds of oligosaccharides, two of the asparagine kind and a third where N-acetylgalactosamine is bound to serine (8). Of the oligosaccharides bound to asparagine, one contained only mannose and N-acetylglucosamine while the other contains in addition galactose, sialic acid and further N-acetylglucosamine residues.

The region proximal to asparagine containing mannose and N-acetylglucosamine residues is called the core region (9). In those cases where the sugar sequence in the core region has been determined, one finds various mannose and some N-acetylglucosamine residues bound to a N,N'-diacetylchitobiose unit, that is N-acetylglucosaminyl- β (1→4)-N-acetylglucosamine. This structure is clearly seen in the oligosaccharides shown in Table 1 and it probably also occurs in various other glycoproteins such as fetuin, ribonuclease B and immunoglobulin A and G. Although exceptions have been reported, it might be that some fundamental oligosaccharide like for instance mannose β-bound to N,N'-diacetylchitobiose, is a common structure to all asparagine-linked glycoproteins.

Studies on the biosynthesis of this type of glycoprotein has been limited to the transfer of the external residues such as sialic acid, galactose and N-acetylglucosamine by techniques similar to those described for other glycoproteins (3).

The biosynthesis of the core part of these oligosaccharides has not been elucidated yet. With the previous experience on the other type of glycoproteins and that of the external sugars of these glycoproteins, it might be just a question of time to solve this problem; that is, to find enzymes for the sequential addition of these internal sugars. However, various authors have remarked that this seems to be a somewhat elusive problem, indicating that there might be some unexpected complication here (1,3,10).

We will try to show how work started with a vague plan related somehow to polysaccharide synthesis in animal tissues, resulted in the detection of a series of compounds and reactions which could perfectly explain how the core region of asparagine-linked glycoproteins is formed.

Papers which appeared around the years 1964-1965 came to our attention, I mean those of the groups of Strominger, Robbins, Osborn and Horecker in which lipid derivatives of sugars seemed to be involved in the biosynthesis of bacterial cell wall polysaccharides (11, 12, 13). However, it was not until one of the members of our Institute, Marcelo Dankert, came back from a very productive period of work in Boston, that we became really aware of the possible importance and implication of these studies. He had worked in Phil Robbins' laboratory and they had described at about the same time as Jack Strominger and his group, the existence of lipid intermediates in the biosynthesis of two different cell wall polysaccharides of bacteria. The lipid portion was shown to be undecaprenol, an eleven isoprene unit alcohol (14,15). The phosphate of this compound acted as sugar acceptor from sugar nucleotides, the first sugar being transferred together with its phosphate thus forming undecaprenol pyrophosphate sugar derivatives. After formation of a sugar repeating unit on the pyrophosphate, these compounds interact with polymerization of the repeating unit on one undecaprenol pyrophosphate residue.

Some time before, Morton and Hemming's group in Liverpool had shown that animal and plant tissues contain long chain isoprene alcohols (16). The polyprenol or prenol isolated from mammals was shown to be a mixture of compounds containing from 17 to 22 isoprene units, where the isoprene which carries the alcohol group is saturated (17). The compound is shown in Fig. 1 and was called dolichol. The predominant prenol in rat is dolichol-18 and in pig tissues dolichol-19.

The term dolichol is used now for all those prenols where the α-isoprene is saturated. Dolichols have been found also in various fungi and in yeasts (17), where they play a role in mannoprotein synthesis (18).

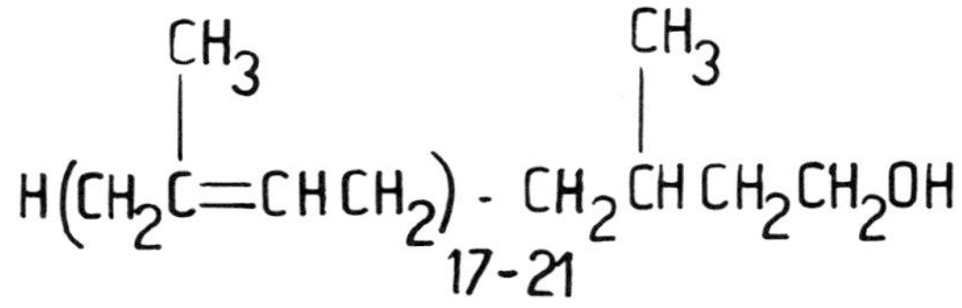

$$H\left(CH_2\overset{\overset{CH_3}{|}}{C}{=}CHCH_2\right)_{17-21}\text{-}CH_2\overset{\overset{CH_3}{|}}{C}HCH_2CH_2OH$$

FIG. 1. Structure of liver dolichol (16).

Dankert convinced us that we should try animal tissues for similar lipid intermediates. We started off with the labelled sugar nucleotide most available to us, uridine diphosphate glucose or UDP-G, and incubated it with a crude microsome preparation from rat liver. After some dubious results, we found a low but reproducible lipid soluble radioactivity which was acid unstable. The water soluble compound liberated was glucose. This acid instability is a very characteristic property of the intermediates in bacteria (14).

The first positive result was so easily obtained, that we thought the structure and the role of the compound would be soon cleared. Both aspects, the demonstration that it is a prenol derivative and its role in the cell have occupied us for the last six years. The difficulties encountered have been such that I think nature played a trick on us luring us into this field with our first experiment.

We tried for some time to obtain the glucolipid in amounts that would permit its chemical detection. For this we purified solvent extracts from liver with radioactive glucolipid added as a marker. However, these efforts never yielded measurable amounts.

Another approach was then sought. It was possible to prepare an enzyme which could be stimulated by the addition of lipid extracts. Fig. 2 shows the

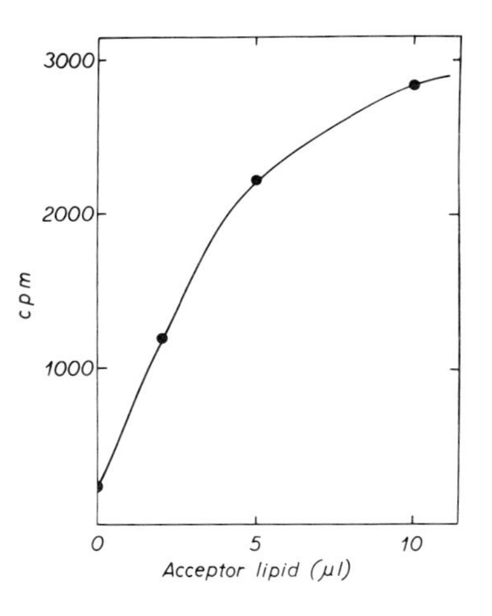

FIG. 2. Stimulation of glucosylation as a function of acceptor lipid concentration (20).

stimulatory effect of these organic extracts. This stimulatory effect was used as an assay for following the purification of the stimulating factor which was thought to be the acceptor lipid for glucose. Although the phosphate in undecaprenol-phosphate had been shown to be very acid labile, due to the allylic structure of the compound, the stimulating activity was found to be stable to mild acid treatment. The acid stability seemed to indicate to us that the compound was not a prenol. This experiment exemplifies how in most cases the obstacles to advance are to be found in human nature. Only after reading a communication to the British Biochemical Society by Hemming (19) did we become aware that dolichols' last isoprene is saturated and should therefore form stable phosphates.

Table 2 shows the purification scheme. The final preparation seemed to be a pure or nearly pure substance and it showed an infrared spectra (Fig. 3) which seemed to be quite similar to that described for dolichol (16) and for bacterial undecaprenol (14). Our compound had a phosphate and this seemed to show also in the spectra.

However, the results were not very clear and at this stage we had all about consumed our ATP reserves in

TABLE 2. Purification of acceptor lipid (20).

	mmoles of organic phosphate	mumoles of glucose incorporated per umole of phosphate	Recovery (%)
Crude extract	44.8	0.008	100
After alkaline and acid treatment	4.2	0.06	70
DEAE-cellulose	0.12	1.9	69
Thin-layer chromatography	0.03	6.8	55

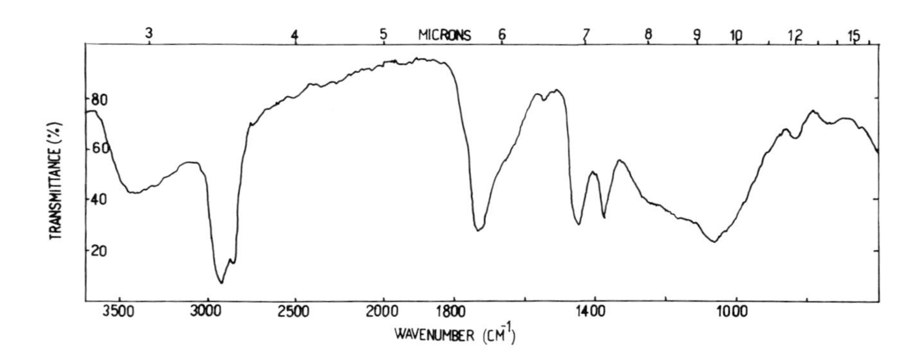

FIG. 3. Infrared spectra of the acceptor lipid. The spectra was measured in a 0.5 mm KBr disk with a Perkin-Elmer Spectrophotometer.

the purification shown in Table 2. We then thought of another approach, to obviate the tedious purification procedures. Morton and Hemming had described a purification method of dolichol and fairly large amounts of pure alcohol are obtained by this technique (16). The compound was chemically phosphorylated by a method described by Cramer (21) and Popjáck (22). Dolichol phosphate prepared by this procedure was added to a reaction mixture and assayed as stimulating factor. It was seen that the compound behaved exactly as the natural acceptor both in its stimulating activity and in the chemical and chromatographic properties of the glucosylated derivative formed (20).

We think the compound formed is dolichol monophosphate glucose or DMP-G. The reaction can be written therefore in the following way:

1. UDP-G + DMP ⟶ DMP-G + UDP,

where DMP stands for dolichol monophosphate.

The purified DMP and the preparation of phosphorylated dolichol have been the only two instances where we have obtained chemically detectable amounts of substance. All further work was carried out with radioactive labelled sugar derivatives, the structural details determined by the properties of their degradation products. This is an unhealthy situation but our endeavours to isolate detectable amounts of DMP-G as well as GEA (see below) have failed. We are still trying.

For a long time we could find no way in which DMP-G would be metabolized. Although the compound was expected to be an intermediate, it was perfectly stable under all conditions tried. By chance we one day forgot to connect the water bath and used by mistake a detergent solution which had concentrated itself by standing uncovered for a long time and found that under those conditions, lower temperature and higher detergent, DMP-G was quickly transformed by liver microsomes (20). The reaction was analyzed by addition of a chloroform-methanol-water mixture where three phases are formed (23). An upper water phase, an interphase consisting of denatured protein precipitate and a lower chloroform-methanol phase containing DMP-G. The radioactivity of DMP-G is transferred to the proteinatious interphase (20,24).

For obvious reasons the compound formed was believed to be a protein. It is insoluble in the two-phase system mentioned and is insoluble in trichloroacetic acid. However it was found that the compound, when dissolved in concentrated piridine acetate and placed on a silica gel G thin layer plate, migrated with polar organic solvent mixtures (25). On paper it migrated with the solvent front when a chloroform-methanol 1:1 mixture was used which contained saturating amounts of water. It was then found that the compound is acid unstable so that it was concluded that we were in the presence of a second lipid intermediate of rather strange solubility properties. We called the compound GEA (Glucosylated Endogenous Acceptor). The reaction detected can be written:

2. $$\text{DMP-G} + \text{EA} \rightarrow \text{GEA} + \text{DMP}$$

Mild acid hydrolysis liberated a water soluble compound which behaved like a neutral oligosaccharide of a molecular weight of about 3,500. That the intermediate contained phosphate was indicated by its absortion to anionic resins. A strong ammonia treatment of GEA yielded the oligosaccharide phosphate as it lost its negative charge by a phosphatase treatment. In fact, the compound seems to be a pyrophosphate since it elutes from DEAE-cellulose together with bacterial pyrophosphate derivatives and is clearly separated from DMP-G and other monophosphates (25).

That GEA really is a dolichol derivative has never been demonstrated as the dolichol moiety is unlabelled and the amounts isolated have not permitted chemical detection. However two different experiments made with GEA seem to validate our proposal.

If GEA is a dolichol derivative, it should liberate DMP on acid treatment and the DMP should stimulate DMP-G formation. Preparative and partially purified extracts of GEA were seen to behave in this way (26). The amount of DMP liberated depended on the intensity of acid treatment, which is interpreted as the result of simultaneous liberation of DMP and DDP (dolichol diphosphate) by mild acid treatment.

A paper which appeared in the Journal of Chemical Education by Herndon, described how bile acids form stoichiometric addition compounds with various organic substances (27). These compounds are called coleic acids. For instance deoxycholate can form complexes with fatty acids which contain, depending on the length of the fatty acid, a fixed amount of even numbers of deoxycholate molecules. These seem to form pairs maintained by hydrogen bonds leaving a tunnel through which the hydrocarbon molecule aligns. It was reasoned that dolichol would bind a fixed number of deoxycholate molecules, independently of the residue bound to it. That is, if GEA were a dolichol derivative, the difference in molecular weight of its coleic acid with that of DMP-G should be the difference in molecular weight of their hydrophilic moieties.

This was tested by molecular weight determination on Sephadex (25). DMP-G was found to bind 28 deoxycholate molecules. Under the same conditions GEA had a molecular weight of about 3,300 Daltons larger. This coincides rather well with the difference in molecular weight between GEA's oligosaccharide pyrophosphate and glucose phosphate (about 3,400 Daltons). GEA formation can thus be written:

2. $\text{DMP-G} + \text{DDP-Ose}_{18} \rightarrow \text{DDP-Ose}_{18}\text{-G} + \text{DMP}$

where Ose_{18} stands for an oligosaccharide of about 18 monosaccharide residues.

At this time, in other laboratories as well as in our own, dolichol derivatives of other monosaccharides were detected. In the presence of DMP, formation of dolichol derivatives of mannose and N-acetylglucosamine was found to take place (24) (28). The following reaction shows dolichol monophosphate mannose formation:

3. $\text{GDP-Man} + \text{DMP} \rightarrow \text{DMP-Man} + \text{GDP}$

It was believed that a similar reaction occured with N-acetylglucosamine. However, Molnar et al. (30) using doubly labelled UDP-N-acetylglucosamine, showed that N-acetylglucosamine-1-phosphate is transferred. The reaction can therefore be written:

4. UDP-GlcNAc + DMP ⟶ DDP-GlcNAc + UMP

Hemming has recently prepared large amounts of DMP-Man and confirmed its structure by various physical methods so that we feel comforted that at least one of these compounds really exists (31). In his study he compared extracted DMP-Man with a sample obtained by organic synthesis from dolichol and mannose by Jeanloz and his group (32).

GEA was shown to transfer its oligosaccharide moiety to an endogenous microsomal protein (33). This glycoprotein is insoluble in water and can be solubilized only under drastic conditions. When exhaustively treated with proteases a glycopeptide is obtained which has a molecular weight similar to that of GEA's oligosaccharide. This fact seems to prove that the whole oligosaccharide has been transferred, as shown in the following reaction:

5. DDP-Ose_{18}-G + protein ⟶ protein-Ose_{18}-G + DDP

This enzyme requires manganese ions, as most glycoprotein transferases do.

Some properties of the oligosaccharide of GEA were studied, following the label in degradative reactions (34). It could be determined that the transferred glucose is not at the reducing end of the oligosaccharide but at the non-reducing end. This was done treating the oligosaccharide obtained by mild acid hydrolysis of GEA with $NaBH_4$. The reduced oligosaccharide was then hydrolyzed completely and it was found that no sorbitol had been formed.

The measurement of the percentage of radioactive label in formic acid after periodate oxidation of GEA's oligosaccharide, indicates that possibly two glucose residues are transferred. Presumably one corresponds to a non-reducing end. Increasing the amount of substrate, more than two glucose residues can be transferred (34).

But more interesting than these results was what happened when the oligosaccharide obtained by methanolysis was subjected to strong alkaline treatment. Fig. 4 shows

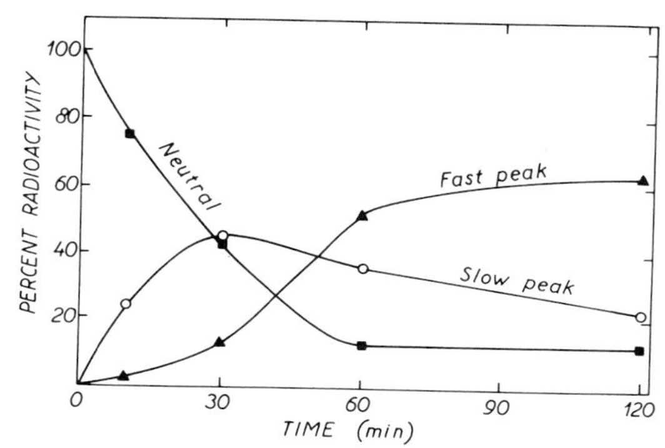

FIG. 4. Alkaline treatment of the GEA oligosaccharide. The oligosaccharide obtained by methanolysis of GEA was heated at 100° in 2M KOH for the times indicated. The samples were neutralized and the compounds separated by paper electrophoresis in 5% formic acid (34).

that in alkali the neutral oligosaccharide disappears rapidly and a positively charged slow moving and afterwards a faster moving compound appears, as determined by paper electrophoresis. This was interpreted as showing that the compound contained two hexosamine moieties, since it could be made neutral again by N-acetylation. Preliminary experiments show that the hexosamines might be at one of the extremes of the oligosaccharide chain, very probably at the reducing end.

This result, which at the beginning seemed to be rather trivial, seems to be rather important. Most of the oligosaccharides bound to asparagine, contain N,N'-diacetylchitobiose bound to the aminoacid and then various mannose and further N-acetylglucosamine residues

(Table 1). If the first two sugars in GEA were N-acetyl-glucosamines and the oligosaccharide grows on a dolichol pyrophosphate molecule, it should be possible to isolate dolichol pyrophosphate N,N'-diacetylchitobiose.

This compound was in fact shown to be formed (35). Under appropiate conditions UDP-GlcNAc transfers the N-acetylglucosamine not only to dolichol monophosphate but also to dolichol diphosphate N-acetylglucosamine (DDP-GlcNAc) to form dolichol diphosphate N,N'-diacetyl-chitobiose. The reaction can be written as follows:

6. UDP-GlcNAc + DDP-GlcNAc $\rightarrow$ DDP-$(GlcNAc)_2$ + UDP

where $(GlcNAc)_2$ stands for N,N'-diacetylchitobiose. The disaccharide was identified by its chromatographic properties. The proposed mechanism was indicated by the pattern of borohydride reduction of the N,N'-diacetyl-chitobiose obtained under different experimental conditions. When the compound was obtained in one step utilizing as acceptor an organic extract which probably contains DDP-GlcNAc, the radioactive label was found in the non-reducing end of the disaccharide. When it was obtained in two steps, incubating only DMP, first with labelled and then with unlabelled UDP-GlcNAc, the label appeared at the reducing end of the disaccharide.

With this promising result we then proceeded to try to add mannose units as would be expected from the known core structures. GDP-Man was incubated with microsomes in the presence of certain lipid fractions of liver. After the incubation the products of mild acid hydrolysis were chromatographed on paper. It appears that oligosaccharides containing labelled mannose are formed, ranging from about 5 to 16 monosaccharide units (36). The fact that this radioactivity is due to oligosaccharides was demonstrated by the pattern obtained from their acetolysis. The results are shown in Fig. 5, where in the upper chromatogram it is seen that in the absence of added acceptors not much oligosaccharides are formed. Treatment of the mannose containing oligosac-charides with alkali liberated positively charged substances as in the case of GEA's oligosaccharide. It is not known if GDP-Man is the sole mannose donor, as

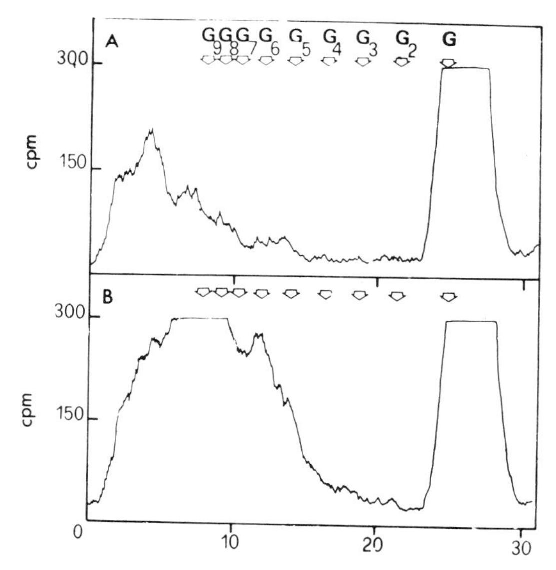

FIG. 5. Formation of Man-EA. Liver microsomes were incubated with ^{14}C labelled GDP-Man in the presence of DMP, buffer, Mg-EDTA and 0.5% Na-deoxycholate. Lipids formed were extracted with the lower phase of a chloroform-methanol- 4mM $MgCl_2$ (3:2:1) mixture, hydrolyzed under mild acid conditions and chromatographed on paper with 1-butanol-pyridine-water (4:3:4) (39). A, formation of Man-EA from GDP-Man; B, the same plus lipid fractions of liver. G, glucose.

DMP-Man is formed during the reaction. Incubation of the dolichol diphosphate bound oligosaccharides with unlabelled DMP-Man increased their size considerably (36). With these results in mind the reaction is tentatively written as follows:

7. GDP-Man and/or DMP-Man + EA $\rightarrow$ Man-EA + GDP and/or DMP

where EA stands again for Endogenous Acceptors. These

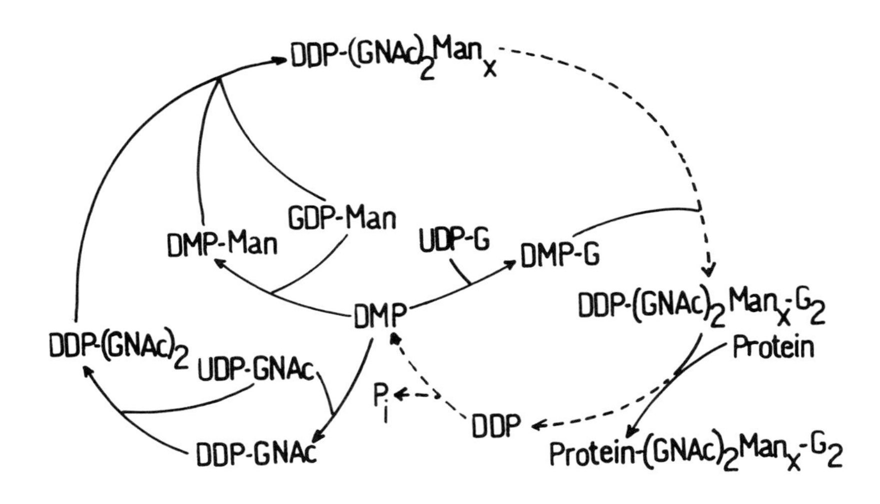

FIG. 6. Biosynthesis of DDP-$(GlcNAc)_2Man_x$-G_2 (GEA) and of a glycoprotein.

seem to be also dolichol diphosphate derivatives by their elution pattern from DEAE-cellulose.

One is tempted to order the reactions detected and suppose that they are involved in glycoprotein synthesis in the sequence depicted in Fig. 6 where DDP-$(GlcNAc)_2$ Man_x stands for the mannosylated endogenous acceptors shown in reaction 7 and one of its higher homologues would be equivalent to the EA shown in reaction 2. DDP-$(GlcNAc)_2$ Man_x would serve not only as precursor of DDP-Ose_{18}-G, but can transfer its oligosaccharide directly to an endogenous protein (36). Hemming and Heath have reported the transfer of mannose from DMP-Man to protein (37,38).

The glycoprotein formed in Fig. 6 is not one of the common ones as glucose is a rather rare carbohydrate in asparagine glycoproteins. It should be pointed out that GEA is formed by all of the mammal tissues tested (34).

At the moment a renewed attempt to isolate detectable amounts of GEA or its oligosaccharide is being made.

Another approach is the study of tissues which specialize in the synthesis of fairly large amounts of glycoproteins of known structure. Promising results have been obtained with hen oviduct. About half of this tissue glycoprotein synthesis is directed to the formation of ovalbumin. As shown in Table 1, this glycoprotein contains only the core sugars. Labelled DDP-$(GlcNAc)_2$ prepared either with liver or oviduct microsomes, was incubated in the presence of unlabelled GDP-Man. Higher oligosaccharides were seen to be formed. Their structure has not been determined but preliminary data seem to indicate that they may be precursors of the complete ovalbumin oligosaccharide.

The results obtained can be summarized as follows. Through the study of new lipid sugar derivatives in animal tissues we have come upon the biosynthesis reactions of a glycoprotein (Fig. 6). The structure of the oligosaccharides resemble the core portion of asparagine glycoproteins. The biosynthesis of this core portion directly from sugar nucleotides has proven most elusive. We propose and will try to demonstrate in further work that this core portion is formed step by step on dolichol diphosphate. Once ready, the core oligosaccharide is transferred as a whole to protein. The external sugars, sialic acids, galactose, N-acetylglucosamine and fucose, might then be added by the classical mechanism.

The work described here was carried out in collaboration with L.F. Leloir, A.J. Parodi, H. Carminatti, R.J. Staneloni, J.A. Levy, E. Tábora and A.I. Cantarella. We thank Dr. B. Frydman and Sirex S.A. for their help in infrared spectra determination and interpretation. Grants from the Consejo Nacional de Investigaciones

Científicas y Técnicas, Universidad de Buenos Aires and National Institutes of Health (USA, No. GM 19808-02) are gratefully acknowledged.

(1) H. Clauser, G. Herman, B. Rossignol and S. Harbon, in: Glycoproteins, Vol. B, ed. A. Gottschalk (Elsevier, Amsterdam, 1972) p. 1151.

(2) R.C. Hughes, Progress Biophys. & Molec. Biol., 26 (1973) 191.

(3) P.J. O'Brien and E.F. Neufeld, in: Glycoproteins, Vol. B, ed. A. Gottschalk (Elsevier, Amsterdam, 1972) p. 1170.

(4) T. Helting and L. Rodén, J. Biol. Chem., 244 (1969) 2799.

(5) A. Gottschalk and A.S. Bhargava, in: Glycoproteins, Vol. B, ed. A. Gottschalk (Elsevier, Amsterdam, 1972) p. 810.

(6) R.G. Spiro, ibid., p. 964.

(7) R.D. Marshall, Ann. Rev. Biochem., 41 (1972) 673.

(8) T. Arima, M.J. Spiro and R.G. Spiro, J. Biol. Chem., 247 (1972) 1825.

(9) R.G. Spiro, Ann. Rev. Biochem., 39 (1970) 599.

(10) H. Nikaido and W.Z. Hassid, Advan. Carbohyd. Chem. Biochem., 26 (1971) 461.

(11) J.S. Anderson, M. Matsuhashi, M.A. Haskin and J.L. Strominger, Proc. Nat. Acad. Sci., USA, 53 (1965) 881.

(12) A. Wright, M.A. Dankert and P.W. Robbins, ibid., 54 (1965) 235.

(13) I.M. Weiner, T. Higuchi, L. Rothfield, Saltmarsh-Andrew, M.J. Osborn and B.L. Horecker, ibid., p. 228.

(14) A. Wright, M.A. Dankert, P. Fennesey and P.W. Robbins, ibid., 57 (1967) 1798.

(15) Y. Higashi, J.L. Strominger and C.C. Sweeley, ibid., p. 1878.

(16) J. Burgos, F.W. Hemming, J.F. Pennock and R.A. Morton, Biochem. J., 88 (1963) 470.

(17) P.J. Dumphy, J.D. Kern, J.F. Pennock, K.J. Whittle and J. Feeney, Biochim. Biophys. Acta, 136 (1967) 136.

(18) P. Babczinski and W. Tanner, Biochem. Biophys. Res. Comm., 54 (1973) 1119.

(19) F.W. Hemming, Biochem. J., 113 (1969) 23P.

(20) N.H. Behrens and L.F. Leloir, Proc. Nat. Acad. Sci. USA, 66 (1970) 153.

(21) F. Cramer and H.J. Bohm, Angew. Chem., 71 (1959) 775.

(22) G. Popjáck, J.W. Cornforth, R.H. Cornforth, R. Ryhage and D.S. Goodman, J. Biol. Chem., 237 (1962) 56.

(23) J. Folch, M. Lees and G.H. Sloane Stanley, ibid., 226 (1957) 497.

(24) N.H. Behrens, A.J. Parodi, L.F. Leloir and C.R. Krisman, Arch. Biochem. Biophys., 143 (1971) 375.

(25) N.H. Behrens, A.J. Parodi and L.F. Leloir, Proc. Nat. Acad. Sci. USA, 68 (1971) 2857.

(26) A.J. Parodi, N.H. Behrens, L.F. Leloir and M. Dankert, Biochim. Biophys. Acta, 270 (1972) 529.

(27) W.C. Herndon, J. Chem. Ed., 44 (1967) 724.

(28) S.S. Alam, R.M. Barr, J.B. Richards and F.W. Hemming, Biochem. J., 121 (1971) 19P.

(29) J.B. Richards, P.J. Evans and F.W. Hemming, ibid., 124 (1971) 957.

(30) J. Molnar, H. Chao and Y. Ikehara, Biochim. Biophys. Acta, 239 (1971) 401.

(31) P.J. Evans and F.W. Hemming, FEBS Lett., 31 (1973) 335.

(32) C.O. Warren and R.W. Jeanloz, ibid., p. 332.

(33) A.J. Parodi, N.H. Behrens, L.F. Leloir and H. Carminatti, Proc. Nat. Acad. Sci. USA, 69 (1972) 3268.

(34) A.J. Parodi, R. Staneloni, A.I. Cantarella, L.F. Leloir, N.H. Behrens, H. Carminatti and J.A. Levy, Carbohyd. Res., 26 (1973) 393.

(35) L.F. Leloir, R. Staneloni, H. Carminatti and N.H. Behrens, Biochem. Biophys. Res. Comm., 52 (1973) 1285.

(36) N.H. Behrens, H. Carminatti, R. Staneloni, L.F. Leloir and A.I. Cantarella, Proc. Nat. Acad. Sci. USA, 70 (1973) 3390.

(37) J.B. Richards and F.W. Hemming, Biochem. J., 130 (1972) 77.

(38) J.W. Baynes, An-Fei Hsu and E.C. Heath, J. Biol. Chem., 248 (1973) 5693.

(39) L.F. Leloir, A.J. Parodi and N.H. Behrens, Revista Soc. argent. Biol., 47 (1971) 108.

Figure 4 is reproduced by permission of the Elsevier Publishing Company; Figures 2 and 5 by permission of the National Academy of Sciences.

DISCUSSION

I. SCHENKEIN: You mentioned gamma-G that is synthesized by plasmacytoma cells, the so-called myeloma protein. There is good evidence that in those proteins, the first sugars, that is the bridge sugars, are added when the protein is still on the ribosomal surface, and that subsequently, during completion and transport from the ribosomal surface towards the exterior (myeloma proteins are export proteins), additional sugars are added, for example in the Golgi apparatus. Have you any information on where the biosynthesis of glycoproteins involving polyprenols takes place in the whole cell?

N.H. BEHRENS: No, we have not. We did determine the activity of the enzymes catalyzing the formation of DMP-G and of DDP-Ose 18-G_2 in the subcellular fractions of liver cells. However, the distribution found did not permit a clear deduction. It is an interesting problem and we should one day study it further.

R.W. JEANLOZ: From our knowledge of the structure of glycoproteins, we expect two types of mannose residue, an α and a β anomer. Have you any evidence of polyprenol intermediates having both structures or is there only one mannose intermediate; this is my first question. My second question concerns the possibility that one mannose residue is being added to the first glucosamine residue before the second glucosamine residue is added. Do you have evidence for the sequence of additions to the first glucosamine residue?

N.H. BEHRENS: To answer the first question, can I have again the last slide (Fig. 6). It is frequently found that the first mannose which is bound to the N,N'-diacetylchitobiose moiety is bound by a β-bond and most of the other mannose residues present in the core are bound by an α-bond. We have no data to support the existence of different donors for the two kinds of mannose bonds found. However, it seems that all the reactions described in Fig. 6 occur with inversion of configuration. Thus, we have evidence indicating that glucose is bound to DMP by a β-bond so that inversion occurs in the transfer of glucose from UDP-G to DMP. Preliminary data indicates that the glucose at the nonreducing end of DDP-Ose 18-G_2 is in the α-configuration so that transfer from DMP-G is also accompanied by inversion. The first N-acetylglucosamine is transferred with its phosphate so that it remains in the α-configuration as in the nucleotide donor, but the second N-acetylglucosamine is transferred with inversion to DDP-glcNAc to form the β-derivative N,N'-diacetylchitobiose. We have not been able to show what the configuration of mannose in DMP-Man is, but if it is β, one could speculate that the α-bound mannose in the core oligosaccharide may be formed by a transfer with inversion from this DMP-intermediate, whereas β-bound mannose residues may be transferred directly from GDP-Man. With regard to your second question, we incubated ^{14}C-labelled DDP-$(glcNAc)_2$ in the presence of unlabelled GDP-Man and showed that higher oligosaccharides are formed on the disaccharide acceptor. This was done with liver and oviduct microsomes. If the oviduct microsome preparation is catalyzing the formation of the ovalbumin oligosaccharide, then this gives some indication that the second N-acetylglucosamine is transferred to form the N,N'-diacetylchitobiose acceptor before the mannose is transferred to the first N-acetylglucosamine residue.

R.W. JEANLOZ: We have synthesized the mannose intermediate with the α configuration. The intermediate that we isolate from the pancreas shows a different behavior. We have not synthesized, as yet, a synthetic β mannosyl phosphate intermediate.

S. ROSEMAN: I don't understand one point about your characterization of the disaccharide of N-acetylglucosamine linked to the dolichol. What did you do to split the dolichol di-N-acetyl-chitobiose? Under what conditions? I presume that you are getting chitobiose by alkaline degradation. My point is that it is very difficult to deacetylate the N-acetyl glycosamines with alkali.

N.H. BEHRENS: We have not done it with N,N'-diacetylchitobiose itself. The procedure was carried out with the oligosaccharide (Ose 18-G_2) and with $GlcNac_2Man_x$. These oligosaccharides were obtained by methanolysis. The reducing ends of the oligosaccharides were thus protected from degradation during the very strong alkaline treatment required for deacetylation (2 M KOH at 100°).

THE ROLE OF MANNOSYL PHOSPHORYL DIHYDRO-POLYISOPRENOL IN THE SYNTHESIS OF MAMMALIAN GLYCOPROTEINS

Edward C. Heath, John W. Baynes and An-Fei Hsu
Department of Biochemistry
University of Pittsburgh School of Medicine

Abstract: A mouse myeloma tumor was used as a model system to study the biochemical steps involved in the incorporation of mannose into glycoproteins. This tumor, MOPC-46B synthesizes a kappa-type immunoglobulin light chain (K-46) which is a glycoprotein with a single oligosaccharide side chain containing mannose as one of its constituent sugars.

MOPC-46B microsomal preparations contain enzymes which transfer mannose from the sugar nucleotide, GDP-mannose, to endogenous lipid and protein acceptors. Formation of the mannolipid proceeds by the reversible transfer of mannose from GDP-mannose to an endogenous phospholipid. The mannolipid was purified and characterized by chemical methods and mass spectrometry as a mannosyl-monophosphoryl-dihydropolyisoprenol (Man-P-DHPI) containing at least 18 isoprene units, one of which is saturated. The mannolipid was implicated as an intermediate in the *in vitro* mannosylation of endogenous protein acceptors by the following observations: (a) Kinetic studies were consistent with the precursor-product relationships between GDP-Man, Man-P-DHPI, and protein; (b) the transfer of mannose from

GDP-Man to both lipid and protein was inhibited by EDTA while the transfer of mannose from Man-P-DHPI to protein was insensitive to EDTA; (c) Pulse-chase experiments were consistent with a precursor-product relationship between Man-P-DHPI and protein; and (d) exogenously supplied Man-P-DHPI served as a donor of mannosyl residues to protein.

During the course of these studies an additional component(s) was observed to contain ^{14}C-Man when either GDP-^{14}C-Man or ^{14}C-Man-P-DHPI was used as substrate. This water-soluble product exhibited an apparent molecular weight of approximately 2000 and upon analysis was shown to be a ^{14}C-mannose-containing oligosaccharide which also contained unlabeled glucosamine and a phosphomonoester residue at its potential reducing terminus. It appears that this oligosaccharide phosphate is derived in crude microsomal incubation mixtures from the degradation of an oligosaccharide phospholipid. Thus, incubation conditions have been defined that permit extraction of the oligosaccharide phospholipid and it was purified to apparent homogeneity by Sephadex, concanavalin A Sepharose, and DEAE-cellulose chromatographic techniques. The ^{14}C-mannose-containing oligosaccharide phospholipid appears to be homogeneous and is clearly separable from all of the other reactants including the oligosaccharide phosphate and Man-P-DHPI by several thin-layer chromatographic systems. Our current experiments support the contention that the oligosaccharide phospholipid is, in fact, an additional intermediate in the overall process of transfer of mannose from GDP-Man to endogenous protein. Thus, kinetic analyses of incubation mixtures starting with either GDP-^{14}C-Man or ^{14}C-Man-P-DHPI suggest the following sequence of reactions:

GDP-^{14}C-Man ⟶ ^{14}C-Man-P-DHPI ⟶
^{14}C-Man-Oligosaccharide-P-(P)-Lipid ⟶
^{14}C-Man-[Oligosaccharide]-Protein

Structural analysis of the ^{14}C-mannose-containing oligosaccharide phospholipid indicated: 5 mannose per 2 N-acetylglucosamine per potential reducing terminus. The compound exhibits the following properties: (a) acid hydrolysis yields ^{14}C-mannose and glucosamine; (b) ^{14}C-mannose is located in non-reducing positions; (c) treatment of the oligosaccharide phospholipid with 10% (v/v) of NH_4OH at 100° for 1 hr results in the release of the oligosaccharide phosphate, suggesting a pyrophosphate linkage between the oligosaccharide and the lipid; (d) treatment of the oligosaccharide phosphate with alkaline phosphatase exposes a reducing terminal N-acetylglucosamine residue. The structure of the lipid moiety of oligosaccharide phospholipid remains to be established.

INTRODUCTION

Many laboratories have contributed a great deal of information over the last decade concerning the mode of biosynthesis and secretion of glycoproteins. The biosynthesis of the external regions of the carbohydrate portion of these proteins proceeds by the addition from nucleotide derivatives of single sugar residues to the growing oligosaccharide side chain. The sequential addition of external N-acetylglucosamine, galactose, sialic acid, and fucose residues by glycosyl transferases of the endoplasmic reticulum and Golgi apparatus has been accomplished using exogenous protein acceptors, generally prepared by removal of sugar residues from native glycoproteins with the appropriate glycosidases (1-3). Investigations on the mode of incorporation of mannose and glucosamine residues which constitute the major sugars of the internal core regions of the oligosaccharide side chains, however, have been hampered by the inability to prepare efficient exogenous acceptors by glycosidase treatments or to find native proteins which can function as acceptors. An additional problem is that most cell systems exhibit a great deal of heterogeneity in endogenous proteins which can serve as acceptors for mannosyl transferase reactions.

We felt that some of these difficulties may be minimized in a cell system that produces primarily a single mannose-containing glycoprotein with one oligosaccharide side chain of defined structure. We have therefore, selected a mouse myeloma tumor, MOPC-46B, as a model system to study the biosynthesis of glycoproteins, particularly with regard to the biosynthesis of the internal core regions of the oligosaccharide side chain. The mouse plasma cell tumor, MOPC-46B, synthesizes a kappa-type immunoglobulin light chain (K-46), a glycoprotein with a molecular weight of approximately 25,000 and which contains a single oligosaccharide side chain attached at asparagine residue 34 in the peptide chain. This mouse tumor was characterized by Dr. Michael Potter of the National Institutes of Health, who kindly made it available to us for these studies. As a result primarily of the efforts of Melchers (4), some general structural features of the oligosaccharide side chain of this protein have been established (Fig. 1); that is, the oligosaccharide is linked to the polypeptide chain by an N-glycosidic bond between GlcNAc and the amide nitrogen of asparagine; the core portion of the oligosaccharide contains N-acetylglucosamine, mannose, and possibly a small amount of galactose; and the external region of the oligosaccharide contains sialic acid, galactose, and fucose. Thus, the general structural features of the carbohydrate side chain of K-46 conform to those of the majority of serum-type glycoproteins. The MOPC-46B tumor is highly directed toward the synthesis of this single species of glycoprotein; incubation of viable tumor cell suspensions _in vitro_ with ^{14}C-glucosamine and

MOPC-46 MOUSE MYELOMA TUMOR
(Dr. Michael Potter, NIH)

Produces kappa-type immunoglobulin light chain (K-46)

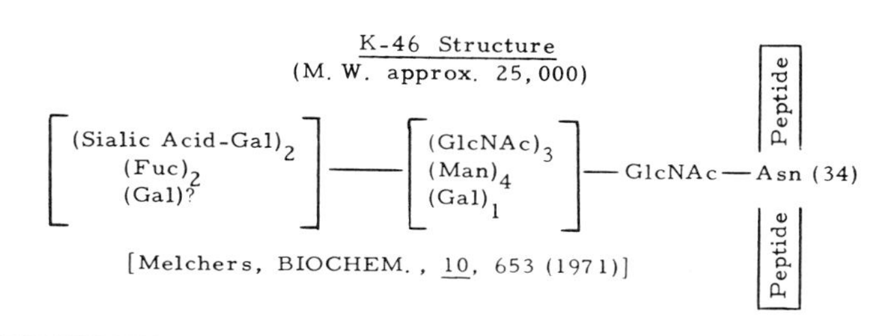

Biosynthesis and Isolation

1. Viable tumor cell suspensions incubated in vitro with ^{14}C-$GlcNH_2$ and ^3H-Leu:
   (a) 40-50% of total radioactivity incorporated into K-46
   (b) >95% of total radioactivity secreted into K-46
2. Tumor-bearing mouse urine contains 1 mg/ml K-46

FIGURE 1

^3H-leucine results in the incorporation of approximately one-half of the total radioactivity into K-46 protein and, in addition, greater than 95% of the total radioactive protein secreted by these cells is K-46. In addition, K-46 may be isolated in reasonably large quantities from the urine of tumor-bearing animals; the urine usually contains approximately 1 mg/ml of the protein.

Using incubation mixtures containing microsomal membrane preparations from MOPC-46B cells and GDP-^{14}C-Man, we observed the biosynthesis of mannosyl-phosphoryl-dihydropolyisoprenol (Man-P-DHPI); the properties of this enzyme system and the characterization of the mannolipid were reported in the Spring of 1972 (5) and subsequently published in detail (6). At that time, we had preliminary evidence suggesting that the manno-lipid serves as a donor for the enzymatic transfer of mannose into endogenous proteins of the microsomal preparations. In this report, we would like to summarize our previous results concerning the biosynthesis of mannosyl-phosphoryl-dihydropolyisoprenol and to present the results of recent experiments which indicate that this mannolipid is an intermediate in the transfer of mannose to glycoprotein. In addition, we will show that mannose- and glucosamine-containing oligosaccharide lipids also participate as intermediates in a proposed sequence of reactions summarized in Fig. 2. Thus, the mannose moiety of Man-P-DHPI is subsequently transferred to an

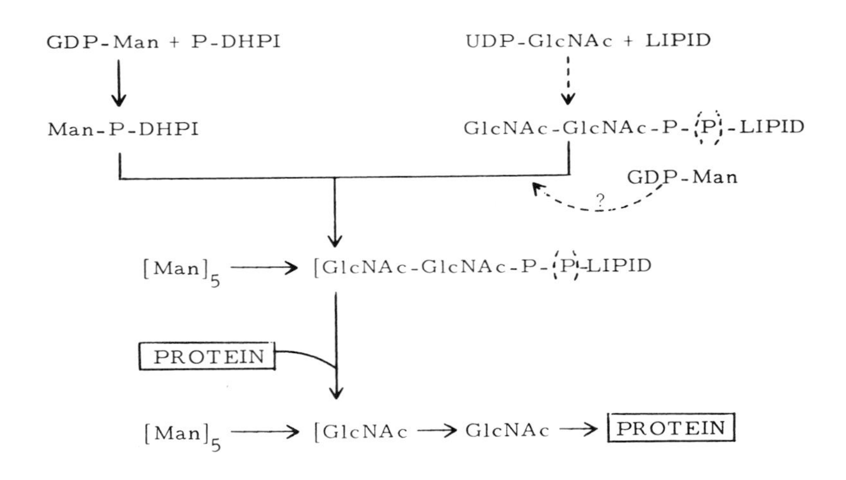

FIGURE 2

endogenous glucosamine- and mannose-containing oligosaccharide phospholipid and, in a subsequent reaction, presumably the entire "preassembled" oligosaccharide is transferred from the "carrier" lipid to endogenous protein acceptor. We have little direct evidence concerning the precise details of the biosynthesis of the glucosamine-containing lipid; however, its structure and properties are consistent with those described by Leloir *et al.* (7) in their report on the biosynthesis of GlcNAc-GlcNAc-P-P-Dolichol.

BIOSYNTHESIS OF MANNOSYL-PHOSPHORYL-DIHYDROPOLYISOPRENOL

Incubation of crude microsomal membrane preparations from MOPC-46B cells with GDP-^{14}C-Man results in the rapid incorporation of radioactive mannose into a butanol-pyridinium acetate-soluble fraction as illustrated in Fig. 3. Kinetic studies indicated that the plateau value of incorporation achieved is a function of the amount of available endogenous lipid acceptor. The mannosyl transferase reaction exhibits the following properties: (a) complete dependence upon divalent cation for activity; (b) the extent of mannose incorporation is unaffected by 10-fold concentrations of unlabeled UDP-Glc, UDP-Gal, UDP-GlcNAc, or CDP-choline; (c) addition of excess GDP to the incubation mixture after mannose incorporation had reached a maximum results in a displacement of radioactivity from the organic solvent-soluble fraction; (d) incorporation of mannose into protein proceeds at a low level after maximum mannolipid has been formed; (e) and pulse-chase experiments suggest that the mannolipid formed in this reaction serves as an intermediate in the subsequent transfer of mannose to protein. Since we had observed an identical reaction using microsomal membrane preparations from bovine liver, we used the liver system in order to prepare sufficient quantities of the mannolipid for structural characterization.

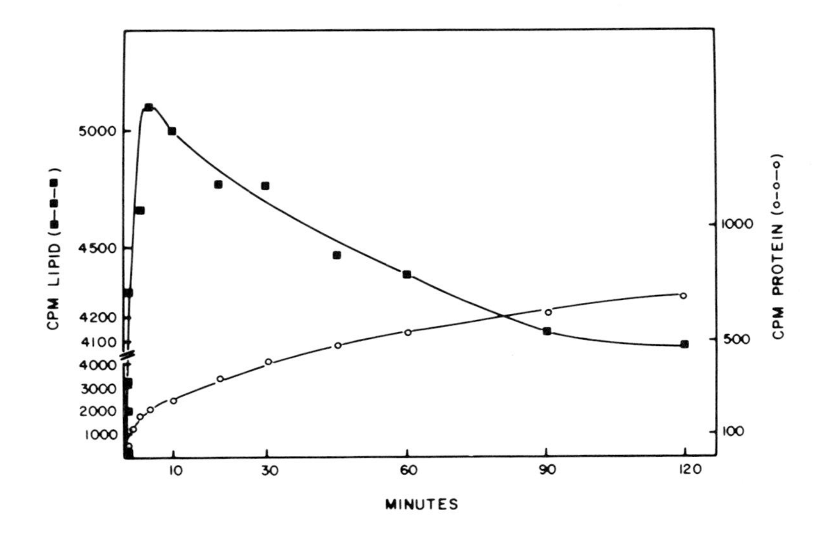

Fig. 3. Incorporation of mannose into lipid and protein. Incubation mixtures contained 4 μM GDP-^{14}C-Man with 1.2 mg of myeloma microsomal protein and 100,000 cpm per 100 μl. Radioactivity in lipid and protein was determined on 100 μl aliquots at the intervals indicated.

As illustrated in Table I, using the phosphorus to mannose ratio as an index of purification, standard fractionation techniques resulted in extensive purification of the mannolipid. The final product, 20,000-fold purified over the crude extract, exhibited an organic phosphorus to acid-labile reducing sugar (after hydrolysis) ratio of 1.2. This product was homogeneous in terms of the distribution of radioactivity and iodine-staining when examined by thin layer chromatography in several solvent systems. The specific activity of mannose in the purified lipid indicated that some endogeneous unlabeled mannolipid was present in the enzyme preparation initially. Analysis of the lipid by mass spectrometry after alkaline hydrolysis indicated that the lipid is a polyisoprenoid

TABLE I

Purification of bovine liver mannolipid

Fraction	Mannose μmoles	Organic phosphorus μmoles	Phosphorus to mannose ratio
Butanol extract	1.8	1.4×10^5	8×10^4
Acetone precipitate	1.5	8.6×10^4	5.7×10^4
Silicic acid eluate	1.5	1.2×10^3	800
DEAE-eluate	1.0		
Sephadex LH-20 eluate	0.8	15	19
Preparative thin layer chromatography	0.4(1.1)	1.4	3.5(1.2)

compound. The mass spectrum revealed a characteristic pattern of triads of fragment ions separated by 68 mass units representing random cleavage between individual isoprene units in each chain. The molecular ion was not visible in the mass spectrum and thus, the precise number of isoprene units in the lipid remains to be established. However, on the basis of the spectral characteristics, it was concluded that the lipid was a dihydropolyisoprenol consisting of at least 18 isoprene units, one of which was saturated. These results suggested that the lipid is a form of dolichol, a C-100 α-saturated polyisoprenol which, as dolichol monophosphate, has been characterized as a glycosyl acceptor in liver microsomes by Behrens and Leloir (8, 9) and Hemming (10) and their coworkers.

We concluded from these studies that microsomal preparations from either bovine liver or mouse myeloma cells catalyze the reaction shown in Fig. 4. Thus, mannose is transferred from GDP-Man to endogenous dihydropolyisoprenol phosphate to form mannosylphosphoryl-dihydropolyisoprenol (Man-P-DHPI) and GDP.

Biosynthesis of Mannosyl Phosphoryl DHPI

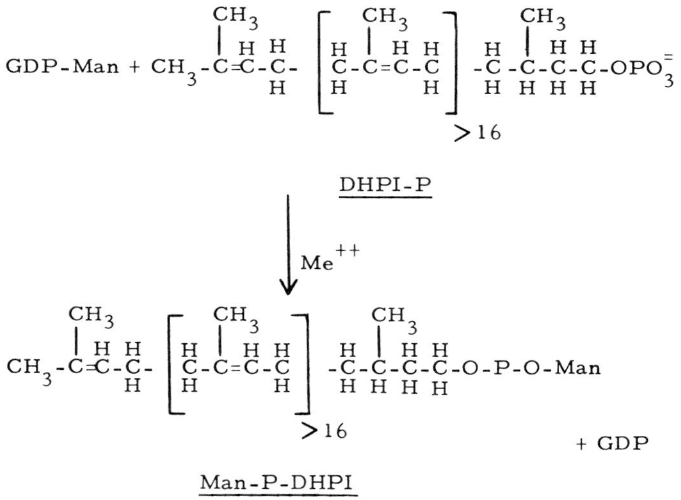

$$\text{GDP-Man} + CH_3\text{-}\overset{CH_3}{\underset{}{C}}{=}\overset{H}{C}\text{-}\overset{H}{\underset{H}{C}}\text{-}\left[\overset{H}{\underset{H}{C}}\text{-}\overset{CH_3}{C}{=}\overset{H}{C}\text{-}\overset{H}{\underset{H}{C}}\right]_{>16}\text{-}\overset{H}{\underset{H}{C}}\text{-}\overset{CH_3}{\underset{H}{C}}\text{-}\overset{H}{\underset{H}{C}}\text{-}\overset{H}{\underset{H}{C}}\text{-}OPO_3^{=}$$

DHPI-P

$\downarrow Me^{++}$

$$CH_3\text{-}\overset{CH_3}{\underset{}{C}}{=}\overset{H}{C}\text{-}\overset{H}{\underset{H}{C}}\text{-}\left[\overset{H}{\underset{H}{C}}\text{-}\overset{CH_3}{C}{=}\overset{H}{C}\text{-}\overset{H}{\underset{H}{C}}\right]_{>16}\text{-}\overset{H}{\underset{H}{C}}\text{-}\overset{CH_3}{\underset{H}{C}}\text{-}\overset{H}{\underset{H}{C}}\text{-}\overset{H}{\underset{H}{C}}\text{-O-P-O-Man} + \text{GDP}$$

Man-P-DHPI

FIGURE 4

By our methods of isolation, approximately 80% of the total radioactivity incorporated into the organic solvent-soluble fraction was recovered as Man-P-DHPI.

BIOSYNTHESIS OF MANNOSE-CONTAINING OLIGOSACCHARIDE LIPID

During the course of our studies assessing the role of Man-P-DHPI in the transfer of mannose residues to protein, the appearance of a water-soluble radioactive compound was detected as shown in Fig. 5. Thus, incubation of pure ^{14}C-Man-P-DHPI with delipidated microsome preparations resulted in the disappearance of radioactivity from the lipid fraction with the concomitant appearance of a small portion of this radioactivity in protein but a larger portion of it in the water-soluble fraction. The

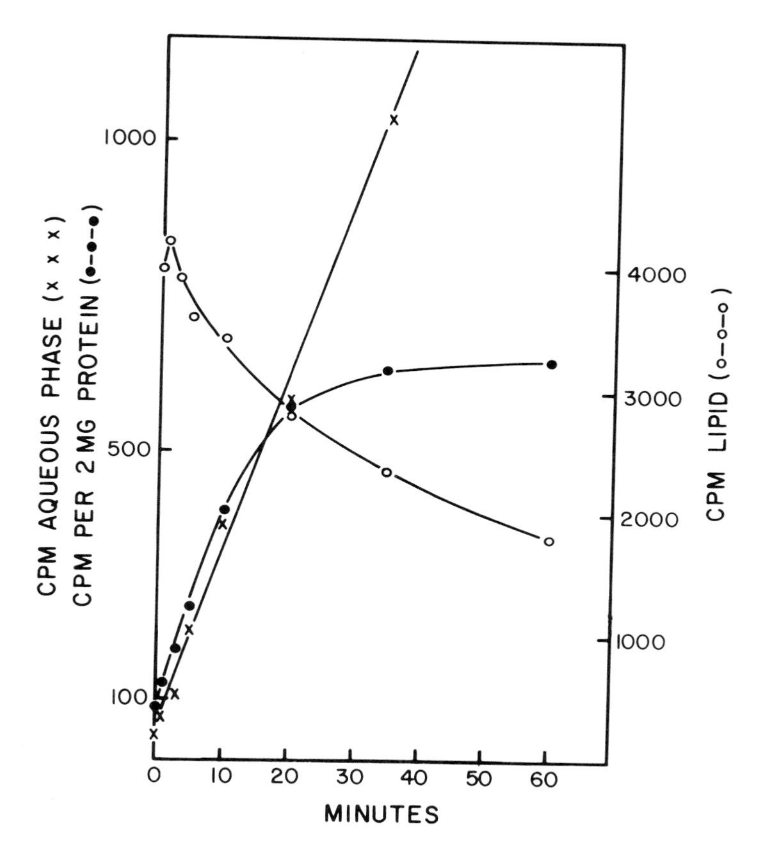

Fig. 5. Direct transfer of mannose from mannolipid to protein. A 1-ml incubation mixture contained 20 mg of delipidated microsomes and 40,000 cpm of partially purified mannolipid. Aliquots (100 μl) were quenched with 300 μl of butanol-pyridinium acetate then diluted with 200 μl of H_2O, and centrifuged. Distinct aqueous and organic phases were collected separately; two additional extractions were performed with 300 μl of butanol-pyridinium acetate and 200 μl of H_2O, and the aqueous and organic extracts were pooled and counted.

water-soluble, radioactive material appeared to be homogeneous when analyzed by gel filtration techniques and by paper chromatography; it exhibited an apparent molecular weight of approximately 2000 and was chromatographed on DEAE-cellulose as illustrated in Fig. 6. The compound eluted from the column at approximately 0.2 molar

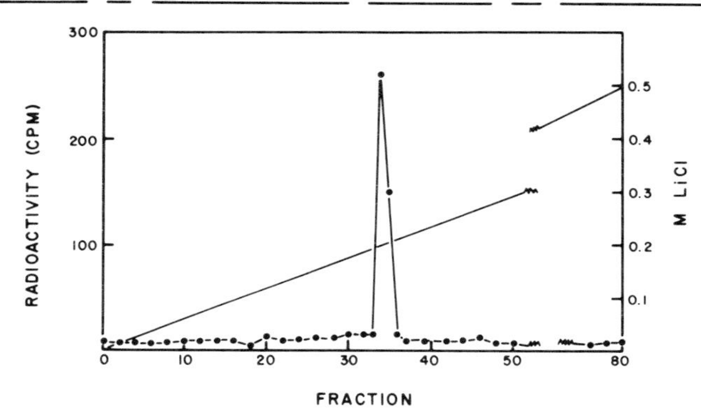

Fig. 6. An aliquot of ^{14}C-Man-oligosaccharide phosphate was applied to a column (1x10 cm) of DEAE-cellulose and eluted with a linear gradient (0-0.5 M) of LiCl.

lithium chloride as a symmetrical peak of radioactivity. The anionic properties of the compound were also manifested when it was subjected to paper electrophoresis as shown in Fig. 7. The compound migrated toward the anode at pH 2 as a single radioactive constituent; the acidic properties of the compound are due to the presence of a phosphomonoester residue as treatment with alkaline phosphatase results in its conversion to a neutral compound.

As illustrated in Fig. 8, on the basis of its partial structural characterization, we have concluded that the water-soluble radioactive compound is a ^{14}C-mannose-

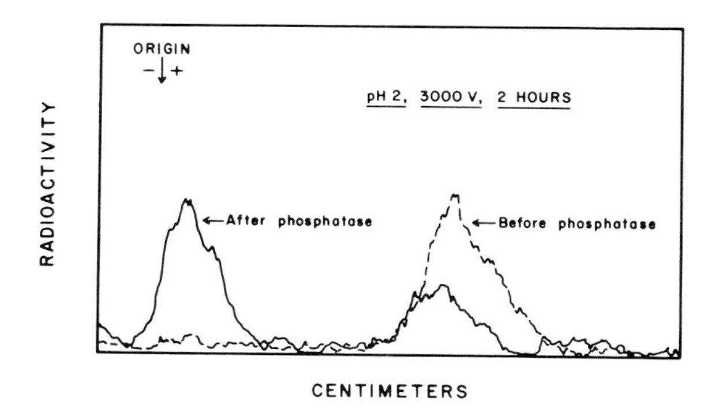

Fig. 7. An aliquot of pure oligosaccharide phosphate (from DEAE-cellulose column) was treated with alkaline phosphatase and then applied to Whatman 3 mm paper and subjected to electrophoresis in acetic acid-formic acid, pH 2, for 2 hours at 3000 volts; a second aliquot was treated identically except without alkaline phosphatase.

---

containing oligosaccharide phosphate. No detectable radioactivity is incorporated into the oligosaccharide phosphate when it is treated directly with sodium borotritide, indicating that the phosphate residue occupies the potential reducing terminus of the oligosaccharide. However, after treatment with alkaline phosphatase, treatment with sodium borotritide, acid hydrolysis, and paper chromatographic analysis indicated that only glucosaminitol contained ^{3}H whereas all of the ^{14}C was present in free mannose. Conversely, after alkaline phosphatase treatment, acid hydrolysis of the oligosaccharide prior to treatment with sodium borotritide and paper chromatographic separation, revealed the presence of ^{3}H-labeled glucosaminitol and mannitol labeled with ^{14}C and ^{3}H. Thus, we have concluded that this compound is an oligosaccharide containing mannose and N-acetylglucosamine with a terminal reducing N-acetylglucosamine residue

PARTIAL STRUCTURAL DETERMINATION OF OLIGOSACCHARIDE PHOSPHATE

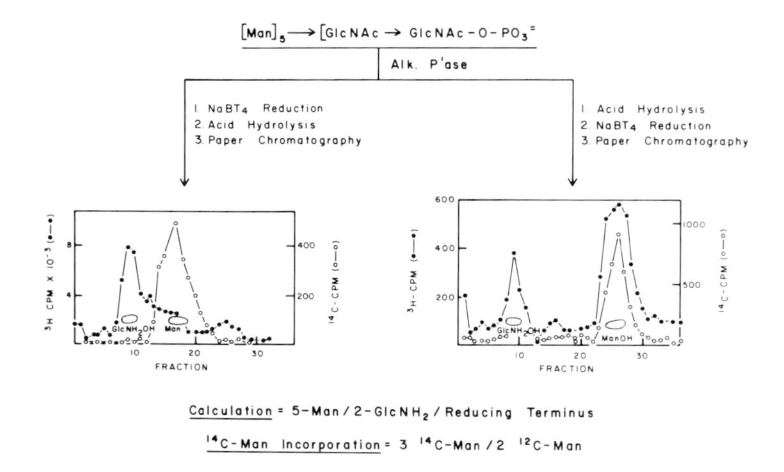

Fig. 8. Structural analysis of oligosaccharide phosphate. Chromatography was conducted on Whatman 3 MM paper in ethyl acetate (18); acetic acid (3); formic acid (1); and water (4) for 15 to 18 hr.

substituted with a phosphomonoester group; from the relative ^3H content of glucosaminitol and mannitol the molar proportions of glucosamine and mannose in the oligosaccharide were estimated to be 5 mannose residues per 2 glucosamine residues per reducing terminus. On the basis of the relative specific activity of ^{14}C-mannose in the oligosaccharide, it was estimated that approximately 3 of the 5 mannose residues had been incorporated into the oligosaccharide from ^{14}C-Man-P-DHPI. Presumably the unlabeled mannose residues were present in the oligosaccharide endogenously. Subsequent studies indicated that the oligosaccharide phosphate is derived from an oligosaccharide phospholipid as shown in Fig. 9. We have been able to isolate a mannose- and glucosamine-containing oligosaccharide lipid which yields the same oligosaccharide phosphate upon hydrolysis with dilute ammonium hydroxide and, on the basis of the rapidity

Formation of Oligosaccharide Phosphate From Lipid

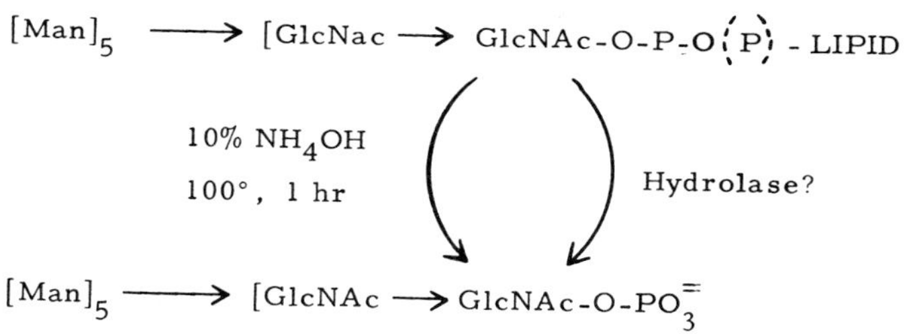

$[Man]_5$ ⟶ [GlcNac ⟶ GlcNAc-O-P-O(P) - LIPID

10% NH_4OH
100°, 1 hr

Hydrolase?

$[Man]_5$ ⟶ [GlcNAc ⟶ GlcNAc-O-$PO_3^=$

FIGURE 9

with which the oligosaccharide phosphate appears in microsomal incubation mixtures, we suspect that enzymatic degradation of the oligosaccharide phospholipid may also occur to yield this product.

ISOLATION OF OLIGOSACCHARIDE PHOSPHOLIPID

Incubation mixtures derived from experiments conducted with GDP-^{14}C-Man as substrate were reexamined for the possible presence of ^{14}C-Man-labeled oligosaccharide phospholipid as shown in Fig. 10. The entire incubation mixture was solubilized in Triton X-100 and applied to a column of Sephadex G-75. The radioactive material which was voided from the gel was pooled and analyzed by thin layer chromatography; two radioactive constituents were present, one of which corresponded to ^{14}C-Man-P-DHPI. The two radioactive constituents were separated quantitatively by fractionation of this mixture on a concanavalin A Sepharose column in Triton X-100 as shown in Fig. 11. Application of the voided material from the Sephadex G-75 column to a concanavalin A Sepharose column resulted in a portion of the radioactivity emerging from the column at the breakthrough volume in a broad peak; addition of α-methylmannoside (as indicated by the arrow) resulted in the elution of the

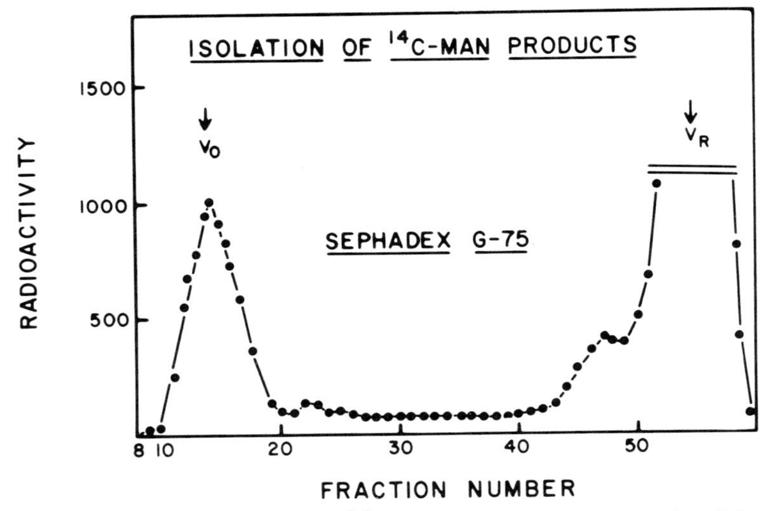

Fig. 10. GDP-^{14}C-Man was incubated with myeloma microsomal protein; the mixture was adjusted to 2% Triton X-100, sonicated, and centrifuged at 40,000 g for 1 hr. The supernate was applied to a column of Sephadex G-75 and eluted with 0.05 M ammonium bicarbonate.

---

remainder of the radioactivity as a single peak. Further analysis of these two radioactive fractions indicated that the material that was not adsorbed to the affinity column is pure ^{14}C-Man-P-DHPI and the material which bound specifically to the gel and which was eluted with α-methylmannoside is the glucosamine- and ^{14}C-mannose-containing oligosaccharide lipid described above. Subsequent evaluation of the solubility properties of these two mannose-containing lipids revealed the basis for not having observed the oligosaccharide lipid in our initial studies; extraction of incubation mixtures with butanol-pyridinium acetate solutions extracts only the Man-P-DHPI and does not extract the oligosaccharide phospholipid, whereas extraction of incubation mixtures with chloroform-methanol-water (1:1:0.3) extracts both of the lipids from the incubation mixtures. The purified mannolipids may

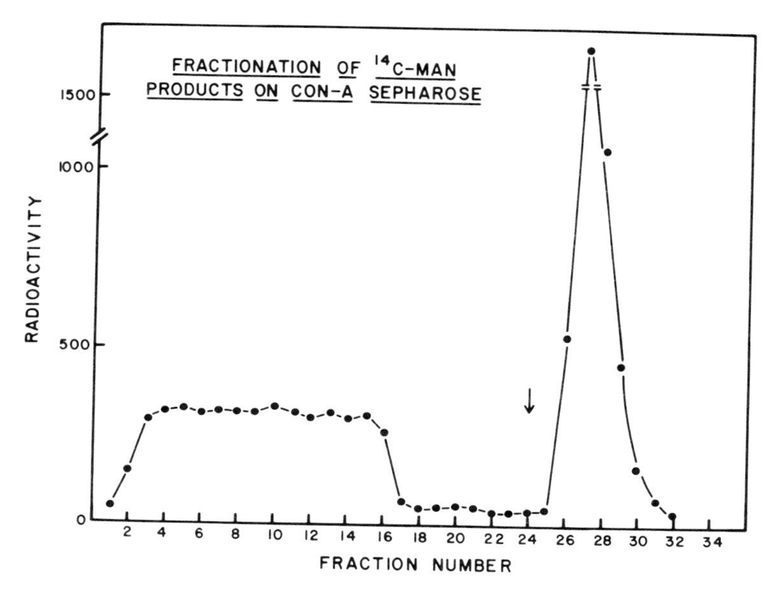

Fig. 11. The pooled radioactive material voided from Sephadex G-75 was applied to a column of concanavalin A Sepharose, equilibrated with 0.5M NaCl, 0.05 M NH_4HCO_3 and 1% Triton X-100. The column was washed with 60 ml of the same buffer; at the point indicated (↓), the buffer solution was adjusted to 10% α-methyl mannoside.

---

also be distinguished from each other on the basis of their fractionation on DEAE-cellulose as shown in Fig. 12. Thus, Man-P-DHPI is eluted from DEAE-cellulose in the chloroform-methanol-water solvent, whereas the oligosaccharide phospholipid is retained on the resin under these conditions and is eluted only when ammonium formate is added to the eluting solvent. The thin-layer chromatographic properties of the two lipids are compared in Fig. 13 indicating that several different solvent systems will distinguish Man-P-DHPI from the mannose-containing oligosaccharide phospholipid.

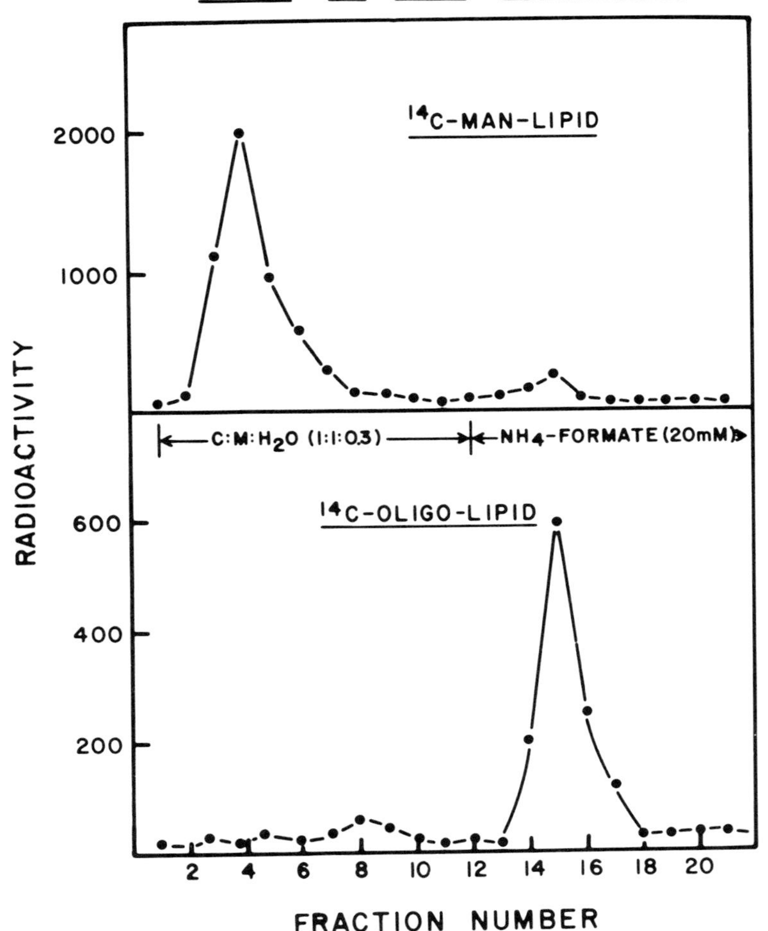

Fig. 12. Each of the radioactive lipid preparations was dissolved in chloroform (1); methanol (1); H_2O(0.3) and applied to a DEAE-cellulose column equilibrated with the same solvent. Each of the columns was eluted as indicated.

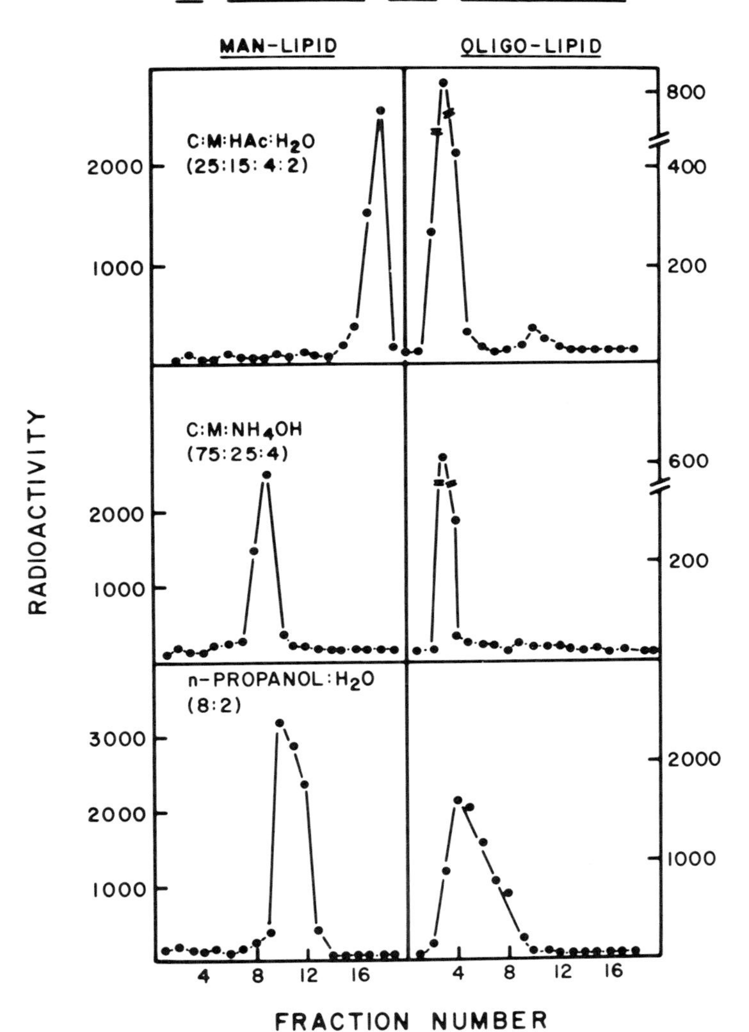

Fig. 13. Thin layer chromatography was conducted with purified preparations of ^{14}C-Man-P-DHPI and ^{14}C-Man-oligosaccharide phospholipid on Silica Gel G in the solvents indicated.

The complete structure of the oligosaccharide phospholipid has not yet been elucidated. The proposed structure presented in Fig. 9 indicates that the linkage between the oligosaccharide and the lipid moieties may involve a pyrophosphate bond and the following properties support this contention: (a) hydrolysis in dilute NH_4OH yields oligosaccharide phosphate; and (b) fractionation on DEAE-cellulose indicates a compound that is more acidic than the mannosyl monophosphoryl lipid. The lipid moiety has not yet been isolated but the properties of the oligosaccharide phospholipid are consistent with the assumption that the lipid is a dolichol as described by Behrens *et al.* (11).

THE ROLE OF MAN-P-DHPI AND OLIGOSACCHARIDE PHOSPHOLIPID IN INCORPORATION OF MANNOSE INTO PROTEIN

As shown in Fig. 14, incubation of microsomal preparations with pure ^{14}C-Man-P-DHPI results in the disappearance of radioactive mannose from the substrate with the concomitant, stoichiometric appearance of ^{14}C-Man in the oligosaccharide phospholipid; quantitatively, only small amounts of radioactivity appear in either the aqueous phase or in protein. Thus, there appears to be a rapid transfer of mannose from the monomannosyl lipid to the oligosaccharide lipid. Conversely as shown in Fig. 15, when pure ^{14}C-Man-oligosaccharide phospholipid is used as substrate with the microsomal membrane preparations, there is a disappearance of radioactivity from the oligosaccharide phospholipid with the concomitant appearance of radioactivity in protein and in the aqueous phase of the fractionated incubation mixture. Kinetically, therefore, the ^{14}C-Man-oligosaccharide phospholipid is synthesized, at least partially, by the transfer of mannose from ^{14}C-Man-P-DHPI and the ^{14}C-Man-oligosaccharide phospholipid appears to be the primary donor of radioactivity in the glycosylation of protein. Again, the radioactivity which appears in the water-soluble

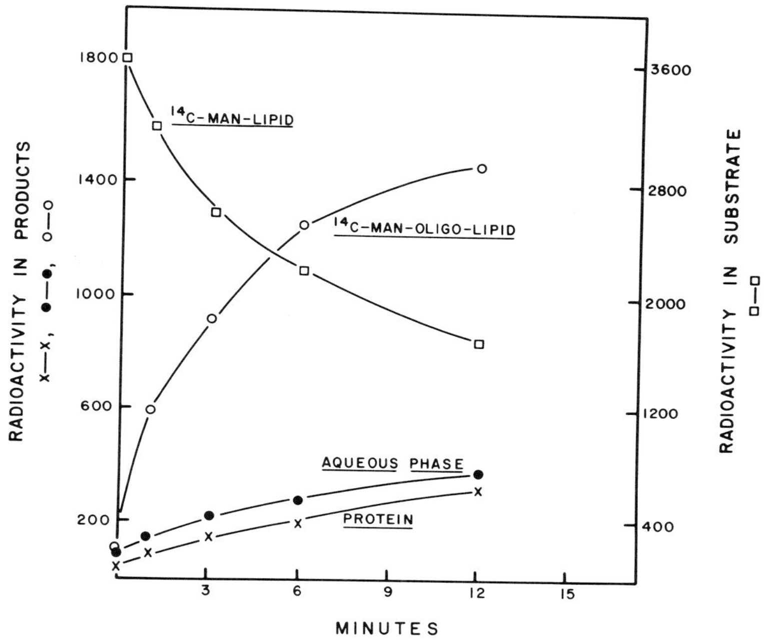

Fig. 14. Transfer of ^{14}C-Man from ^{14}C-Man-P-DHPI to oligosaccharide phospholipid by myeloma microsome preparation. Aliquots of the incubation mixture were adjusted to 1.5% NaCl and then extracted with chloroform-methanol (3:2); the organic phase contained ^{14}C-Man-P-DHPI. The insoluble interphase was extracted with chloroform-methanol-H_2O (1:1:0.3); the soluble fraction contained ^{14}C-Man-oligosaccharide phospholipid and the insoluble fraction contained protein. The protein fraction was washed several times in 10% trichloroacetic acid.

---

component of the incubation mixture under these conditions was isolated and characterized as ^{14}C-Man-oligosaccharide phosphate as described previously. Because of the rapidity with which the oligosaccharide phospholipid is decomposed in this system, it seems likely that

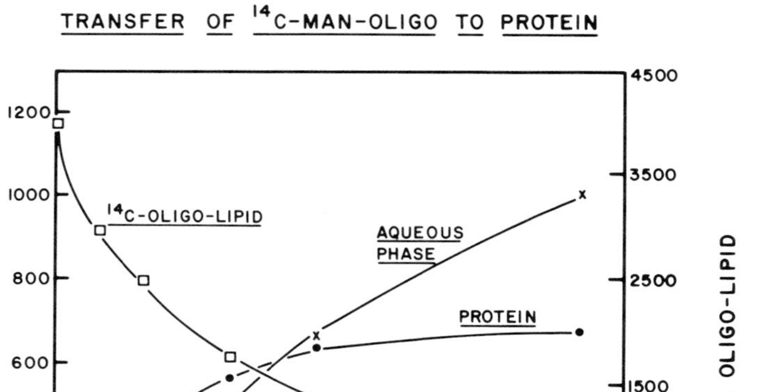

Fig. 15. ^{14}C-Man-oligosaccharide phospholipid was incubated with myeloma microsome preparation. Aliquots were treated as described in the legend to Fig. 14 in order to determine the radioactive content of each constituent.

enzymatic hydrolysis of the compound is occurring.

The identity of the endogenous protein(s) that serve as acceptor for ^{14}C-mannose in the sequence of reactions described above remains to be established. However, it should be pointed out that the family of proteins labeled by this sequence of events is distinct for two types of tissues studied. Thus, as illustrated in Fig. 16, the Triton X-100-solubilized proteins derived from the mouse myeloma tumor microsomal system exhibited the profile of radioactivity shown in Panel A. Regardless of whether the radioactive protein was derived

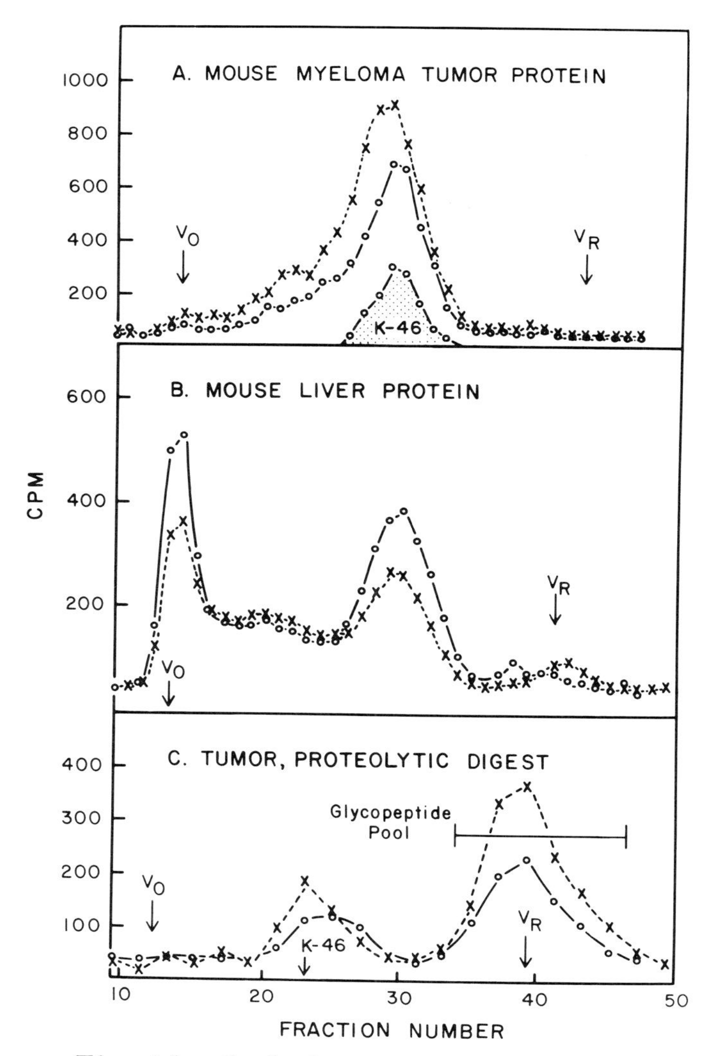

Fig. 16. Sephadex G-150 profiles of sodium dodecyl sulfate-solubilized protein from incubations using GDP-mannose (X---X) or mannolipid (O---O) substrates. Protein recovered from incubations was solubilized in 2.5% sodium dodecyl sulfate, 1% mercaptoethanol, 0.1 M EDTA, 0.05 M sodium phosphate, pH 8.0, applied to a column (1 x 40 cm)

of Sephadex G-150, and eluted with the same buffer containing 1% sodium dodecyl sulfate. A, mannose-containing protein from incubations with mouse myeloma microsomes. B, mannose-containing protein from incubations with mouse liver. C, tumor protein after proteolysis by Pronase and subtilisin.

---

from an incubation mixture with ^{14}C-Man-P-DHPI as substrate, or an incubation mixture with GDP-^{14}C-Man as substrate, a similar radioactive profile of the solubilized proteins was obtained. It is also of interest to note that the migration of the bulk of the radioactive proteins corresponded to the same relative position of migration of pure K-46 protein. The gel filtration profile of radioactive proteins obtained from an incubation using mouse liver preparations indicated (Panel B) that at least a portion of the proteins labeled under these conditions were markedly distinguished from those labeled in the tumor microsome preparations. In comparable incubation mixtures with microsome preparations from both tissues using ^{14}C-Man-oligosaccharide phospholipid as substrate, identical radioactivity profiles were obtained for the solubilized proteins.

The covalent attachment of the radioactive carbohydrate constituents to the polypeptide chain was indicated by digestion of the radioactive tumor proteins with proteolytic enzymes as shown in Panel C. Regardless of the substrate (GDP-^{14}C-Man, ^{14}C-Man-P-DHPI, or ^{14}C-Man-oligosaccharide phospholipid) used for biosynthesis of the labeled proteins, 70-80% of the radioactivity was converted to low molecular weight species under the conditions of proteolysis used in these studies. The low molecular weight products obtained by these procedures exhibited electrophoretic properties consistent with those that one would predict for glycopeptides derived from glycoproteins by proteolytic digestion.

In the myeloma tumor system , the role of the mannolipid in biosynthesis of K-46 protein remains to be firmly established. Recent experiments conducted in our laboratory indicate that 10-20% of the radioactive protein synthesized by the myeloma microsome system is immunoprecipitable with specific anti-K-46 antibody. Whether only this portion of the labeled protein products are, in fact, K-46 or whether a portion of the radioactivity is in a larger protein which is a precursor of K-46 is not known at this time. It is of interest to note that Mach *et al.* (12) recently suggested that the kappa-type light chain biosynthetic sequence may involve proteolytic cleavage of a larger precursor molecule to yield the light chain.

The observations that a mannose- and glucosamine-containing oligosaccharide phospholipid may be an intermediate in the biosynthesis of the core region of the oligosaccharide side chains of glycoproteins is of particular significance. If this "pre-assembled" oligosaccharide lipid proves to be an obligatory intermediate in the biosynthesis of the core region of oligosaccharide side chains of glycoproteins, it would establish an interesting analogy with oligosaccharide-polyisoprenol intermediates that function in the biosynthesis of bacterial polymers (13). It must be emphasized however, that the quantitative significance of the oligosaccharide phospholipid intermediates in mammalian glycoprotein biosynthesis remains to be established and it is entirely possible that the bulk of the core portion of the carbohydrate side chains of glycoproteins are synthesized by a series of individual glycosyl transferases functioning in sequence as has been clearly demonstrated to be the case with regard to the addition of distal sugar residues in the oligosaccharide side chains. On the other hand, the core region of the carbohydrate side chains of glycoproteins appears to be a relatively uniform structure regardless of the source of glycoprotein and, in fact, the general structural features of the oligosaccharide portion of the phospholipid (Fig. 9) are consistent with the general structural features that have been proposed for the core portions of several glycoproteins as shown in Fig. 17. Thus, the common core structures, proposed by Kornfeld

Proposed Glycoprotein Core Structures

Man-1 $\xrightarrow{\beta}$ 4-GlcNAc-1 $\xrightarrow{\beta}$ 4-GlcNAc $\xrightarrow{\beta}$ Asn

Ovalbumin, α-amylase, bromelain

Lee and Scocca, J. Biol. Chem., 247, 5753 (1972).

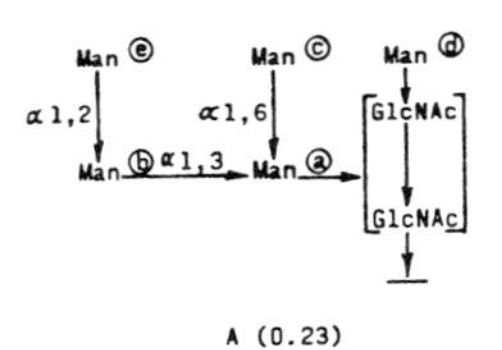

γ - M-Immunoglobulin

Hickman, et al., J. Biol. Chem., 247, 2156 (1972).

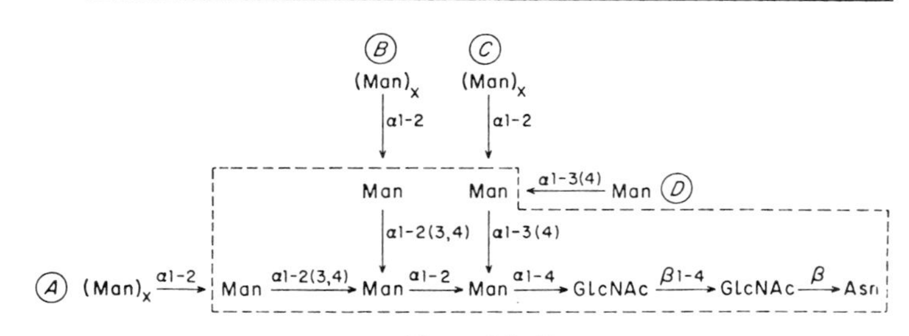

Thyroglobulin

Arima and Spiro, J. Biol. Chem., 247, 1836 (1972).

FIGURE 17

(14) and Spiro (15) and their coworkers, consisting of an oligosaccharide containing five mannose residues and two N-acetylglucosamine residues, with an N-acetylglucosamine residue at the reducing terminus to provide a point of attachment through an N-glycosidic linkage to asparagine are completely consistent with the structural analysis of the mannose- and N-acetylglucosamine-containing oligosaccharide phospholipid described above. Lee and Scocca (16) have firmly established the structure, Man-1 → 4-GlcNAc-1 → 4-GlcNAc → Asn, as the common peptide linkage region in several glycoproteins.

Of prime consideration in our further studies are the enzymatic details of attachment of the N-acetylglucosaminyl residue occupying the reducing terminus of the oligosaccharide that may form the N-glycosidic bond to the asparagine residue in the polypeptide chain. The mode of biosynthesis of the GlcNAc-asparagine linkage site has not been elucidated (2). This aspect of glycoprotein biosynthesis may be resolved if it is possible to demonstrate that an oligosaccharide lipid of the type described in these studies serves as the immediate precursor to this linkage region.

REFERENCES

(1) S. Roseman. In: Biochemistry of glycoproteins and related substances: Cystic Fibrosis, Part II, ed. E. Rossi and E. Stahl (S. Karger, New York, 1968) p. 244.

(2) R. G. Spiro. Ann. Rev. Biochem., 39 (1970) 599.

(3) P. J. O'Brien and E. F. Neufeld. In: Glycoproteins, 2nd Edition, Part B, ed. A. Gottschalk (American Elsevier Publishing Co., New York, 1972) p. 1170.

(4) F. Melchers. Biochemistry, 10 (1971) 653.

(5) J. W. Baynes and E. C. Heath. Fed. Proc., 31 (1972) 437.

(6) J. W. Baynes, A. -F. Hsu, and E. C. Heath. J. Biol. Chem., 248 (1973) 5693.

(7) L. F. Leloir, R. J. Staneloni, H. Carminatti, and N. H. Behrens. Biochem. Biophys. Res. Communs., 52 (1973) 1285.

(8) N. H. Behrens and L. F. Leloir. Proc. Nat. Acad. Sci. U. S. A., 66 (1970) 153.

(9) N. H. Behrens, A. J. Parodi, L. F. Leloir, and C. R. Krisman. Arch. Biochem. Biophys., 143 (1971) 375.

(10) J. B. Richards and F. W. Hemming. Biochem. J., 130 (1972) 77.

(11) N. H. Behrens, H. Carminatti, R. J. Staneloni, L. F. Leloir, and A. I. Cantarella. Proc. Nat. Acad. Sci. U. S. A., 70 (1973) 3390.

(12) B. Mach, C. Faust, and P. Vassalli. Proc. Nat. Acad. Sci. U. S. A., 70 (1973) 451.

(13) W. J. Lennarz and M. G. Scher. Biochim. et Biophys. Acta, 265 (1972) 417.

(14) S. Hickman, R. Kornfeld, C. K. Osterland, and S. Kornfeld. J. Biol. Chem., 247 (1972) 2156.

(15) T. Arima and R. G. Spiro. J. Biol. Chem., 247 (1972) 1836.

(16) Y. C. Lee and J. R. Scocca. J. Biol. Chem., 247 (1972) 5753.

DISCUSSION

L. DELUCA: Do you have any idea whether glycoproteins other than secreted ones involve the same lipid intermediates that you have described for secreted mannose-glycoproteins?

E.C. HEATH: No - until we prove that this is really involved in the secreted one, I think it is a moot point at this stage.

L. DELUCA: Do you have any ATP requirements for the *in vitro* synthesis of mannolipid and do you have the same inhibition that Frank Hemming finds with ADP?

E.C. HEATH: There is no requirement for ATP nor inhibition by ADP but we do get inhibition with GDP.

S.A. KORNFELD: Have you gone back and examined the secreted light chain to see if there is a high molecular weight mannose oligosaccharide on it that was missed?

E.C. HEATH: We are doing that right now, that is to isolate the glycopeptide. We do not have good analysis on it of this protein from the work by Melchers.

S. ROSEMAN: Did you try incorporating the water soluble oligosaccharide phosphate into the protein in the presence and absence of nucleotides?

E.C. HEATH: No.

S. ROSEMAN: Do you get transfer of sialic acid or other sugars into the lipid?

E.C. HEATH: No. We tried several sugars, sialic acid, galactose and fucose, but none were incorporated into the lipid in our *in vitro* system.

S. ROSEMAN: Do you get any acceleration or increase in the rate of formation of the water soluble oligosaccharide phosphate from the oligosaccharide lipid when you add GMP or GDP: in other words is it reversible?

E.C. HEATH: No, we observed no change in the rate of formation with either of those nucleotides. That really has not been done exhaustively, though.

S. ROSEMAN: What is your feeling about retinol sugars in biosynthesis?

E.C. HEATH: I believe that retinol-mannose is a real thing but I think its more appropriate for Dr. DeLuca to comment on its role in biosynthesis.

L. DELUCA: We have isolated retinol phosphate-mannose *in vitro*. We have a TLC system (chloroform: methanol: water: 60:25:4) which enables us to isolate two mannolipids from an *in vitro* system using an enzyme preparation from normal rats. One mannolipid has an R_F of 0.25 and the other has an R_F of 0.42 with slight variations. If we use an enzyme preparation from retinol deficient animals the mannolipid with R_F 0.25 disappears. This mannolipid is not synthesized using the membrane fraction of retinol deficient rats nor when the deficient animals are injected with ^{14}C-mannose and the lipids extracted. However, if chemically synthesized retinol-phosphate is added to the system containing enzyme from retinol deficient animals, the R_F 0.25 mannolipid reappears.

S. ROSEMAN: Are you looking at a non-specific glycosylation of a compound somewhat similar to dolichol or is retinol-phosphate-mannose a real intermediate in the glycose transfer reactions?

L. DELUCA: Hemming has shown that the different chain length polyisoprenals stimulated synthesis of mannolipid nonspecifically. The *in vivo* existence of the two mannolipids thus is very important. Twenty minutes after intraperitoneal injection of ^{14}C-mannose into rats, we isolated labeled glycophospholipids possessing R_Fs of 0.25 and 0.42, as obtained for the *in vitro* system - the ratio of radioactivity in the two is approximately 1:1. Vitamin A deficiency reduces the R_F 0.25 glycophospholipid specifically. However, the two mannolipids synthesized *in vivo* have not yet been characterized and may differ from the corresponding mannolipids synthesized *in vitro*. We have no information regarding the function of the 0.25 R_F

glycophospholipid.

R.W. JEANLOZ: What is your evidence for the homogeneity of your oligosaccharide phosphate intermediate? Have you tried sequential enzymic degradation?

E.C. HEATH: The evidence for homogeneity is based on fractionation on DEAE cellulose chromatography and electrophoresis. We have not tried any enzymatic degradation of the oligosaccharide to obtain further structural details.

A.D. ELBEIN: In the synthesis of the lipid-linked oligosaccharides, I was wondering whether you knew how many mannoses you were adding? That is, since the immunoglobin oligosaccharide is a branched structure, is it possible that the dolichol-phosphoryl mannose is adding terminal mannoses in the same way that the ficaprenyl-phosphoryl-mannose described by Lennarz adds terminal mannoses to the micrococcal mannan?

E.C. HEATH: On the basis of the specific activity of the ^{14}C-mannose in the product, we calculated that an average of three to five mannoses came from the radioactive substrate.

A.D. ELBEIN: Have you tried Smith degradation on the oligosaccharide to determine whether the ^{14}C-mannoses are terminal?

E.C. HEATH: No, we have not done any further structural analysis on this material. The difficult part is to get enough of this compound to do the things you and Dr. Jeanloz suggested.

CHEMICAL COMPONENTS OF SURFACE MEMBRANES RELATED TO BIOLOGICAL PROPERTIES

MARY CATHERINE GLICK
Department of Pediatrics
University of Pennsylvania School of Medicine

Abstract: Specific glycoproteins are expressed reproducibly on the cell surface in a variety of conditions. Similar groups of glycopeptides are found in the membranes after virus transformation and during mitosis and tumorigenesis. In cell differentiation, as measured by neurite formation in clones of mouse neuroblastoma, another group of glycopeptides is predominantly present. Observations of several kinds are compared and lead to suggestions concerning the arrangement of surface membrane components in the four conditions. A simplified model is presented suggesting that a particular group of glycoproteins is loosely associated with the membrane while another group extends through the membrane as well as into the external environment. It is possible that in each cell the variety of functions observed are mainly associated with one high molecular weight group of glycoproteins which differ in the arrangement and amount of monosaccharides. This group of glycoproteins and the loosely associated group confer specificities which correlate with the biological properties of the cell.

INTRODUCTION

The participation of surface membrane glycoproteins in biological functions is suggested by demonstrating that specific glycoproteins are expressed reproducibly on the cell surface in a variety of conditions. The presence or absence of the glycoproteins may alter the cell environment and markedly change cell-cell interactions, communication, transport, and any antibody complexing or hormone-binding reactions which the cell might have. It is also possible that the membrane components play some as yet unknown role in cell division.

One way of approaching this problem is to examine the components of the surface membrane under defined conditions and look for deviations from the normal or other comparative states. This approach is possible because of the availability of cells in culture which can be manipulated to respond to certain conditions or which through their genetic complement respond with particular phenotypic expressions. These cells in culture can be extracted for specific membrane components. In addition, whole surface membranes can be isolated from the cells. Thus by combining these parameters with the use of the radioactive technique of dual labeling, a variety of comparisons can be interpreted to suggest relationships of chemical components to biological properties.

Other approaches to this problem are presented in this Symposium and therefore will not be discussed here.

EXPERIMENTAL

The detailed procedures for most of the experiments described have been published so only a few relevant details will be given. It is important to note that since changes occur in the glycoproteins of the cells and surface membranes which are growth dependent (1,2) the cells for all experiments were harvested while growing exponentially unless otherwise stated.

The basic experimental procedure is outlined schematically in Figure 1.

Any kinds of comparisons can be substituted into this scheme. In all experiments discussed here, the cells were labeled with radioactive L-fucose. Minor differences were obtained with the use of radioactive D-glucosamine or D-glucose, correlating with the chemical composition of the glycoproteins. The glycoproteins from the membrane can be analyzed by chromotography on Sephadex G-200 or the oligosaccharide portion with a few amino acids attached can be examined on Sephadex G-50 or DEAE cellulose after

Pronase digestion of the polypeptides. The latter two procedures have been used most extensively. Thus far cells have been examined for four different functions: transformation, tumorigenesis, mitosis and differentiation.

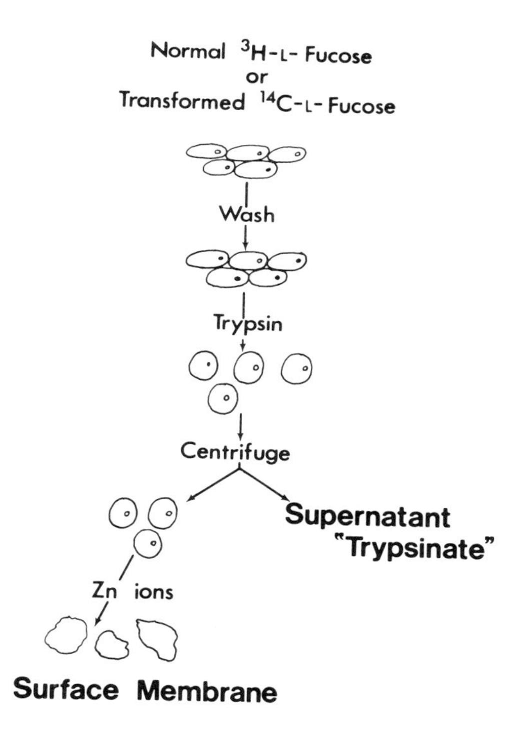

Fig. 1. Schematic outline of the basic protocol used for preparation of surface membranes and trypsinates.

RESULTS AND DISCUSSION

Transformation and tumorigenesis. Using the technique outlined in Figure 1 the radioactively labeled trypsinates and surface membranes from virus transformed cells were shown to contain a specific group of glycopeptides

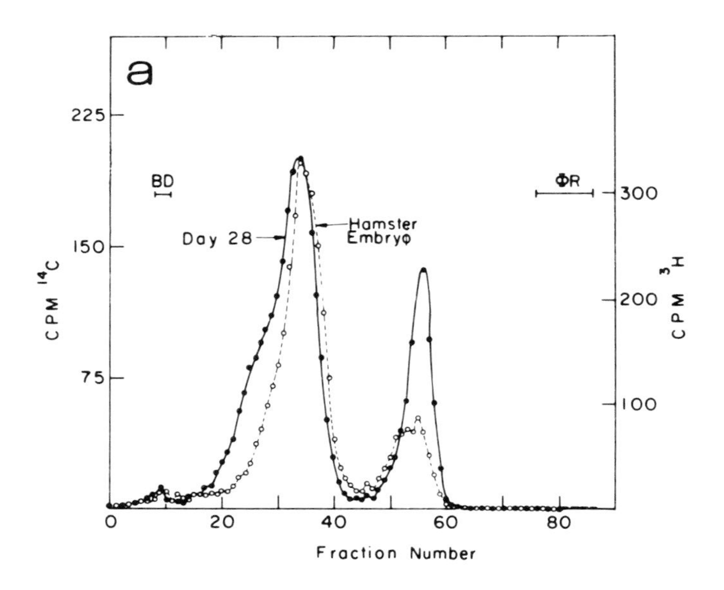
a
225
150
75
CPM ^{14}C
BD
ΦR
300
200
100
CPM ^{3}H
Day 28
Hamster
Embryo
0
20
40
60
80
Fraction Number

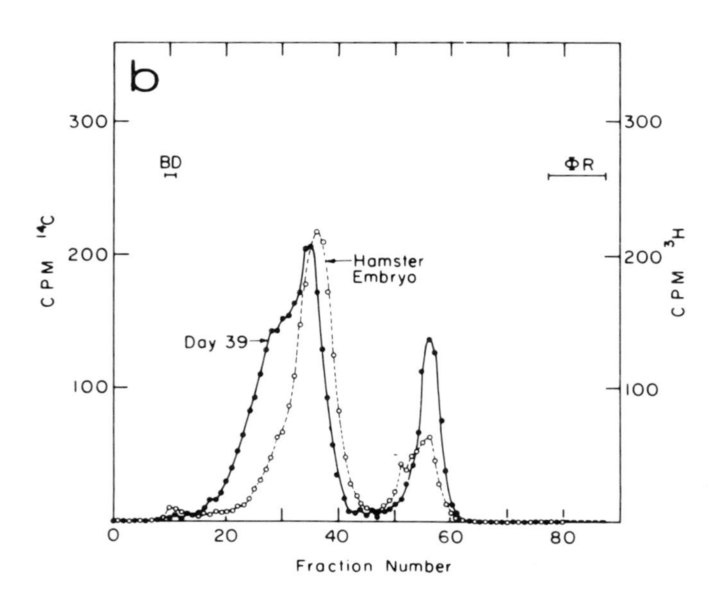
b
300
200
100
CPM ^{14}C
BD
ΦR
300
200
100
CPM ^{3}H
Day 39
Hamster
Embryo
0
20
40
60
80
Fraction Number

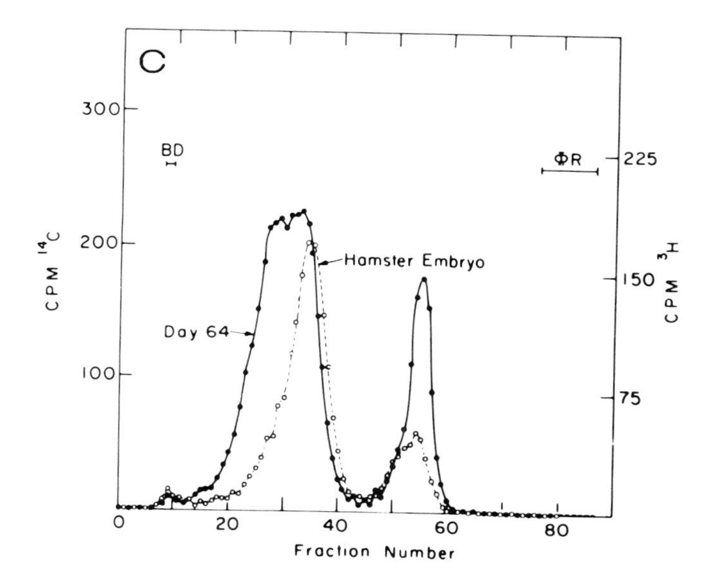

Fig. 2. Sephadex G-50 profile of surface membrane glycopeptides from control hamster embryo cells and cells infected with polyoma virus and subsequently examined after (a) 28; (b) 39; and (c) 64 days in culture. The radioactive glycopeptides were obtained from cells grown in the presence of L-[^{14}C] or [^{3}H]fucose, (O--O, ●—●, respectively). The trypsinates from the cells to be compared, were combined, digested with Pronase and chromatographed on Sephadex G-50. BD and ϕR are the fractions in which Blue Dextran 2000 and phenol red were eluted.

which was not present to the same extent in the normal counterpart. A clone of baby hamster kidney fibroblasts (BHK_{21}/C_{13}) and cells from this clone transformed by Rous sarcoma virus (C_{13}/B_4) were originally compared by filtration of the Pronase-digested glycopeptides through Sephadex G-50. A similar group of glycopeptides was found in other cells transformed by RNA or DNA viruses (4).

These studies were extended to follow the appearance of this specific group of glycopeptides after infection of hamster embryo cells with polyoma virus (5). Only as the cells began to form progressively growing tumors in adult hamsters was the appearance of this specific group of glycopeptides noted. Other phenotypic properties considered to be characteristic of transformation such as saturation density did not correlate with the appearance of these glycopeptides. Twenty-eight days after infection of the hamster embryo cells with polyoma virus, the glycopeptides appeared more like the uninfected hamster embryo cells (Fig. 2a) than 39 days after infection (Fig. 2b). By 64 days after infection the cells were highly tumorigenic and large amounts of the group of glycopeptides characteristic of virus transformation appeared (fractions 20-30, Fig.2c). In fact, a linear relationship was found with the amount of this specific group of glycopeptides present and the number of cells required to form tumors. That is, as the tumorigenecity increased, requiring fewer cells to form tumors, the amount of this group of glycopeptides increased. Similar results were obtained by analyzing hamster embryo cells which underwent rapid or slow transformation (5).

The correlation of this specific group of membrane glycopeptides with tumorigenesis was shown in another way (6). A hamster embryo cell line which was obtained after treatment by the chemical, dimethylnitrosamine (7) and a series of cloned variants of these cells were examined. The parent cell line gave phenotypic properties of transformed cells including tumor formation in adult hamsters. Examination of these cells in culture by the techniques described did not reveal this specific group of membrane glycopeptides. Figure 3a shows that the glycopeptides from the transformed cell line (DMNA) were similar to those of the untransformed hamster embryo cells. However, when cells derived from tumors formed by these

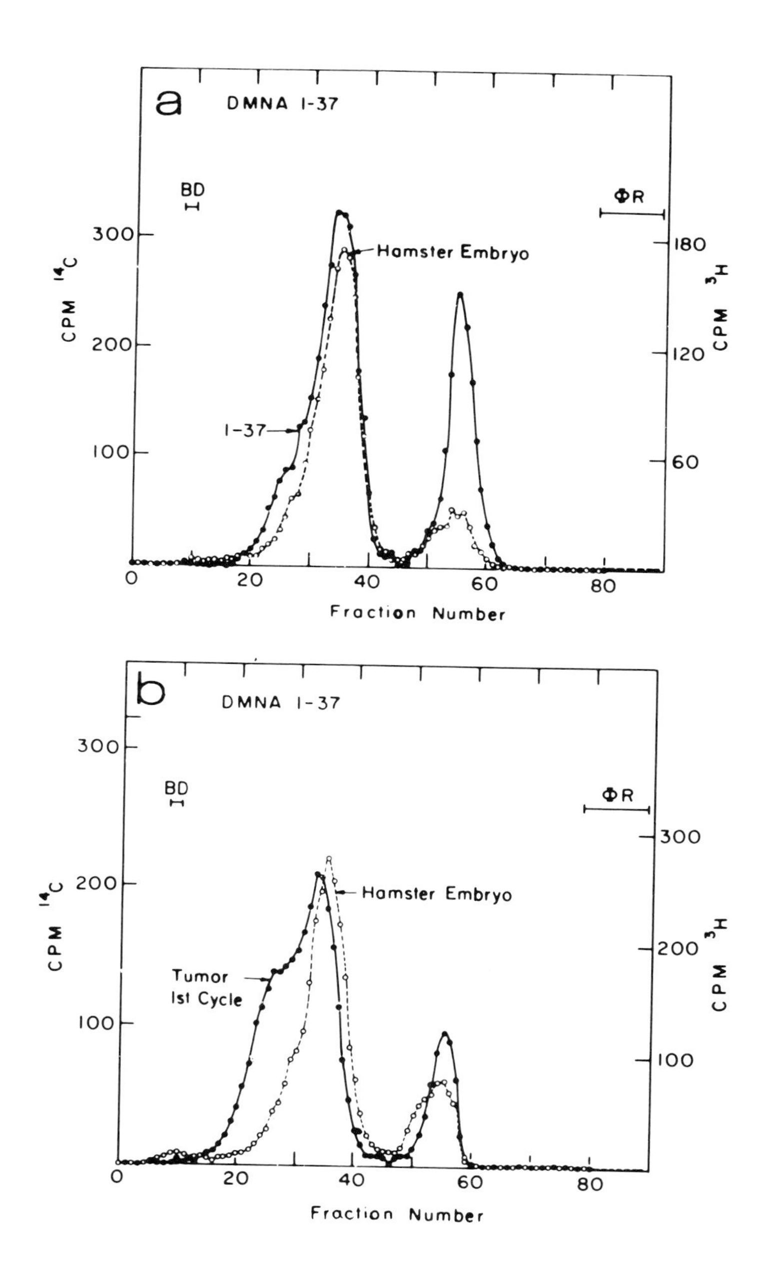
a
DMNA 1-37
BD
ΦR
Hamster Embryo
1-37
CPM ^{14}C
CPM ^{3}H
300
200
100
180
120
60
0
20
40
60
80
Fraction Number
b
DMNA 1-37
BD
ΦR
Hamster Embryo
Tumor
1st Cycle
CPM ^{14}C
CPM ^{3}H
300
200
100
300
200
100
0
20
40
60
80
Fraction Number

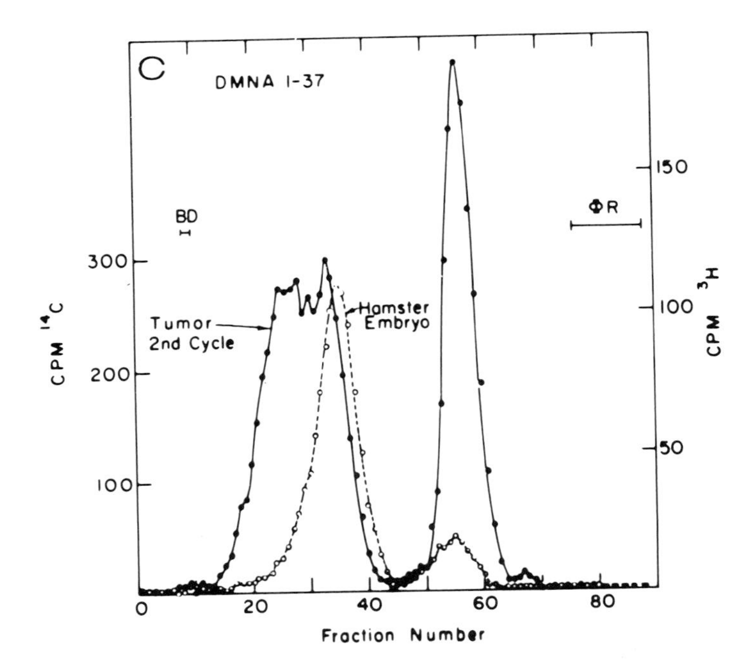

Fig. 3. Sephadex G-50 profile of trypsinates from cells (a) transformed (1-37) after treatment with dimethylnitrosamine (DMNA); (b) derived from a tumor of the DMNA cells shown in Figure 3a (Tumor 1st cycle); and (c) derived from the tumor cells shown in Figure 3b (Tumor, 2nd cycle). Secondary hamster embryo cells served as a control (Hamster embryo). All cells were made radioactive by growth in the presence of L-[^{14}C]- or [^{3}H]fucose, (O--O and ●–●, respectively). The trypsinates were combined and digested with Pronase prior to gel filtration. Further details are described in legend to Figure 2 and ref. (6).

cells were examined, these glycopeptides were prevalent (fractions 20-30, Fig. 3b). When these tumor cells were reinnoculated into adult hamsters, these glycopeptides were even more abundant (fractions 18-30, Fig. 3c). A cloned variant of the parental transformed line was reverted to lower tumorigenecity and loss of other phenotypic properties of transformation. Again these cells in culture did not show the specific glycopeptides. Small tumors were formed by these cells and examination of the tumor cells revealed the presence of the particular membrane glycopeptides. Reinnoculation of these tumor cells into adult hamsters greatly increased the tumorigenecity and also increased the amount of specific glycopeptides present in the membrane. Another cloned variant derived from the DMNA cells was more tumorigenic than the parental cell line. Again the cells in culture showed very little of these membrane glycopeptides and when the tumor cells were examined, large amounts of these glycopeptides were present. In all of the cells examined, other properties characteristic of the transformed phenotype were not consistant with either the membrane glycopeptides or with tumorigenesis (6).

In addition to these transformed cells, one other type of transformation did not show the presence of this particular group of glycopeptides. That is, mouse lymphocytes which were stimulated by Concanavalin A or by antibodies did not show a shift in glycopeptide profile which differed significantly from the untransformed lymphocytes. However, a marked increase in the incorporation of radioactive precursors into the glycoproteins was noted after stimulation (8).

Thus during two aberrant functions of the cell, virus transformation and tumorigenesis, the surface membrane glycopeptides differ from those present before these functions are acquired. Transformation after treatment by a carcinogen or lymphocyte stimulation does not produce this altered cell surface.

<u>Mitosis</u>. BHK_{21}/C_{13} fibroblasts are arrested in metaphase after treatment with vinblastin sulfate. Two populations of cells both grown in the presence of vinblastin sulfate can be obtained. One population, arrested in metaphase will come off the monolayer and can be harvested

directly. The second population not yet in metaphase remains attached to the monolayer and can be harvested by brief trypsinization and serve as a control.

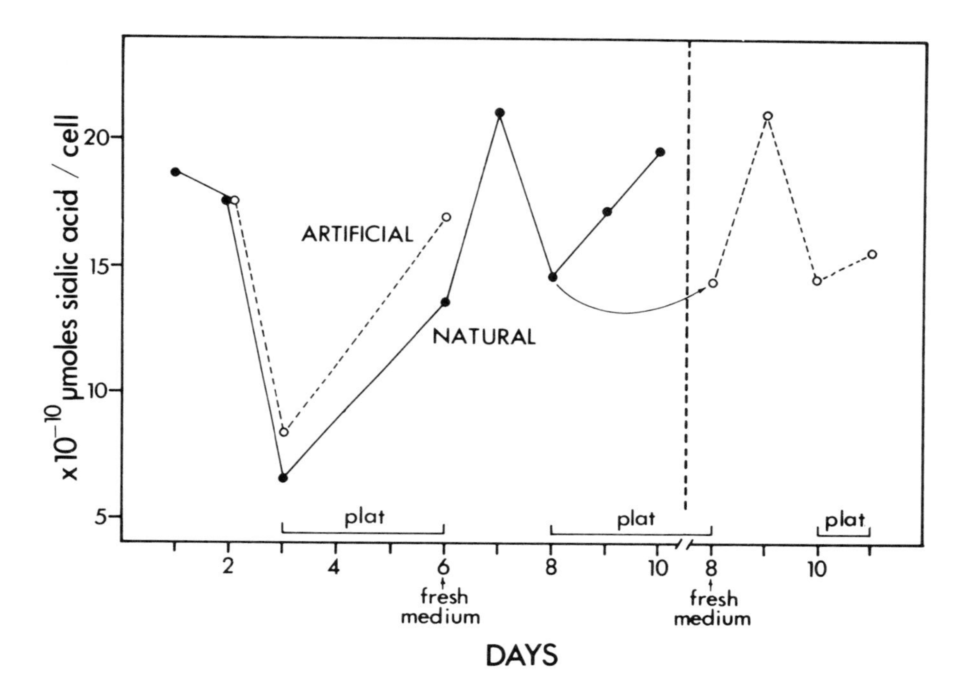

Fig. 4. Sialic acid content of L cells in exponential and confluent growth. L cells in suspension culture, growing exponentially (2-7 x 10^5 cells/ml of medium) on all days with the exception of the days designated "plat". On these days the cells were artificially concentrated (0--0) or naturally grown (●—●) to 10-20 x 10^5 cells/ml of medium. Duplicate aliquots were removed from the cultures each day to determine the cell number and the sialic acid and protein contents of the cells. The generation time of the cells when growing exponentially was 20 hours.

When comparisons of the surface membrane glycopeptides from these two populations were made as outlined in Figure 1 and examined on Sephadex G-50 after Pronase digestion, the cells in metaphase showed small amounts of a specific group of glycopeptides not present to the same extent in the control cells (9). Thus fibroblasts in mitosis express reproducible changes in the membrane glycoproteins.

Several other observations have suggested the relationship of the surface membrane glycoproteins to cell division. 1) These observed alterations or shift in the glycopeptide profile were observed to be growth dependent (2). 2) Changes in the monosaccharide content of KB cells were observed throughout the cell cycle (1). 3) A dramatic drop in sialic acid content was observed when L cells, mouse fibroblasts, enter the plateau or stationary phase of growth. This is represented in Figure 4, where the sialic acid content of the cells is plotted against the numbers of days in culture. The stationary phase of growth was determined by the cell number and is marked on the graph. The sialic acid content of the L cells was approximately 1.7 nmoles per cell. These cells, artificially concentrated to confluency or grown to confluency, showed a decrease in sialic acid content when compared to the cells growing exponentially. However, on Day 2, 3 and 4 of confluency, the sialic acid content increased to the content of sialic acid of the exponentially growing cells. The surface membrane reflected this change in sialic acid content.

Although these experiments suggest a correlation of the surface membrane glycoproteins with cell division, they unfortunately do not delineate any mechanism of their envolvement.

Differentiation. As a model system for cell differentiation, clones of mouse neuroblastoma, C-1300, were examined. Differentiation is expressed in these clonal lines by axon or dendrite formation and the expression of enzymes leading to the synthesis of neurotransmitters. The variety of clonal lines which have been isolated and well characterized (10) make it possible to compare the composition of the surface membranes to a number of functions.

Table 1 lists the clones of mouse neuroblastoma cells which were examined and summarizes their characteristics (11). Three of these clonal lines have the ability to form axons or neurites while two do not extend processes under any known conditions. The five clones differ in their ability to synthesize various neurotransmitters as shown in Table 1 by the enzyme activities. There are also differences in the ability to respond to electrical stimuli. All of these cell lines were obtained from Dr. Marshall Nirenberg, N.I.H.

TABLE 1

Properties of clones of mouse neuroblastoma C-1300

Clone	Tyrosine hydroxylase	Choline acetyl-transferase	Acetyl-choline-esterase	Neurites
	pmoles product formed/min/mg protein			
NIE-115	980	0.1	256,000	+
NS-20	0	490	48,000	+
N-18	2	2	105,000	+
N-1	70	22	23,000	-
NIA-103	0	2	19,000	-

A comparison was made of the clonal lines which form axons to those lines which do not form axons by the experimental procedure similar to that outlined in Figure 1 (14). The comparisons in this series of experiments were made with confluent cells as many of the properties of the differentiated state are expressed to a greater extent in confluency (12, 13). Sequential removal of the membrane glycopeptides was achieved by a two-step trypsinization procedure and the sum of the glycopeptides which were removed was similar to those removed by the trypsinization procedure used with the other studies. This sequential

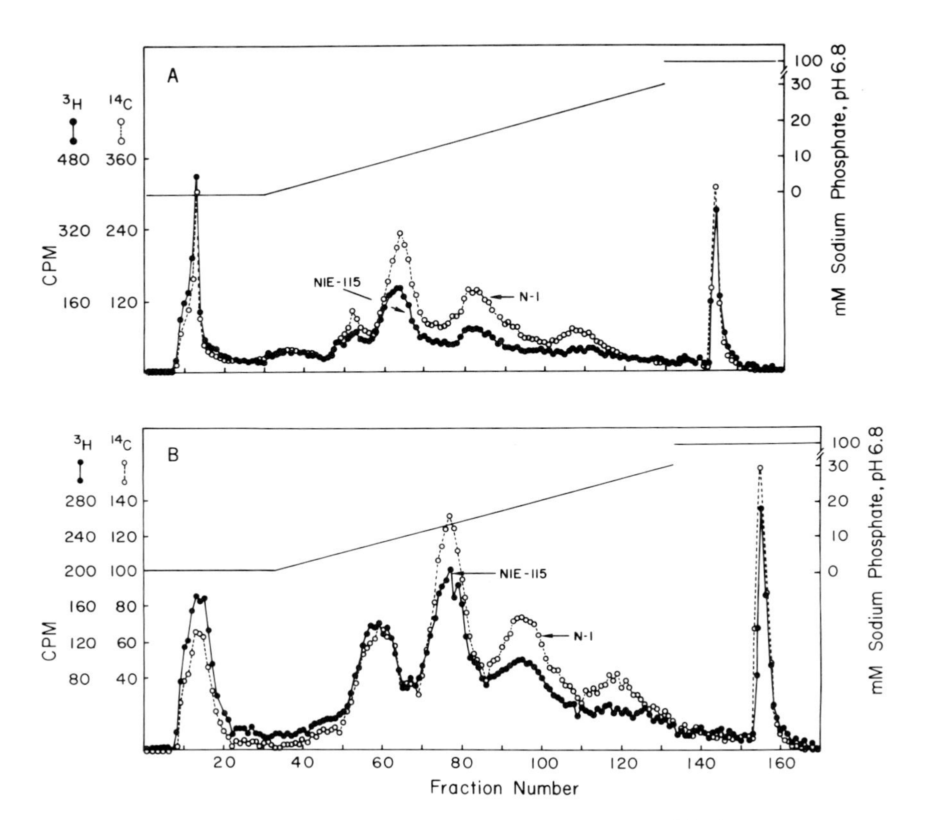

Fig. 5. Chromatography on DEAE-cellulose of Pronase-digested trypsinates from clones of mouse neuroblastoma, C-1300. Cells were grown, labeled, harvested, trypsinized, and were combined and digested with Pronase as described (14). After precipitation of the mixture with 5% trichloroacetic acid, followed by dialysis, the radioactive glycopeptides were separated on a column of DEAE-cellulose (16 x 1 cm) with sodium phosphate buffer. (a)Trypsinates A and (b) trypsinates B of clone N-1 labeled with L-[^{14}C]fucose (0--0); and clone NIE-115 labeled with L-[^{3}H]fucose (●—●), with Y. Kimhi and U. Z. Littauer.

treatment with trypsin appears to separate certain glycopeptides. The glycopeptides were examined by filtration through Sephadex G-50 after Pronase digestion. Among the five clonal lines examined, the membrane glycopeptides from the axon-forming cells were similar and differed from those of the axon-minus cells (14). The difference was also seen after chromatography of the glycopeptides on DEAE-cellulose. Figure 5 shows the differences observed between clone NIE-115, axon-forming cells, and clone N-1, axon-minus cells. Trypsinate A represents the glycopeptides which were most readily removed by trypsin while B represents those removed by the second trypsinization. The membrane glycopeptides from the non-axon-forming cells appeared more like those derived from the virus transformed and tumor cells than those from the axon-forming cells which appear more similar to non-transformed cells. There was no obvious correlation of the glycopeptides to the expression of other differentiated functions such as the enzymes involved in neurotransmitter synthesis (14).

It may be possible that the positive correlation which exists between the glycopeptides and axon formation may be the expression of the carbohydrate configuration which is necessary for the membrane to extend into processes or aid in performing the specialized nerve cell functions.

Chemical composition. In order to determine the chemical nature of the reproducible differences in the membrane glycopeptides which were observed with the variety of cell functions, the Pronase-digested glycopeptides were fractionated by chromatography on DEAE cellulose. Figure 6 shows this separation of the membrane glycopeptides from non transformed (BHK_{21}/C_{13}) and virus transformed fibroblasts (C_{13}/B_4). By collecting fractions from the Sephadex G-50 columns and subsequently purifying them on DEAE cellulose it was found that Sephadex G-50 fractions 20-30 (see Fig. 2) represented areas IV-VII of the DEAE cellulose column (Fig. 6). Fractions 30-40 (see Fig. 2) represented areas I-III with some overlap into area VII. Indeed each glycopeptide area of the DEAE cellulose column was accounted for in an inverse positioned relationship by dividing the Sephadex column fractions

20-45 into seven fractions. Sephadex column fractions 50-60 (see Fig. 2) were found to be dialyzable and contain in addition to free fucose, other small carbohydrate complexes which are currently being studied.

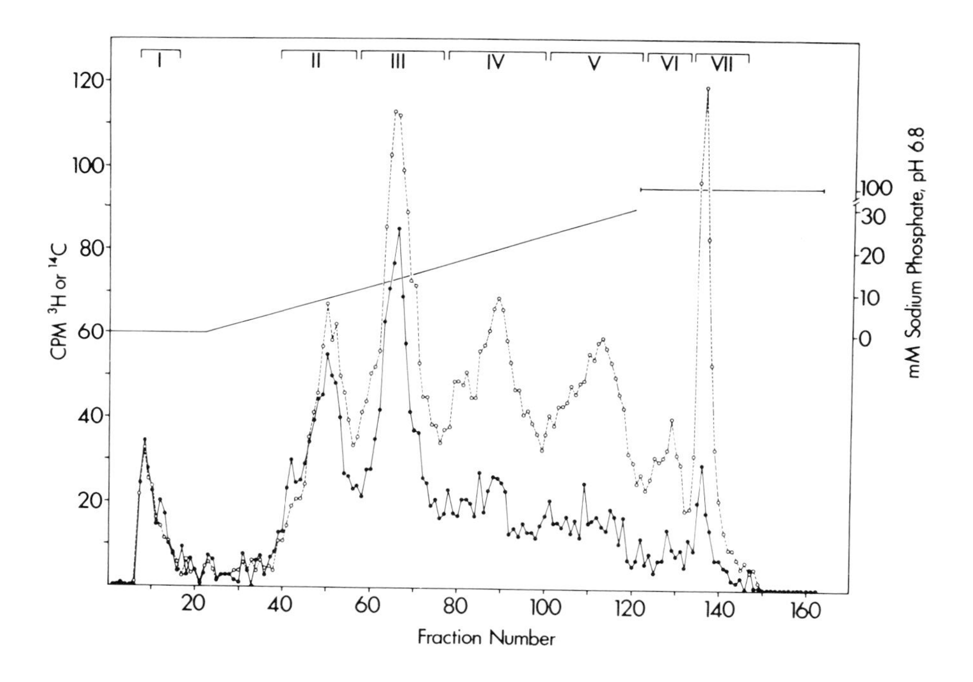

Fig. 6. Cochromatography on DEAE-cellulose of Pronase-digested trypsinates from BHK_{21}/C_{13} and C_{13}/B_4 fibroblasts grown in the presence of L-[^{14}C]- and [^{3}H]fucose (0--0, ●—●, respectively). The Roman numerals designate the fractions which were combined for further characterization. See legend to Figure 5.

Areas IV-VII (Fig. 6) are those glycopeptides which are characteristic of the virus transformed or tumorigenic cell. The glycopeptides of areas IV-VII are found also in the normal membranes when large amounts of glycoproteins are processed although the proportion to areas I-III is much less than in the virus transformed cells as shown in Figure 6.

TABLE 2

Comparison of the sialic acid content of membrane glycopeptides

DEAE cellulose fractions	C_{13}	C_{13}/B_4	$\frac{C_{13}/B_4}{C_{13}}$
	Sialic Acid		
	μmoles per 10^9 cells		molar ratio
I-III	0.16	0.17	1.06
IV-VII	0.28	0.68	2.43

Analysis of the sialic acid content of each of these areas are summarized in Table 2 to point out the differences between the glycopeptides which are characteristic of the virus transformed cells (areas IV-VII) and their normal couterpart (areas I-III). When expressed as μmoles per 10^9 cells the amount of sialic acid in both cell types was similar in areas I-III. In contrast the amount of sialic acid found in the glycopeptides of areas IV-VII of the transformed cells was more than twice that of the C_{13} cells as was suggested by the increased amounts of radioactive fucose. In addition, the glycopeptides of areas IV-VII of both cells showed increasing amounts of sialic acid in the individual glycopeptides when compared to glycopeptides I-III.

Fractions IV-VII represent larger glycopeptide fragments than the fractions I-III. That is, the glycopeptides of these fractions contained increased amounts of galactose and mannose as well as sialic acid and fucose. This is suggested by a similar molar ratio of fucose: galactose: mannose: sialic acid (0.5:2:1:1) in these glycopeptides.

Although the glycopeptides of areas I-VII which have been analyzed are by no criteria completely individual glycopeptides some information can be obtained. The analytical results thus far suggest that the glycopeptides characteristic of the virus transformed cell (areas IV-VII, Fig. 6) are larger and more branched than those found in both the transformed cell and the normal counterpart (areas I-III, Fig. 6). It is anticipated that the detailed structural analyses of the more purified glycopeptides will reveal more subtle differences as well as possible differences in the carbohydrate to amino acid linkages.

Functional glycoproteins. The possibility exists that the most accessible proteins and glycoproteins on the cell periphery at a given time are those which participate not only in interactions envolving other cells but also in the cells response to the external environment. It is therefore important to examine the components which are on the outside of the membrane. Observations of several different kinds are summarized and interpreted to suggest an arrangement of the membrane glycoproteins.

The first observation suggests that only one group of high molecular weight proteins is exposed to the external environment to a significant extent (15). This was demonstrated by the use of the technique of lactoperoxidase catalyzed iodination of the proteins of the intact cells with radioactive iodine. Only the proteins on the outer surface are iodinated as lactoperoxidase does not enter the cells. The surface membranes isolated from the iodinated whole cells showed only one protein to contain significant radioactivity. When the surface membranes were iodinated after isolation all but one group of proteins were iodinated suggesting that either most of the proteins are on the cytoplasmic side of the membrane or that the proteins undergo rearrangement during the isolation procedures (15). The high molecular weight protein which is iodinated stains with periodic acid-Schiff's reagent on polyacrylamide gels, suggesting that it is a glycoprotein. These results were obtained using mouse fibroblasts grown in suspension culture and baby hamster kidney fibroblasts grown on monolayers.

A second observation suggests that the groups of glycopeptides seen after Pronase digestion of the glycoproteins can be found on one high molecular weight glycoprotein. Extraction of IMR-32, a cell line derived from human neuroblastoma cells, with Triton X100 yields a high molecular weight component. Digestion of this extract with Pronase produces glycopeptides similar to those removed from the surface of these cells with trypsin (16).

The third observation concerns the two-step trypsinization procedure used to remove sequentially, the glycopeptides from the clones of mouse neuroblastoma (14). The first trypsinization removes preferentially one group of glycopeptides suggesting that these glycopeptides are in a more accessible location on the cell surface.

The fourth observation, the extraction of the cells with EDTA also suggests that this group of glycoproteins is more readily removed than the other glycoproteins. SV40-transformed hamster embryo cells were grown in the presence of L-[^{14}C]- or [^{3}H] fucose, washed and divided into two groups. A trypsinate was obtained from one half of the flasks of radioactive cells as outlined in Figure 1. The cells were removed from the remaining flasks with 0.5mM EDTA. The cell treated with trypsin were extracted subsequently with EDTA and the cells removed with EDTA were trypsinized subsequently. Thus four fractions were processed, combining those to be compared for Pronase digestion and gel filtration. Figure 7a shows the profile of membrane glycopeptides after filtration on Sephadex G-50 of the trypsinates from the SV40 transformed hamster cells compared with secondary hamster embryo cells. The profile of the SV40-transformed cells is similar to that seen in Figure 2 for the virus transformed or tumor cells showing the shift to more rapidly migrating glycopeptides (fractions 20-30) compared to that of the normal counterpart. Figure 7b compares the profiles of the EDTA extract of these cells with the cells treated with EDTA following removal of the trypsinate. The latter serves as a control

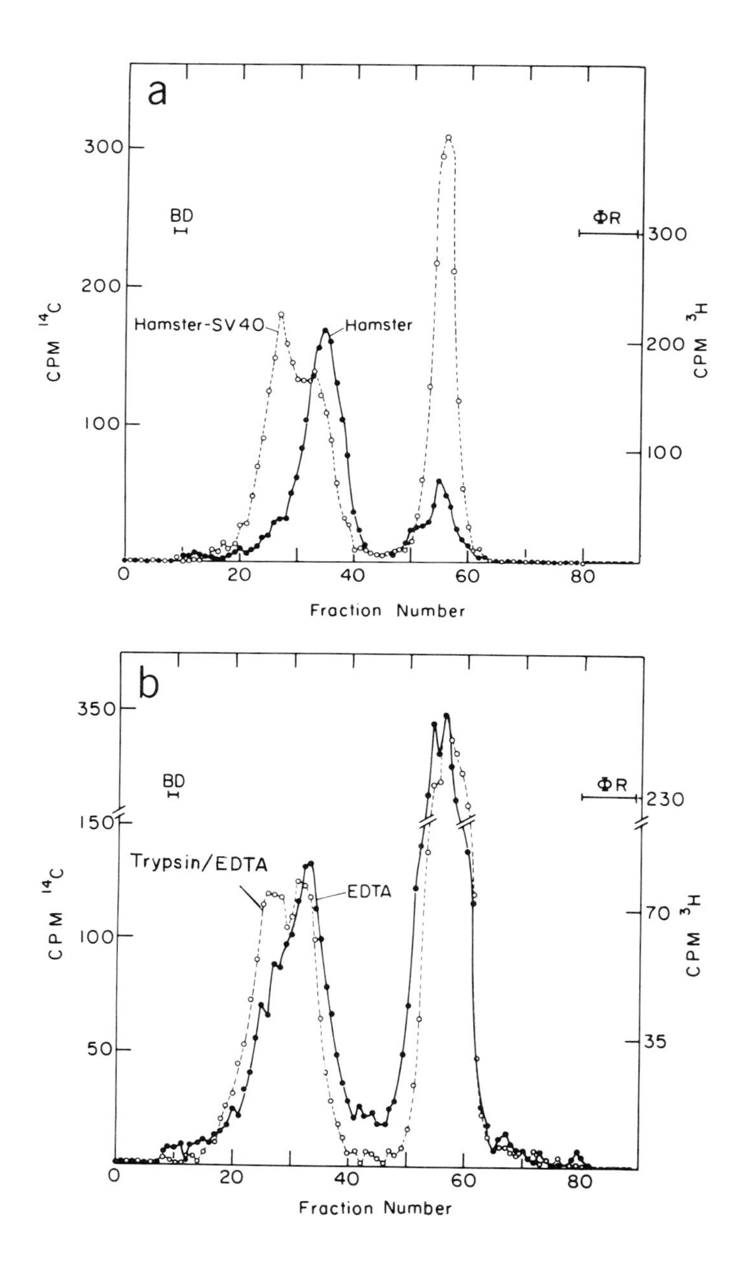
a
300
200
100
CPM ^{14}C
BD
ΦR
300
200
100
CPM ^{3}H
Hamster-SV40
Hamster
0
20
40
60
80
Fraction Number
b
350
150
100
50
CPM ^{14}C
BD
ΦR
230
70
35
CPM ^{3}H
Trypsin/EDTA
EDTA
0
20
40
60
80
Fraction Number

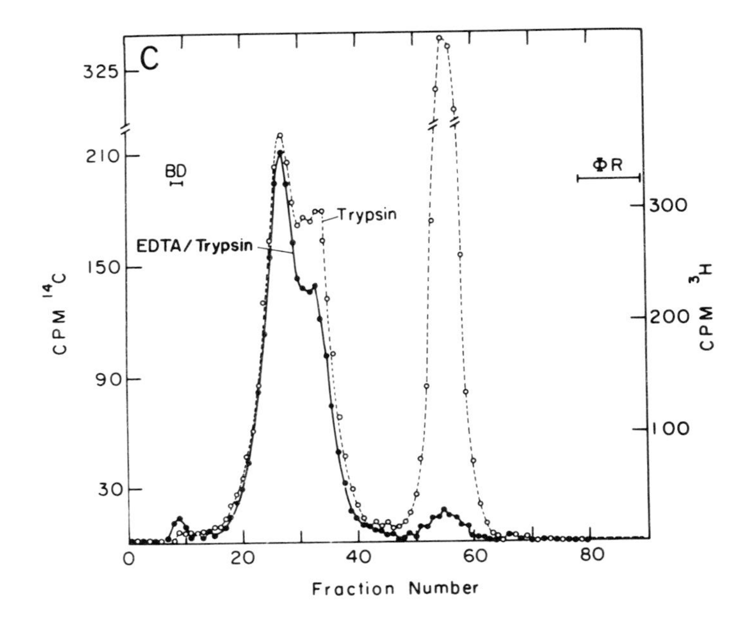

Fig. 7. Sephadex G-50 profile of fucose-containing glycopeptides from the surface of SV40-transformed hamster embryo cells. (a) Trypsinate from the transformed cells (Hamster-SV40) compared with a trypsinate from secondary hamster embryo cells (Hamster); (b) extracts of the transformed cells after treatment with 0.5 mM EDTA. Trypsin/EDTA, the cells of Figure 7a after removal of the trypsinate were subsequently treated with EDTA for 10 minutes; EDTA, cells treated with EDTA for 20 minutes without trypsinization; (c) trypsinates from the transformed cells. EDTA/Trypsin, the cells of Figure 7b (EDTA) after treatment with EDTA were subsequently trypsinized; Trypsin, trypsinate of the transformed cells. All fractions which were compared were combined and digested with Pronase before gel filtration. Details are in the legends of Figures 2 and 3. ^{3}H, ●—●; ^{14}C, O--O.

showing that the group of glycopeptides removed by EDTA extraction is not just a procedural artifact since both groups of glycopeptides are seen when there is a prior trypsinization. EDTA removes predominantly one group of glycopeptides, fractions 28-40 , when the cells are extracted before trypsinization. Figure 7c shows the profile of the glycopeptides removed by trypsin after a prior EDTA treatment. Again showing that EDTA removed the specific glycopeptides found in fractions 28-38 as a decrease is noted when compared to the regular trypsinate. EDTA appeared to generate larger amounts of low molecular weight material (fractions 50-60, Fig. 7b) than the trypsin treatment. When the cells were treated with EDTA following trypsinization, large amounts of this material was still observed (Fig. 7b) whereas the reverse, trypsin treatment after EDTA extraction yielded negligible amounts of this material (fractions 50-60, Fig. 7c).

The fifth observation suggests that a group of glycopeptides similar to that which is removed by EDTA or sequential trypsinization is found in the media in large amounts. It is possible that this group of glycopeptides represents those which are "shed" from the cell surface. These glycopeptides may be for this reason only loosely associated with the membrane and readily accessible for removal.

The final observation is that these loosely associated glycoproteins are those which contain less sialic acid (Table 2 and Fig. 6). They are not the group of glycopeptides which are observed as characteristic of mitosis, virus transformation and tumorigenesis. Cell differentiation, defined in terms of neurite formation, produces an excess of these glycopeptides.

All of the above observations may be summarized and schematic drawings can be used to depict the arrangement of membrane glycoproteins (Fig. 8). The major glycoprotein extends through the membrane and into the external environment where it is loosely associated with another glycoprotein of similar monosaccharide composition to that found in the normal cell or axon-forming cells. The black areas represent the portion of the glycoprotein where the monosaccharide composition resembles that which is predominantly seen as characteristic of virus transformation and

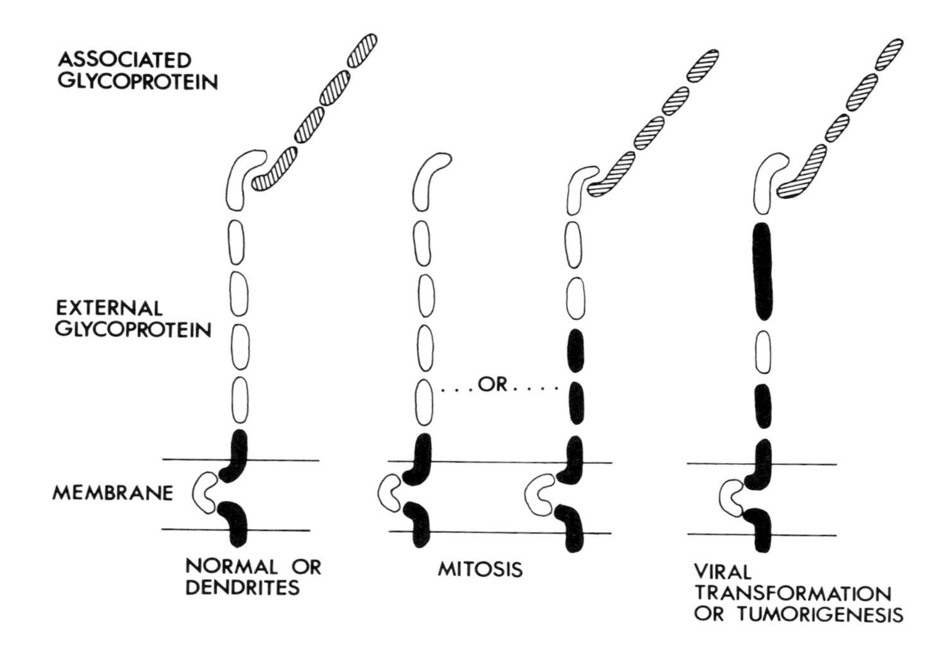

Fig. 8. Schematic arrangement of glycoproteins in the surface membranes from cells with different biological properties.

tumorigenesis. Mitosis can be represented as lack of the associated glycoprotein or additional amounts of the "transformed" glycopeptides. From the chemical studies of the Pronase-digested glycopeptides a scheme can be drawn showing more detail of each segment of Figure 8. The polypeptide backbone is represented by a staff with the variety of monosaccharides attached, represented as circles (Fig. 9). In virus transformation and tumorigenesis (represented by the closed circles) the monosaccharides units of the membrane are larger and more branched than in the "normal" membrane or than found in the associated glycoprotein (open circles, Fig. 9). Each oligosaccharide unit represents a Pronase-digested fragment.

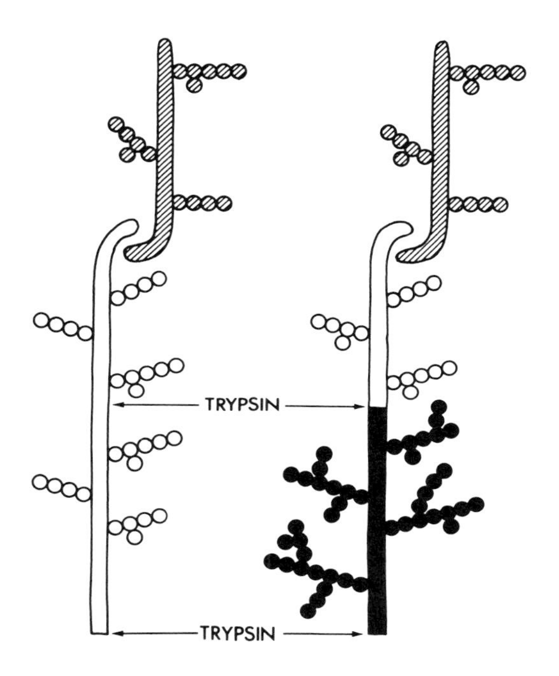

Fig. 9. Simplified arrangement of membrane glycoproteins. The circles represent monosaccharides attached to a polypeptide backbone, each unit representing a Pronase fragment. The trypsin fragments are indicated. The black areas are the glycopeptides characteristic of virus transformed cells.

Obviously both of these schemes are oversimplified and many alternate arrangements are possible. However, even from these simplified concepts one can see many possibilities for variation in specific arrangements. As is well known from studies of the blood group substances (17) a change of one monosaccharide unit can confer a different specificity. Alterations in these surface components could lead to the cell acquiring alternate functions and biological properties. Indeed, the four different biological properties described have reproducible chemical differences in the membrane glycoproteins. More detailed analyses are required to determine the full extent of the differences.

REFERENCES

(1) M.C. Glick, E.W. Gerner and L. Warren, J. Cell. Physiol. 77 (1971) 1.

(2) C.A. Buck, M.C. Glick and L. Warren, Biochemistry 10 (1971) 2176.

(3) C.A. Buck, M.C. Glick and L. Warren, Biochemistry 9 (1970) 4567.

(4) C.A. Buck, M.C. Glick and L. Warren, Science 172 (1971) 169.

(5) M.C. Glick, Z. Rabinowitz and L. Sachs, Submitted for publication.

(6) M.C. Glick, Z. Rabinowitz and L. Sachs, Biochemistry 12 (1973) 4864.

(7) E. Huberman, S. Salzberg and L. Sachs, Proc. Nat. Acad. Sci. USA 59 (1968) 77.

(8) L. Stavy and M.C. Glick, unpublished observations.

(9) M.C. Glick and C.A. Buck, Biochemistry 12 (1973) 85.

(10) T. Amano, E. Richelson and M. Nirenberg, Proc. Nat. Acad. Sci. USA 69 (1972) 258.

(11) B.K. Schrier, S.H. Wilson and M. Nirenberg, Methods in Enzymology, in press

(12) A. Blume, F. Gilbert, S. Wilson, J. Farber, R. Rosenberg, Jr. and M. Nirenberg, Proc. Nat. Acad. Sci. USA 67 (1970) 786.

(13) D. Schubert, H. Tarikas, A.J. Harris and S. Heinemann, Nature New Biol. 233 (1971) 79.

(14) M.C. Glick, Y. Kimhi and U.Z. Littauer, Proc. Nat. Acad. Sci. USA 70 (1973) 1682.

(15) J.F. Poduslo, C.S. Greenberg and M.C.Glick, Biochemistry 11 (1972) 2616.

(16) M.C. Glick, H. Schlessinger and K. Hummeler, unpublished observations.

(17) V. Ginsberg and A. Kobata, in: Structure and Function of Biological Membranes, ed. L.I. Rothfield (Academic Press, N. Y., 1971) p. 439.

Supported in part by American Cancer Society Grants PRA-68 and BC109.

DISCUSSION

S. ROSENBERG: We know that new antigens appear on the surface of virally transformed and tumorigenic cells. Can you correlate in any way the appearance of your glycopeptides in virally transformed cells with new antigens which appear on malignant or transformed cells? Have you tested the antigenicity of these glycopeptides?

M.C. GLICK: No, I have not tested the antigenicity of the animal cell glycopeptides but I am in the process of doing so with glycopeptides from some human cell lines.

R. LONGTON: Have you noticed any glycopeptide changes in consecutive passages of cells?

M.C. GLICK: I looked at some polyoma transformed cells which had been in culture in Dr. Sach's laboratory for more than a year. The patterns of the membrane glycopeptides from these cells showed slightly less of the "transformed" glycopeptides than the clones which had been cultured for a much shorter time - however, this is only one system. I have made the restriction of not using cells which have been in culture beyond 12 passages because we don't know what happens in long term culture. It also cuts down on the possibility of Mycoplasma contamination.

R. LONGTON: In our work we have noticed some media changes in serial passages. The content of neutral sugars and hexosamine decreased as the number of serial passages increased. Has this any correlation with your work?

M.C. GLICK: I haven't looked in that sort of detail since I have not been involved with cells of long-term passage. I should point out again, though, that the glycopeptides do show some growth-dependent changes.

C.W. CHRISTOPHER: I have a quick comment which might turn out to be trivial or not so trivial. We have found that with your gel filtration technique, phenyl red has a peculiar property of being retained on the column.

M.C. GLICK: Yes, if you noticed, it always eluted after the free monosaccharides.

C.W. CHRISTOPHER: We used a number of different compounds (free amino acids, nucleotides) and even varied the pH to determine the true internal volume, V_i. My point is that you focused our attention on the left-hand portion of most of your column chromatograms, but on the right-hand portion there were larger differences which often (but not always) appeared to be related to one of these four different biological conditions of the cells. If this small molecular weight region contains material other than single monosaccharides, could differences in this region represent abortive or incomplete glycosylations related to any one of the four biological conditions you studied?

M.C. GLICK: Yes it could but we have never been able to make any correlation between changes in the amount of material in this region and the appearance or disappearance of the other glycopeptides. It is found in the fraction which is most readily removed by mild trypsinization and in the EDTA extracts. It is present in all EDTA extracts, even after trypsinization. It is never present in the surface membranes. We are currently examining it in detail and we know that it contains more than free fucose.

D. GARDNER: Did you look at the transformed glycopeptides in any tumors derived from the clones of the C-1300 neuroblastoma?

M.C. GLICK: Yes I did, and they formed a "tumor" pattern. The interesting thing is that the clones which do not form axons, that is, those that are less differentiated, give the more "tumor"-like pattern from all the neuroblastoma clones examined.

D. GARDNER: Did you study the effect of cyclic AMP on all cultures to determine whether the glycopeptide pattern is shifted when axon formation is stimulated?

M.C. GLICK: Yes, cAMP or dbcAMP will stimulate axons formation, although we have worked mostly with aminopterin. In each case, the glycopeptide pattern is shifted with axon formation.

M. RIEBER: You show that of about 20 membrane proteins, only one high molecular weight protein is iodinated. In this regard I'd like to point out a remarkable similarity to our own work on the phosphorylation of membrane proteins. Using essentially your procedure, after separating plasma membrane proteins from hamster cells, we also got 20 protein peaks, and only 1 or 2 high molecular weight peaks. It is these high molecular weight proteins which are phosphorylated (M. Rieber and J. Bacalao, Biochem. J. *132* (1973) 641). This may be somehow related to some role of such high molecular weight proteins. I was somewhat confused by the fact that you get essentially the same protein iodinated when you use normal cells, transformed cells, or even cells in suspension like L-cells. Can you provide any explanation as to why you always get the identical protein iodinated under various conditions?

M.C. GLICK: I am proposing that there is a main high molecular weight glycoprotein in all cells and that it is the oligosaccharide structure of this protein which is different under various conditions of the cell, such as virus transoformations.

I. SCHENKEIN: Did you ever try the effect of nerve growth factor on your dendrite or non-dendrite producing neuroblastoma cells?

M.C. GLICK: I have not looked at any effects of nerve growth factor. Others have found that it stimulates some human neuroblastoma cell lines but probably not the mouse lines.

C. BOONE: We found that if you radioiodinate the animal serum protein in tissue culture medium, some of the proteins bind very tightly to cultured cells to the extent of about a tenth to a half of a microgram per million cells and can't be washed off. Have you related any of these proteins to the ones that you've been working with?

M.C. GLICK: There is a similarity between the glycoproteins which are loosely associated with the cells and the radioactive glycoproteins found in the media. However, since these glycoproteins are radioactive, it is more likely that they are cell products than components of the media.

GLYCOPROTEINS AT THE CELL SURFACE OF SUBLINES OF THE TA3 TUMOR

R.W. JEANLOZ and J.F. CODINGTON
Laboratory for Carbohydrate Research
Harvard Medical School and Massachusetts General Hospital

Abstract: The two ascites sublines, TA3-Ha and TA3-St, were originally strain specific, but the TA3-Ha subline lost strain specificity and only a fraction of the H-2 antigens at its surface are expressed. Treatment with proteases released from the TA3-Ha cells a high-molecular weight glycoprotein that was absent from the TA3-St cells. This glycoprotein, which strongly inhibits the agglutinin of *Vicia graminea*, has a high content of threonine and serine residues, most of which are linked to the carbohydrate chains through *N*-acetylgalactosamine residues. The glycoprotein was fractionated into two components of about 100,000 and 450,000 daltons, which appeared on electron micrographs as rods with average lengths of 90 and 350 nm, respectively. A chemical structure of the carbohydrate side-chains of this glycoprotein is proposed.

INTRODUCTION

The two ascites sublines, TA3-Ha and TA3-St, of the TA3 mammary adenocarcinoma that arose spontaneously in the strain A/HeHa mouse in Dr. T. S. Hauschka's laboratory at Roswell Park Memorial Institute were originally strain specific (1,2). After numerous (more than 200) transfers in the strain of origin, the TA3-Ha subline (maintained at Roswell Park) became hyperdiploid and lost strain specificity, whereas the heteroploid TA3-St subline (maintained at the Karolinska Institutet) retained strain specificity (2). Sanford (3) reported that treatment of the TA3-Ha cells with neuraminidase reduced their transplantability in the allogeneic C3H mouse, and Gasic and Gasic (4) have described the heavy coat of "sialomucin" that coats the surface of these cells. Furthermore, Sanford _et al_. (5) have reported recently that the non-strain-specific subline, TA3-Ha, and the strain-specific subline, TA3-St, possess approximately the same number of H-2 antigen sites at their surface, but that not all sites are exposed at the surface of the TA3-Ha subline. In view of the important roles suggested for glycoproteins localized at the surfaces of mammalian cells, such as active transport through the membrane (6), contact inhibition (7), histotypical recognition (8), transplant rejection (9), and suppression of antigenic activity (8,10), it became of interest to investigate the glycoproteins at the surface of the cells of both sublines and to compare their chemical and physical properties.

ISOLATION OF GLYCOPROTEINS FROM THE CELL SURFACE

Information on the chemical structure of the carbohydrate chains of the glycoproteins located at the surface of the cell may be obtained (a) by immunochemical procedures based on agglutination by or adsorption of plant agglutinins, or of antisera, of known specificity, (b) by isolation of the plasma membrane, followed by solubilization and fractionation of the surface glycoproteins, and (c) by direct extraction of the glycoproteins or of fragments of these macromolecules from the living cell by enzymic cleavage or by solubilization with salts or detergent solutions. The immunochemical procedures give important information on the nature of some biological phenomena taking place at the surface of the cell (11-13), but give relatively little information on cell-surface chemical structures or physical parameters. The isolation of plasma membranes has generally resulted, in the hands of other investigators (14,15), in major losses of cell surface components, although it has proved useful in the study of certain membrane proteins. For these reasons, we selected the third method and investigated various procedures of enzymic degradation and direct extraction of the TA3 cells. Most of the preliminary experiments were performed with the TA3-Ha subline, since it was directly available to us and showed the presence of a relatively large amount of surface glycoprotein material (4); in further experiments, some of these procedures were extended to the TA3-St subline. In a comparison of various isolation procedures, it is important to keep in mind that the cells investigated are grown in ascites form and, consequently, that cell characteristics from one batch to another may vary slightly and also that the cells may change with an increasing number of generations.

Action of neuraminidase on TA3-Ha and TA3-St cells

The proportion of sialic acid in the components removed from the cells extracted was considered an indication of the success of the extractions, since sialic acid residues have always been found at the external extremeties of the carbohydrate chains of glycoproteins and, in view of their very polar properties, are assumed to be the components farthest from the membrane. Extensive treatment of the TA3-Ha cells with *Vibrio cholerae* neuraminidase in phosphate buffered saline solution at 25°C liberated about 75% of the sialic acid component (16) and a similar proportion was released from the TA3-St subline (17). These proportions are based on the totality of the sialic acid residues detected by neuraminidase treatment, after extensive proteolytic digestion of the cells has released all the proteins. No protein material, in addition to that lost by the cells in the absence of the enzyme, seems to be released from either subline by neuraminidase at 25°C (16,17). The absolute amount of sialic acid residues released from the more specific TA3-St subline was about 60% of that released from the TA3-Ha subline. This difference between both sublines is more striking if the cell diameters of both types of cells are considered; if one assumes similar topographies for the two cells, the concentration of sialic acid at the surface of the TA3-Ha cell can be estimated at 2.3 times that at the surface of the TA3-St cell. Differences in the relative proportion of *N*-acetyl and *N*-glycolyl residues between the two sublines were also observed (17). Although the presence of sialic acid residues at the surface of the cells has been linked to the degree of oncogenicity of the TA3 cell

(18,19), results obtained with other cell lines by other investigators (20,21) show clearly that it is not possible, at the present time, to relate the degree of oncogenicity with proportion, type, or nature of the linkage of the sialic acid residues, either in cancer cells or in transformed cells relative to normal cells (22).

Action of proteolytic enzymes

The action of trypsin after modification by treatment with L-(1-tosylamido-2-phenyl)ethyl chloromethyl ketone (TPCK-trypsin) to remove chymotrypsin activity, papain, and Pronase was investigated. As expected, the degradation of the protein chain was far less extensive with the two first-named enzymes, and consequently the purification of the resulting fragments was more convenient than that of the fragments obtained from a Pronase digest (16); subsequent work was performed either with TPCK-trypsin or with papain. Both enzymes were active at 4°C and at low enzyme concentrations and five or six 20-min successive incubations were performed. Under these conditions, the viability of the cells remained higher than 95% during the entire experiment, and very little cytoplasmic material was released (16). After five incubations of 20 min each, almost no additional material containing sialic acid could be released. The material obtained by the action of papain contained a total amount of sialic acid as great as that released by the direct action of neuraminidase on the cells at 23°C (23). The content of sialic acid in the material obtained by tryptic digestion was approximately three-quarters of that released by papain, or slightly less than half the content of sialic acid present in the cell (16,23).

The amount of carbohydrate and protein material released by TPCK-trypsin from the TA3-Ha cell can reach about 6% of the dry weight of the cell, but the proportion of protein is greater in the material released from the TA3-St cell than in that released from the TA3-Ha cell. The amino acid composition of the peptide material released from both sublines was similar, except that a greater proportion of alanine and glycine and a lower proportion of aspartic and glutamic acids were present in the TA3-Ha material (17). These data suggest that glycoproteins of the serine and threonine (mucin) type are preponderant in this material and that, as compared to the TA3-Ha cell, the TA3-St cell is attacked more extensively by proteolytic digestion, although its viability is not impeded to a greater degree.

Except for a larger proportion of sialic acid released by papain, the carbohydrate compositions of the crude material released by TPCK-trypsin and papain were quite similar and consisted of residues of galactose, glucose, mannose, N-acetylgalactosamine, N-acetylglucosamine, and N-acetylneuraminic acid (16,17).

The large proportion of carbohydrate-containing material obtained by protease digestion without impairment of the viability of the cell, and the general similarity in composition between the materials obtained by trypsin and papain suggest that this method of removal of surface material can give valid, valuable information on the chemical structure of the surface. This method depends, of course, upon the presence of peptide linkages susceptible to enzyme action, and molecules not cleaved by the enzyme would remain undetected on the cell residue. Nevertheless, the information thus gained, coupled with that obtained by immunochemical procedures, seems to be the best means, at the present time, to investigate the chemical structure of glycoproteins at the cell surface.

Action of chemical reagents

A preliminary investigation of the release of material by treatment of the living cells with a buffered, balanced salt solution, with chelating agents (citrate ions and ethylenediamine tetraacetate), and with lithium diiosalicylate was performed (16,23).

Incubation of TA3-Ha cells in media devoid of calcium and magnesium ions had no effect on the viability of the cell, but resulted in the release of an increased proportion of protein. The material released by the cells in the presence of citrate ions did not differ significantly from that obtained in a balanced salt solution (16). Ethylenediamine tetraacetate, however, released large amounts of material from the cells (16,23), but the proportion of sialic acid in this material was small, thus indicating that a relatively small proportion of glycoproteins was released. In view of the emphasis placed in our investigation on carbohydrate structures, these experiments were not encouraging and were discontinued, but it is clear that they may give important clues on the total assembly of the cell membrane.

Lithium diiodosalicylate, a reagent known to extract glycoproteins from the erythrocyte membrane (24), greatly affected the morphology of the TA3 cells, which resulted in early death of the cells at a 0.3 M concentration of reagent. At a lower concentration (3 mM), the viability was about 80% after 60 min, but the proportion of sialic acid-containing material released remained low, and the possible presence of macromolecules originating from the interior of the cell complicated further purification (23).

FRACTIONATION OF CELL SURFACE GLYCOPROTEINS

Fractionation of the glycoprotein material obtained from either the TA3-Ha or the TA3-St cells presents numerous difficulties (see 25), because the enzymic attack may produce molecules with various protein chain-lengths. In addition, heterogeneity of the molecules located at the cell surface is expected, as well as a structural microheterogeneity (26). Thus, in the past, only a few glycoproteins have been obtained in a relatively pure state from the surface of a eukaryotic cell, for example, a glycoprotein fragment from Meth-A cells possessing H-2 antigenicity (27). To avoid the effect of structural microheterogeneity, methods based on ion-exchange properties were not selected and our scheme of fractionation depended solely on molecular weight (28).

Passage of the material, obtained by TCPK-tryptic action on the TA3-Ha cell, successively through columns of Bio-Gel P-4, P-30, and P-100 resins gave, in the void volume of the P-100 column effluent, a high-molecular-weight fraction which represented about 10% of the total amount of protein and carbohydrate components and about 30% of the total amount of the sialic acid released from the cell (28).

In striking contrast, the material obtained from the TA3-St cell surface by TPCK-trypsin digestion did not contain any detectable proportion of this high-molecular-weight component, and the elution profiles from the P-30 gel column of the materials obtained from the two sublines were very different (Fig. 1).

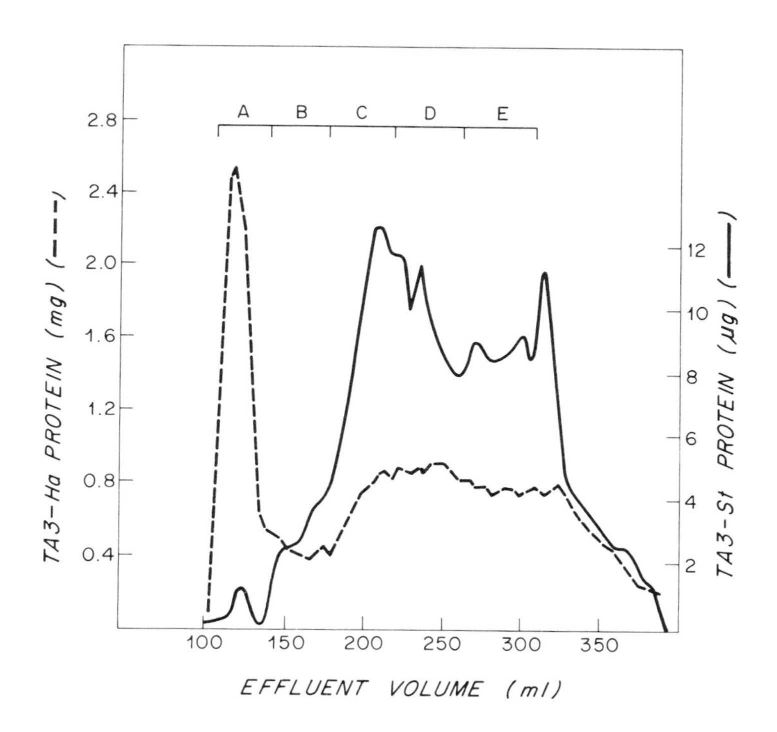

Fig. 1. Fractionation of material, obtained in the void volume of a Bio-Gel P-4 column, on a column of Bio-Gel P-30 (100-200 mesh) at 4°C. Eluent: 0.05 M pyridine acetate, pH 5.3. Material from 2.0 x 10^{10} TA3-Ha cells on a column 2.7 x 87 cm (---). Material from 5 x 10^{9} TA3-St cells on a column 2.8 x 73 cm (——) (17).

The absence of a significant amount of the high-molecular-weight component at the surface of the TA3-St cell was confirmed by immunochemical examination (17).

Similar fractionation of the material released by papain from the TA3-Ha cell showed a greater proportion (about 50% more) of high-molecular-weight component than was observed in the TPCK-trypsin-released material and, in addition, the content of sialic acid was also higher in the macromolecules released by papain. Thus, the higher content of sialic acid observed in the crude material released by papain as compared to that released by TPCK-trypsin is due to the liberation, by papain, of additional high-molecular-weight glycoproteins rich in sialic acid. This difference in behavior of the two proteolytic enzymes may result from some inhibition of trypsin by sialic acid residues (see 29,30) or from the absence of trypsin-labile linkages in some sialic acid-rich glycoproteins.

PROPERTIES OF THE HIGH-MOLECULAR-WEIGHT COMPONENT OF TA3-Ha CELL SURFACE

Chemical composition and structure

The high-molecular-weight component released by TPCK-trypsin action on the TA3-Ha cell surface contains 40-50% of the total residues of sialic acid located at the surface of the cell and is composed of about 30% of protein; the remaining material consists almost entirely of four carbohydrate residues: galactose, *N*-acetylgalactosamine, *N*-acetylglucosamine, and *N*-acetylneuraminic acid in the approximate molar proportions of 4:2:1:1. Approximately two-thirds of the amino acid residues consists of serine and threonine; the other third includes mainly proline, alanine, and glycine in nearly equal proportions. This chemical composition differs from that of other glycoproteins obtained from cell surfaces (9,25,31-34), and only a glycopeptide of low molecular-weight obtained from erythrocyte stroma by tryptic digestion possesses the same carbohydrate components, but in different proportions (35, 36). The amino acid composition is qualitatively similar to that of glycoproteins isolated from mucous secretions (37), but the total content of serine and threonine is greater than that of other glycoproteins investigated up to the present time.

The high content of sialic acid suggests the presence of either numerous straight chains or a few highly ramified carbohydrate chains, since sialic acid residues have always been found to be located at the nonreducing extremity of the carbohydrate chains in glycoproteins. Preliminary experiments of alkaline reduction with sodium borohydride indicated that about two-thirds of the serine and threonine residues, i.e. about one-third of the total amino acid residues, are involved in the carbohydrate-protein linkage. These serine and threonine residues are bound to almost all residues of *N*-acetylgalactosamine (38), and the number of reducing end-groups suggests an average carbohydrate chain-length of four residues. Degradation by methanolysis confirmed the *N*-acetylgalactosamine to serine (threonine) linkages, and periodate oxidation before and after neuraminidase treatment indicated that each sialic acid residue is linked to a galactose residue (38). These results are best explained by a structure composed mainly of two different chains (A and B), as follows:

```
                                                  |
(A) NANA→Gal→(Gal, GNAc)→Gal→GalNAc→Ser→(Thr)
                                                  :
                             (B) Gal→GalNAc→Ser→(Thr)
                                                  |
```

The high content of sialic acid found in the high-molecular-weight component released by papain suggests that, in this material, sialic acid may also be attached to the (B) chain (23), but whether other difference of fine structure exists between the materials released by the TPCK-trypsin and papain will require further investigation.

Homogeneity, size, and configuration

Elution of the purified high-molecular-weight component, obtained by TPCK-trypsin action on TA3-Ha cells, from a Bio-Gel A-5m column with 6 M guanidine hydrochloride in the presence of 1 mM cysteine gave two peaks with apparent molecular weights of approximately 90,000 and 180,000. Determination of the molecular weight by sedimentation equilibrium in the presence of 6 M guanidine hydrochloride and 10 mM ethylenediamine tetraacetate gave values 1.6-1.7 times greater than those obtained by gel filtration (28).

Preparative fractionation (39) on a column of Sepharose 4B gel in 0.05 M pyridine acetate solution gave a single peak with a shoulder at a lower effluent volume. The column effluent was arbitrarily divided into three fractions. Sedimentation equilibrium analysis of the last-eluted fraction indicated homogeneity and a molecular weight of 100,000. The first-eluted fraction was found to have a molecular weight of about 450,000 and the plot of sedimentation equilibrium data indicated monodispersity. The three fractions were examined by electron microscopy after being shadowed with platinum. All fragments appeared as long rods, about 2.5 nm in diameter, with a number-average length of 94 nm, 170, and 320 nm, respectively, corresponding to approximate molecular weights of 100,000 for the smallest component and of 300,000 to 500,000 for the largest component. These results suggest that the native glycoprotein possesses a molecular weight higher than 500,000 and it is probable that the tryptic digestion has attacked the peptide chain at various locations. One of these points may be near the membrane, and, as suggested by Marchesi *et al.* (40) for the glycoprotein at the surface of erythrocytes, the possibility that a hydrophobic part of the protein may be embedded in the membrane itself must be considered.

When a similar preparative fractionation on Sepharose 4B was applied to the material released by papain, a similar elution pattern was observed and the fractions showed apparent molecular weights similar to those of the fractions obtained from TPCK-trypsin-released material. Examination by electron microscopy of one of the fractions showed a similar rod-like structure, as just described (23).

The proportion of the carbohydrate components in the various fractions obtained from Sepharose 4B suggests that the fractionation was also influenced by the content of sialic acid, in addition to the molecular size. All fractions obtained either from TPCK-trypsin or from papain-released material showed the same four major components, namely, galactose, *N*-acetylgalactosamine, *N*-acetylglucosamine, and *N*-acetylneuraminic acid, but in varying proportions (Table 1) (23).

TABLE 1

Carbohydrate composition in molar proportions of the fractions eluted from Sepharose 4B gel (23)

Fraction	Gal	GalNAc	GNAc	NANA
Released by TPCK-trypsin				
Higher mol. wt.	3.8	2.3	1.0	1.2
Lower mol. wt.	5.3	2.1	1.0	1.6
Released by papain				
Higher mol. wt.	3.0	3.9	1.0	0.7
Lower mol. wt.	5.1	4.3	1.0	1.7

The different compositions of the fractions may correspond to various proportions of the (A) and (B) chains just described, and may be related to the concept of microheterogeneity of carbohydrate chains in glycoprotein molecules (26).

IMMUNOCHEMICAL PROPERTIES

The high-molecular-weight glycoprotein was found to be as active as the most potent N-blood group substance in the specific inhibition of the agglutinination of human blood-group NN erythrocytes by antihuman blood-group N specfic lectin of *Vicia graminea* (41). The antigenic determinant of blood-group N specificity is constituted by a ramified carbohydrate structure including nonreducing β-D-galactosyl and α-N-acetylneuraminosyl terminal residues, but the *V. graminea* lectin has been shown to react only with the β-D-galactopyranosyl residue (42). The TA3 glycoprotein exhibited no N or M blood-group specificity (41). The structure proposed for the (B) chains of the glycoprotein fully account for these immunochemical properties and is similar to a structure previously proposed for the *V. graminea* receptor site (43). The lack of agglutinination of the TA3-Ha cells by the _V. graminea_ extract (41) may possibly be explained by unfavorable molecular conformation or spacing of the high-molecular-weight glycoproteins. The sensitivity and specificity of the inhibition test makes it the method of choice for the determination of minute amounts of the macromolecular component in biological fluids (45).

CONCLUSION

The presence, at the surface of the cells of the non-strain-specific TA3-Ha subline, of approximately 3×10^6 molecules of a macromolecular glycoprotein that is absent in the strain-specific TA3-St subline represents a major difference between the two sublines. There is no evidence for the presence, at the surface of the TA3-St cell, of similar macromolecules having an increased lability toward trypsin or papain, which would result in low-molecular-weight fragments of the same general structure after proteolysis, since no fraction, released from this cell and adsorbed on the P-30 resin, showed _V. graminea_ activity (17).

It seems reasonable to suggest that the high number of these macromolecules at the surface of the TA3-Ha cell, where they represent more than 0.5% of the weight of the lyophilized cells, may be a major factor in the difference between the biological activities of the TA3-Ha and TA3-St cells. Although the size of the native molecules at the surface of the cell is not known, it can be assumed that their molecular weight is higher than 500,000. Electron micrographs suggest that the cell surface is approximately 2.5 times that of a perfect sphere (16). On this basis, the molecules, in an extended configuration as suggested by Winzler (36), would stretch outward from the cell surface for at least 5000 Å. They would be separated at the surface by an average distance of 220Å (46).

The presence of these glycoproteins extending far from the surface of the cell suggests that they may be involved in the masking of the histocompatibility (H-2) antigens (5), and thus explains the low strain-specificity of the TA3-Ha cells, as compared to the TA3-St cells. Disruption of this surface, macromolecular network by lyophilization

(5,47), which results in a large increase in the adsorption of H-2 antisera by the TA3-Ha cell but not by the TA3-St cell, supports this hypothesis, as does the finding that the H-2 antigens of the TA3-Ha cells, partially masked (5), are restricted to the cell membrane and become accessible after homogenization (48). Elucidation of the mechanism of biosynthesis of this high-molecular-weight compound and of its possible role in tumor formation are logically the next steps of this investigation.

REFERENCES

(1) T.S. Hauschka, Trans. N.Y. Acad., 16 (1973) 64.

(2) T.S. Hauschka, L. Weiss, B.A. Holdrige, T.L. Cudney, M. Zumpft and J.A. Planinsek, J. Nat. Cancer Inst., 47 (1971) 343.

(3) B.H. Sanford, Transplantation, 5 (1967) 1273.

(4) G. Gasic and T. Gasic, Nature, 196 (1962) 170.

(5) B.H. Sanford, J.F. Codington, R.W. Jeanloz and P.D. Palmer, J. Immunol., 110 (1973) 1233.

(6) P.M. Kraemer, in: Biomembranes, Vol. 1, ed. L.A. Manson (Plenum, New York, 1961) p. 67.

(7) M. Abercrombie and E.J. Ambrose, Cancer Res., 22 (1962) 525.

(8) A. Moscona, Proc. Soc. Exp. Biol. Med., 92 (1956) 10.

(9) A.R. Sanderson, P. Cresswell and K.I. Welsh, Nature New Biol., 930 (1971) 8.

(10) C.A. Appfel and J.H. Peters, J. Theor. Biol., 26 (1970) 47.

(11) H. Lis and N. Sharon, Science, 177 (1972) 9119.

(12) M.M. Burger, Proc. Nat. Acad. Sci. U.S., 62 (1969) 994.

(13) M. Inbar and L. Sachs, Nature, 223 (1969) 710.

(14) D.F.H. Wallach, in: The Specificity of Cell Surfaces, eds. B.D. Davis and L. Warren (Prentice-Hall, Englewood Cliffs, 1967) p. 129.

(15) H.B. Bosmann, A. Hagopian and E.H. Eylar, Arch. Biochem. Biophys., 128 (1968) 51.

(16) J.F. Codington, B.H. Sanford and R.W. Jeanloz, J. Nat. Cancer Inst., 45 (1970) 637.

(17) J.F. Codington, B.H. Sanford and R.W. Jeanloz, J. Nat. Cancer Inst., 51 (1973) 585.

(18) B.H. Sanford and J.F. Codington, Tissue Antigens, 1 (1971) 153.

(19) R.L. Simmons, A. Rios, G. Lundgren, R.K. Ray and C.F. McKhann, Surgery, 70 (1971) 38.

(20) H. Kalant, W. Mons and M. Guttman, Can. J. Physiol. Pharmacol., 42 (1964) 25.

(21) W.J. Grimes, Biochemistry, 12 (1973) 990.

(22) N. Ohta, A.B. Pardee, B.K. McAuslan and M.M. Burger, Biochim. Biophys. Acta, 158 (1968) 98.

(23) J.F. Codington, B. Tuttle and R.W. Jeanloz, Proc. 2nd Intern. Symp. Glycoconjugates, Lille, June 1973, in press.

(24) V.T. Marchesi and E.P. Andrews, Science, 174 (1971) 1247.

(25) E.F. Walborg, Jr., R.S. Lantz and V.P. Wray, Cancer Res., 29 (1969) 2034.

(26) R. Montgomery, in: Glycoproteins, ed. A. Gottschalk (Elsevier Pub., Amsterdam, 1972) p. 518.

(27) T. Muramatsu and S.G. Nathenson, Fed. Proc., 29 (1970) 910.

(28) J.F. Codington, B.H. Sanford and R.W. Jeanloz, Biochemistry, 11 (1972) 2559.

(29) I. Yamashina, Acta Chem. Scand., 10 (1956) 1666.

(30) A. Gottschalk and S. Fazekas de St. Groth, Biochim. Biophys. Acta, 43 (1960) 513.

(31) S.G. Nathenson, A. Shimada, K. Yamane, T. Muramatsu, S. Cullen, D.L. Mann, J.L. Fackey and R. Graf, Fed. Proc., 29 (1970) 2026.

(32) D.A.L. Davies, in: Blood and Tissue Antigens, ed. D. Aminoff (Academic Press, New York, 1970) p. 101.

(33) L. Shen and V. Ginsburg, in: Biological Properties of the Mammalian Cell Surface, ed. L.A. Manson (Wistar Institute, Philadelphia, 1968) p. 67.

(34) O. K. Langley and E. Ambrose, Biochem. J., 102 (1967) 367.

(35) R.J. Winzler, E.D. Harris, D.J. Pekas, C.A. Johnson and P. Weber, Biochemistry, 6 (1967) 2195.

(36) R.J. Winzler, in: Blood and Tissue Antigens, ed. D. Aminoff (Academic Press, New York, 1970) p. 117.

(37) M. Bertolini and F. Bettelheim, in: The Carbohydrates, Vol. II, eds. W. Pigman and D. Horton (Academic Press, New York, 1970) p. 677.

(38) J.F. Codington, B.H. Sanford and R.W. Jeanloz, Fed. Proc., 31 (1972) 465.

(39) H.S. Slayter and J.F. Codington, J. Biol. Chem., 248 (1973) 3405.

(40) W.T. Marchesi, T.W. Tillack, R.L. Jackson, J.P. Segrest and R.E. Scott, Proc. Nat. Acad. Sci. U.S., 69 (1972) 1445.

(41) G.F. Springer, J.F. Codington and R. W. Jeanloz, J. Nat. Cancer Inst., 49 (1972) 1469.

(42) G.F. Springer, Y. Nagai and H. Tegtmeyer, Biochemistry, 5 (1966) 3254.

(43) G. Uhlenbruck and W. Dahr, Vox Sang., 21 (1971) 338.

(44) J.F. Codington and A.G. Cooper, unpublished data.

(45) A.G. Cooper, J.F. Codington and M.C. Brown, Proc. Nat. Acad. Sci. U.S., in press.

(46) J.F. Codington and R.W. Jeanloz, 2nd Eur. Symp. Connective Tissue Res., Hanover, Sept. 1970; Z Klin. Chem. Klin. Biochem., 9 (1971) 61.

(47) S. Friberg, Jr. and B. Lilliehöök, Nature New Biol., 108 (1973) 112.

(48) J. Molnar, G. Klein and S. Friberg, Jr., Transplantation, 16 (1973) 93.

DISCUSSION

H. HEMPLING: What functional attributes of the membrane still remain after successive trypsinization? My reason for asking that is that several years ago, Dr. Gasic and I showed that as little as 1 mM ATP added to the TA 3 cells will produce very striking increases in the permeability of these membranes to electrolytes. I know that you use trypan blue exclusion and ^{51}Cr release, but I would like to know about whether you studied more sensitive criteria of membrane function such as permeability to water and electrolytes and the like.

R.W. JEANLOZ: No, we have not done those studies.

G.A. JAMIESON: I would like to give a word of caution about determining the viability of cells solely by trypan blue exclusion. Dr. Jett in our laboratory has recently carried out some extensive studies on RAJI cells in which she has compared the products of the tryptic digestion of intact cells with those from isolated membranes. (M.J. Jett and G.A. Jamieson, Biochem. Biophys. Res. Comm. 55: 1125 (1973)). In these experiments, the trypan blue exclusion was retained, but there was evidence that high molecular weight components from the cell cytoplasm, molecular weight of about 200,000, were being released. The true membrane components are, in fact, of two size classes of molecular weights about 10,000 and 50,000. There is no component on the membrane corresponding to that released from the cytoplasm.

CARBOHYDRATE ANTIGENS OF CELL SURFACES

DAVID A. ZOPF AND VICTOR GINSBURG
National Institute of Arthritis, Metabolism, and Digestive Diseases
National Institutes of Health

Abstract: Sugar sequences occurring in glycolipids and glycoproteins of cell surfaces are also found in the free oligosaccharides of human milk. We have been isolating and characterizing these oligosaccharides for use in serologic studies on carbohydrate antigens of cell surfaces. The studies include (a) hapten inhibition of antibodies directed against cell surface carbohydrates; (b) binding studies on the antibodies using oligosaccharides labeled by reduction with sodium borotritide; and (c) production of antibodies against specific sugar sequences by immunization of animals with synthetic antigens prepared by coupling the oligosaccharides to carriers.

The carbohydrate antigens of cell surfaces are glycoproteins and glycolipids whose serologic specificity in general is determined by sugar sequences at the nonreducing ends of their carbohydrate chains. In most cases this specificity is probably independent of whether the chains are attached to lipids or to proteins. Many of the sugar sequences that occur in glycolipids and glycoproteins are also found in the free oligosaccharides of human milk (1).

We have been isolating and characterizing these oligosaccharides for use in serologic studies on the carbohydrate antigens of cell surfaces. They are much easier to isolate than their counterparts in membranes and can be obtained in relatively large amounts (2).

Some important sugars of milk are shown in Table 1. Of these sugars, lactose at 50 g per liter is the most abundant by 2-4 orders of magnitude.

TABLE 1

Important oligosaccharides of human milk

Name	Structure
Lactose (3)	Galβ1-4Glc
Lacto-*N*-tetraose (4)	Galβ1-3GlcNAcβ1-3Galβ1-4Glc
Lacto-*N*-*neo*tetraose (5)	Galβ1-4GlcNAcβ1-3Galβ1-4Glc
Lacto-*N*-hexaose (6)	Galβ1-4GlcNAcβ1 6 Galβ1-3GlcNAcβ1-3Galβ1-4Glc
Lacto-*N*-*neo*hexaose (7)	Galβ1-4GlcNAcβ1 6 Galβ1-4GlcNAcβ1-3Galβ1-4Glc

Most of the remaining known oligosaccharides of milk are fucosyl and sialyl derivatives of the five sugars of Table 1. The fucosyl derivatives mentioned in this paper are given in Table 2.

Fucose is attached in at least four linkages: α1-2 to galactose, α1-3 and α1-4 to *N*-acetylglucosamine and α1-3 to glucose. Of the four, the linkage to glucose is quantitatively the least important and possibly the enzyme which forms this linkage is the same one that forms the α1-3 linkage to *N*-acetylglucosamine (10). Sialic acid is

TABLE 2

Some oligosaccharides containing fucose

Name	Structure
Lacto-*N*-fucopentaose I (8)	Galβ1-3GlcNAcβ1-3Galβ1-4Glc 2 Fucα1
Lacto-*N*-fucopentaose II (9)	Galβ1-3GlcNAcβ1-3Galβ1-4Glc 4 Fucα1
Lacto-*N*-fucopentaose III (10)	Galβ1-4GlcNAcβ1-3Galβ1-4Glc 3 Fucα1
Lacto-*N*-difucohexaose I (11)	Galβ1-3GlcNAcβ1-3Galβ1-4Glc 2 4 Fucα1 Fucα1
Lacto-*N*-difucohexaose II (11)	Galβ1-3GlcNAcβ1-3Galβ1-4Glc 4 3 Fucα1 Fucα1

attached in at least three linkages: α2-3 and α2-6 to galactose and α2-6 to *N*-acetylglucosamine. Because of these different linkages many isomeric milk oligosaccharides are possible. With lacto-*N*-tetraose there are theoretically possible 4 monofucosyl derivatives, 6 difucosyl derivatives and 12 monofucosylmonosialyl derivatives; with lacto-*N*-*neo*hexaose, the corresponding numbers are 6, 15 and 36. While all possible isomers for steric reasons probably do not exist, there is evidence that some isolated oligosaccharides are complex mixtures of isomers. For example, the serologic properties of samples of difucosyl derivatives of lacto-*N*-hexaose and lacto-*N*-*neo*hexaose vary with the blood type of the donor from which they were obtained (7). Difucosyl derivatives from donors belonging to the Lewis (a-b+) blood group (all of whom are "secretors") inhibit the agglutination of H, Le^a and Le^b-active erythrocytes by the appropriate antiserum; difucosyl derivatives from donors belonging to the Lewis (a+b-) blood group (all of whom are "nonsecretors") inhibit only the agglutination of Le^a-active cells; and difucosyl derivatives from donors belonging to the Lewis (a-b-) blood group (who happen to be "secretors") inhibit only the agglutination of H-active cells. The variation of isomers with blood type is caused by the fact that the glycosyltransferases involved in the synthesis of the free oligosaccharides in milk are specified by the same genes as those involved in the synthesis of the carbohydrate antigens of cell surfaces. Possession of a secretor gene (*Se* gene) is necessary for the presence in secretory organs of the fucosyltransferase that forms Fucα1-2Gal structures (12) and possession of a Lewis gene (*Le* gene) is necessary for the formation of the fucosyltransferase that forms Fucα1-4GlcNAc structures (13). Thus, the formation of H-active oligosaccharides (those with the structure, Fucα1-2Gal...) requires only the possession of a secretor gene; the formation of Le^a-active oligosaccharides (those with the structure, Galβ1-3GlcNAc...) requires only the

```
Galβ1-3GlcNAc...
        4
        |
      Fucα1
```

possession of a Lewis gene; and the formation of Le^b-active oligosaccharides (those with the structure, Gal 1-3GlcNAc...) requires the possession of both genes.

```
   Gal 1-3GlcNAc...
   2        4
   |        |
Fucα1    Fucα1
```

The variation of oligosaccharides with blood type is of practical importance for the isolation of certain sugars. For example, we have recently been fractionating the oligosaccharides in milk from a donor with the relatively rare genotype, seselele (a nonsecretor belonging to the Le(a-b-) blood group). The milk lacks oligosaccharides containing Fucα1-2Gal or Fucα1-4GlcNAc structures and is a rich source of lacto-*N*-fucopentaose III, the sugar found in sphingoglycolipids of adenocarcinomas (14). A comparison of the sugars from this donor and the sugars from donors with other genotypes is shown in Fig. 1.

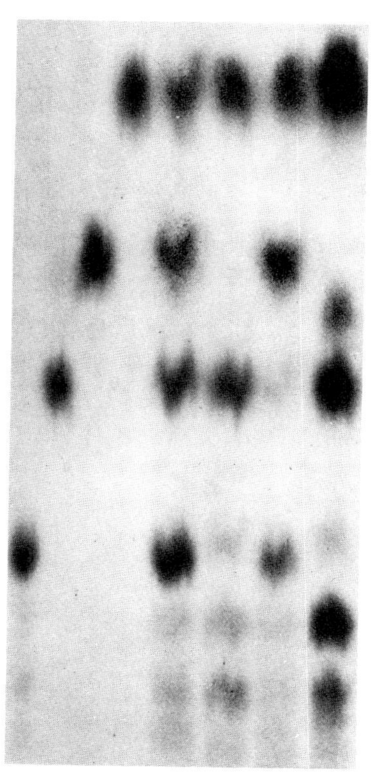

Fig. 1. Paper chromatography of the higher neutral oligosaccharides obtained by Sephadex gel filtration and deionization (2) from milk of donors with different genotypes as follows: lane A, *LeSe*; lane B, *Lesese*; lane C, *leleSe*; and lane D, *lelesese*.

Milk from donors with both a *Se* gene and a *Le* gene (Fig. 1, lane A) contain lacto-*N*-tetraose, lacto-*N*-fucopentaose I, lacto-*N*-fucopentaose II and lacto-*N*-difucohexaose I. Milk from donors with a *Le* gene but not a *Se* gene (Fig. 1, lane B) lacks sugars with Fucα1-2Gal linkages

(12) and consequently is missing lacto-*N*-fucopentaose I and lacto-*N*-difucohexaose II (the residual sugar visible in the lacto-*N*-difucohexaose I area is mainly lacto-*N*-difucohexaose II, lacto-*N*-hexaose and lacto-*N*-*neo*hexaose). Milk from donors with a *Se* gene but not a *Le* gene (Fig. 1, lane C) lacks sugars with Fucα1-4GlcNAc linkages (13) and consequently is missing lacto-*N*-fucopentaose II (the residual sugar is lacto-*N*-fucopentaose III) and lacto-*N*-difucohexaose I (the residual sugar is lacto-*N*-hexaose, lacto-*N*-*neo*hexaose and a hexasaccharide isomer of lacto-*N*-difucohexaose I and II). Milk from the donor with neither the *Le* gene nor the *Se* gene is shown in Fig. 1, lane D. The sample lacks lacto-*N*-fucopentaose I and II and lacto-*N*-difucohexaose I and II. The sugar in the lacto-*N*-fucopentaose II area is pure lacto-*N*-fucopentaose III as judged by the fact that it liberates fucose on mild acid hydrolysis and gives rise to a tetrasaccharide that is negative in the Elson-Morgan test (10). The sugar running slightly ahead of lacto-*N*-fucopentaose III, not seen in the other samples, has been tentatively identified as

```
Galβ1-3GlcNAcβ1-3Galβ1-4Glc.
                         3
                         |
                     Fucα1
```

The slow-running sugars in all the samples include mono-, di- and trifucosyl derivatives of lacto-*N*-hexaose and lacto-*N*-*neo*hexaose. There are more of these derivatives in the milk from the donor with the genotype lelesese than in milk from other genotypes and, unlike the other samples, their fucose residues are presumably attached exclusively in α1-3 linkages. From this sample of milk we can therefore isolate sugars not obtainable from other samples. The trifucosyl derivative, for example, should be exclusively that of lacto-*N*-*neo*hexaose (there are not enough places for three 3-linked fucosyl residues in lacto-*N*-hexaose) and have the following structure:

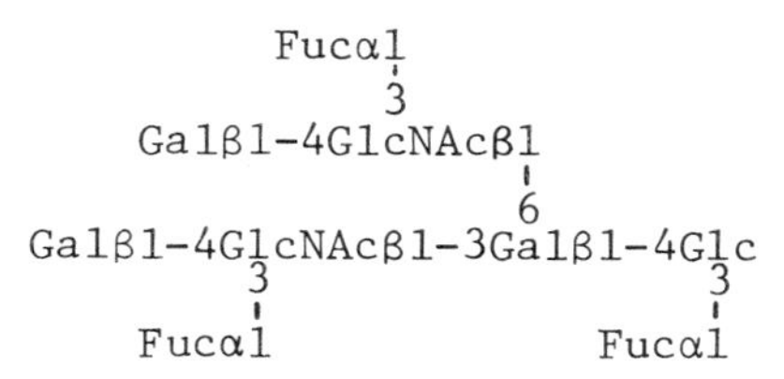

The pattern shown in Fig. 1, lane D, was not observed in milk obtained from another nonsecretor belonging to the Le(a-b-) blood group analyzed previously (15).

Oligosaccharides from milk were important in elucidating the structures of the H, Le^a and Le^b determinants by being used as hapten inhibitors (16). A related approach is to use them in equillibrium binding experiments as shown in Table 3. They are easily labeled with ^3H by reduction of the glucose at their reducing ends to sorbitol with NaB^3H_4.

TABLE 3

Binding of ^3H-labeled, reduced oligosaccharides by anti-Le^a and anti-Le^b goat sera

Initially, one side of an equillibrium dialysis cell contained 70 µl of a 40% ammonium sulfate concentrate of goat anti-Le^a and anti-Le^b sera (kindly supplied by Dr. D. M. Marcus (17)) containing 0.56 and 0.63 mg/ml of precipitable antibody protein, respectively. The opposite side contained 70 µl of phosphate-buffered saline, pH 7.4, containing approximately 0.01 µg of reduced oligosaccharide. Equillibrium was reached after 3 days at 2°.

Antiserum	Oligosaccharide	Ratio $\frac{\text{CPM serum side}}{\text{CPM saline side}}$
Anti-Le^a	Lacto-*N*-fucopentaose I	1.0
	Lacto-*N*-fucopentaose II	4.3
	Lacto-*N*-fucopentaose III	1.1
	Lacto-N-difucohexaose I	2.3
Anti-Le^b	Lacto-*N*-fucopentaose I	1.0
	Lacto-*N*-fucopentaose II	1.4
	Lacto-*N*-fucopentaose III	1.1
	Lacto-*N*-difucohexaose I	11.3

As expected from inhibition studies (17), the anti-Le^a serum bound lacto-*N*-fucopentaose II strongly and lacto-*N*-difucohexaose I weakly while the reverse was true for the anti-Le^b serum. Neither lacto-*N*-fucopentaose I nor lacto-*N*-fucopentaose III were significantly bound by either antiserum (at least 2-3 orders less than the other two sugars).

Equillibrium dialysis is particularly useful in cases where the availability of sugars is limited. For example, if lacto-*N*-hexaose is labeled by reduction with NaB^3H_4 and partially hydrolyzed most of the labeled fragments can be separated from each other by paper chromatography as shown in Fig. 2. Peak 1 is unhydrolyzed lacto-*N*-hexaose; peak 2 contains two labeled pentasaccharides which are not separable by chromatography; and peak 3 contains two labeled tetrasaccharides which can be separated from each other by rechromatography (Fig. 2C).

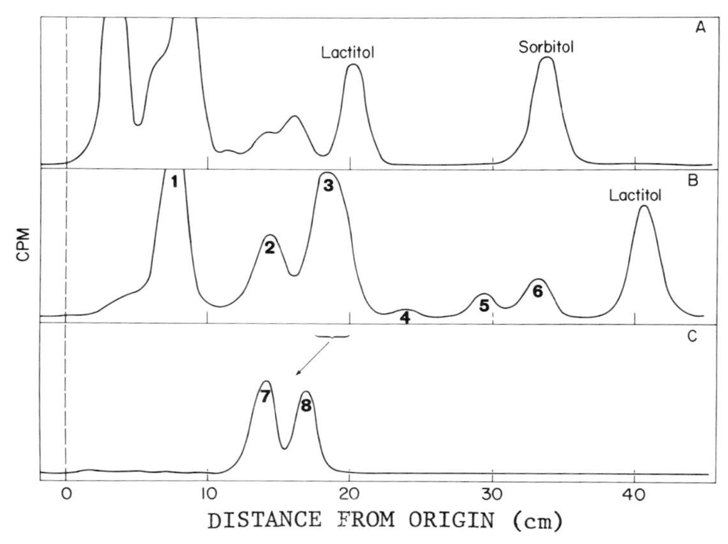

Fig. 2. Partial hydrolysis of reduced lacto-*N*-hexaose.

Reduced lacto-N-hexaose, 1.5 μmole, labeled with ^3H (3.5 x 10^7 CPM) was hydrolyzed in 0.5 ml of

0.1N HCl for 40 min at 100°. After deionization by passage through an ion-exchange column (0.5 x 3 cm) containing Amberlite AG 50(H+) and AG 3(OH⁻), the hydrolysate was subjected to paper chromatography using ethyl acetate-pyridine-acetic acid - H_2O (5:5:1:3) as the developer. After 18 hours (Fig. 2A) and 42 hours (Fig. 2B) of development the chromatogram was scanned for radioactivity. Peak 3 in Fig. 2B was eluted with H_2O and chromatographed for 4 days using ethyl acetate-pyridine - H_2O (12:5:4) as the developer. The resulting chromatogram was scanned for radioactivity (Fig. 2C). The identities of the sugars in the numbered peaks are given in the text and in Table 4.

The identity and yields of labeled hydrolysis products are given in Table 4.

TABLE 4

Identity and yield of oligosaccharides shown in Fig. 2.

Peak	Oligosaccharide	Yield	
		cpm	%
2	Galβ1-4GlcNAcβ1 | 6 GlcNAcβ1-3Galβ1-4Sorbitol and GlcNAcβ1 | 6 Galβ1-3GlcNAcβ1-3Galβ1-4Sorbitol	1.8×10^6	5
4	GlcNAcβ1 | 6 GlcNAcβ1-3Galβ1-4Sorbitol	4×10^5	1
7	Galβ1-4GlcNAcβ1-6Galβ1-4Sorbitol	2.7×10^6	8
8	Galβ1-3GlcNAcβ1-3Galβ1-4Sorbitol	1.5×10^6	4
5	GlcNAcβ1-6Galβ1-4Sorbitol	9×10^5	3
6	GlcNAcβ1-3Galβ1-4Sorbitol	1.2×10^6	4

Although the yields are not high, more than enough material is obtained for binding studies. Sugars such as those in Table 4 should be especially useful in testing for binding to antibodies like anti-CEA (18) which are believed to be directed against the "incomplete" carbohydrate chains of blood group determinants. The branched portion of lacto-*N*-hexaose is identical to the branched portion of the carbohydrate chains of soluble blood group substance (19).

Under the conditions of Table 3 binding constants of less than 10^3 would not be detected. For example, agglutination of red cells by a particular Anti-I cold agglutinin (called "Ma") is inhibited by sugars containing the structure Galβ1-4GlcNAcβ1-6Gal (20), including lacto-*N*-hexaose. However, using conditions shown in Table 3 we could not demonstrate binding of this antibody with lacto-*N*-hexaose or any of the fragments derived from it shown in Table 4, presumably because of a low hapten binding constant.

Sugars of the isomaltose series have been oxidized to aldonic acid derivatives and coupled to bovine serum albumin by the mixed anhydride reaction to produce effective antigens (21). Using a modification of this method, we have coupled milk oligosaccharides to polylysine and produced molecules capable of precipitating specific antibodies. For example, the polylysine-lacto-*N*-fucopentaose II conjugate gives quantitative precipitin curves with goat anti-Le^a serum similar to those obtained with human Le^a-active soluble blood group substance and precipitation of antibody by both the conjugate and the soluble blood group substance is specifically inhibited by lacto-*N*-fucopentaose II. Milk sugars coupled to polylysine should be useful for the production of antibodies directed towards specific sugars which can then be used for the detection of the sugars on cell surfaces.

REFERENCES

1. V. Ginsburg, in: Advances in Enzymology, Vol. 36, ed. A. Meister (John Wiley & Sons, New York, 1972) p. 131.

2. A. Kobata, in: Methods in Enzymology, Vol. 28, ed. V. Ginsburg (Academic Press, New York, 1972) p. 262.

3. E. O. Whittier, Chem. Rev. 2 (1925-1926) 84.

4. R. Kuhn and H. H. Baer, Chem. Ber. 89 (1956) 504.

5. R. Kuhn and A. Gauhe, Chem. Ber. 95 (1962) 518.

6. A. Kobata and V. Ginsburg, J. Biol. Chem. 247 (1972) 1525.

7. A. Kobata and V. Ginsburg, Arch. Biochem. Biophys. 150 (1972) 273.

8. R. Kuhn, H. H. Baer and A. Gauhe, Chem. Ber. 89 (1956) 2514.

9. R. Kuhn, H. H. Baer and A. Gauhe, Chem. Ber. 91 (1958) 364.

10. A. Kobata and V. Ginsburg, J. Biol. Chem. 244 (1969) 5496.

11. R. Kuhn and A. Gauhe, Chem. Ber. 93 (1960) 647.

12. L. Shen, E. F. Grollman and V. Ginsburg, Proc. Nat. Acad. Sci. U. S. 59 (1968) 224.

13. E. F. Grollman, A. Kobata and V. Ginsburg, J. Clin. Invest. 48 (1969) 1489.

14. H. Yang and S. Hakomori, J. Biol. Chem. 246 (1971) 1192.

15. A. Kobata, E. F. Grollman, B. F. Torain and V. Ginsburg, in: Blood and Tissue Antigens, ed. D. Aminoff (Academic Press, New York, 1970) p. 497.

16. W. M. Watkins, in: Glycoproteins, ed. A. Gottschalk (Elsevier, London-New York, 1966) p. 462.

17. D. M. Marcus and A. P. Grollman, J. Immunol. 97 (1966) 867.

18. D. A. R. Simmons and P. Perlman, Cancer Research 33 (1973) 313.

19. K. O. Lloyd and E. A. Kabat, Proc. Nat. Acad. Sci. U. S. 61 (1968) 1470.

20. T. Feizi, E. A. Kabat, G. Vicari, B. Anderson and W. L. Marsh, J. Immunol. 106 (1971) 1578.

21. Y. Arakatsu, G. Ashwell and E. A. Kabat, J. Immunol. 97 (1966) 858.

DISCUSSION

S. ROSENBERG: When you were talking about your glycosyltransferase studies you stated that the quantity of the glycosyltransferase would determine how many fucoses were attached and thus determine whether Le^a or Le^b substance is made. Does that mean that in individuals who are homozygous you get a different Lewis structure than in people who are heterozygous for the Lewis gene?

V. GINSBURG: I would think so.

S. ROSENBERG: Aside from family studies, is there any way one can serologically determine whether an individual is homozygous or heterozygous for the Lewis gene?

V. GINSBURG: No.

S. ROSENBERG: Do you feel that with your kind of study you could determine the genotype based on a qualitative difference in the carbohydrate pattern?

V. GINSBURG: I don't think so.

M. HOROWITZ: On one of your slides it was indicated that long chain oligosaccharides are found in glycoprotein and glycolipids. Have any long chain oligosaccharides been found in milk? In other words, is there any evidence for high molecular weight oligosaccharides larger than a hexa- or heptasaccharide?

V. GINSBURG: Milk contains large oligosaccharides. From their electrophoretic mobility after oxidation to aldonic acid derivatives, I would estimate that some of the neutral oligosaccharides have 15-20 sugar residues.

V. MONTEZ DE GOMEZ: Do you know if the differences in animal blood types is reflected by differences in these milk oligosaccharides containing galactose? - in different animals, not only humans.

V. GINSBURG: Milk from some animals is very rich in oligosaccharides but I know of no study to correlate the presence or absence of individual oligosaccharides without the blood type of the animals.

MOLECULAR ORIENTATION OF ERYTHROCYTE MEMBRANE GLYCOPROTEINS

V.T. MARCHESI, H. FURTHMAYR, and M. TOMITA
Department of Pathology
Yale University School of Medicine

Abstract: Membrane glycoproteins seem to be amphipathic molecules which are oriented so that their sugar moieties reside entirely external to the lipid barrier of the membrane. These glycoproteins, which make up the bulk of the integral membrane proteins, are bound tightly to the lipid matrix of the membrane. Parts of these molecules may also extend across the lipid bilayer so that segments of their polypeptide chains associate with proteins attached to the inner surface of the membrane. Previous studies suggested that glycophorin, the major sialoglycoprotein of the erythrocyte membrane, was associated with macromolecular structures [called intramembranous particles (IMP)] which are situated within the lipid matrix of the membrane. More recent evidence indicates that essentially <u>all</u> of the major glycoproteins of the red cell membrane are oriented asymmetrically in the membrane. Parts of the glycophorin may be linked to spectrin a fibrous protein which forms filamentous threads along the inner surface of the membrane. Both proteins, along with segments of the Band III polypeptides are specifically phosphorylated under appropriate conditions.

On the basis of common solubility properties and other features it is reasonable to suggest that the major glycoproteins of the human RBC membrane (glycophorin, Band III) are associated together as a functional complex which may correspond to the intramembranous particles.

MEMBRANE GLYCOPROTEINS ARE AMPHIPATHIC MOLECULES

Many investigators have found that the glycoproteins of the red cell membrane are bound tightly to the lipid elements of the membrane. The interactions between the polypeptide segments of these glycoproteins and the lipids of the membrane is probably of a non-covalent nature, but the bonding between the two species is difficult to disrupt with the usual protein solvents. On this basis it was suggested by Singer and Nicolson (1) that proteins of this type could be considered "integral" to the structure of the membrane. Within the last few years several methods have been developed for the isolation of these molecules (2-5) and as a result we now have enough preliminary information on the properties of the major sialoglycoprotein to begin to formulate some ideas as to where and how this molecule is situated in the membrane.

Recent studies of the major sialoglycoprotein, called glycophorin (6), confirm the early suggestions (7,8) that it was bound to the membrane through hydrophobic associations between a lipophilic segment of the molecule and the lipids of the membrane. A schematic model of the molecular features of the glycophorin molecule is shown in Figure 1. This model, which must still be considered provisional, shows that all the covalently-linked sugar residues are attached to the amino acids located in the N-terminal half of the polypeptide chain. Since the oligosaccharide residues form various antigenic determinants and binding-sites for lectins, this part of the molecule has been designated the receptor segment.

Another part of the molecule has been designated the "hydrophobic region" because of its high content of nonpolar amino acids. This has been determined by isolating and analyzing a tryptic peptide and a cyanogen bromide-derived fragment from this part of the molecule. Portions of these peptides have also been subjected to amino acid sequence analysis (9), and the results are consistent

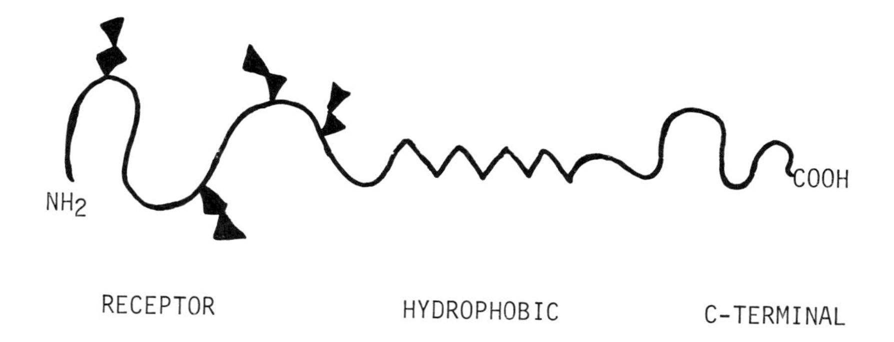

Figure 1. Linear arrangement of the glycophorin molecule. Peptides produced by cyanogen bromide cleavage and by tryptic digestion have been aligned on the basis of amino acid and carbohydrate compositions and partial amino acid sequence as illustrated in this provisional model. Carbohydrate residues are located at the N-terminal end (receptor region) of the molecule. The segment linking the receptor region with the C-terminal segment is composed largely of nonpolar amino acids, and this has been designated the hydrophobic region of the molecule. The solid triangles represent the approximate locations of oligosaccharide chains; however, the total number of such chains and their precise locations have not yet been determined.

with the idea that they are derived from a "hydrophobic" segment of the molecule which is capable of interacting with the aliphatic portions of phospholipids.

Glycophorin also seems to have a polypeptide segment which extends from the C-terminal end of the hydrophobic segment (designated the C-terminal portion of Figure 1) which contains a number of charged amino acids. Thus this molecule is really made up of three segments two of which would be most likely to reside in aqueous media.

At present we do not have enough information about the other membrane bound glycoproteins to decide whether the anatomy of the glycophorin molecule is representative of this class of molecules as a whole. The polypeptides in the Band III region (to be described below) show many of the properties of glycophorin (e.g. degree of binding to the membrane, asymmetrical orientation, etc.) and it is likely that they have similar molecular features as well, but this has not yet been established.

ORIENTATION OF GLYCOPHORIN IN THE MEMBRANES

A glycoprotein which is amphipathic or polarized so that one end is primarily hydrophilic and the other primarily hydrophobic would most likely sit in the membrane with the segment containing charged groups (e.g. sialic acids, acidic and basic amino acids) extending into an aqueous compartment, either inside or outside the cell, while its hydrophobic segment would most likely insert into the lipid bilayer. Such an orientation has been suggested on many grounds and seems to be the most likely mode for glycophorin. However, since this molecule also has a hydrophilic C-terminal segment of its polypeptide chain, it is not possible to decide, on *a priori* grounds, where this segment is likely to reside. This segment could be located outside the membrane, inside the cell, or even possibly in a hydrophilic pocket or channel *within* the membrane. Many attempts have been made to decide this question using differential labeling techniques, and the most persuasive evidence favors the idea that the C-terminal segment of glycophorin is located inside the cell (10,11). However, most of these studies rely on the differential labeling of membrane proteins of osmotically-lysed ghosts vs. those of intact cells, and these must be interpreted with caution since we do not know whether subtle structural rearrangements take place in the membrane during osmotic shock.

Further support for the idea that glycophorin and probably all the major glycoproteins of the red cell membrane are oriented asymmetrically at the membrane surface has been provided by the extensive studies of Steck and co-workers. They have found that glycophorin is readily digested by proteolytic enzymes when intact cells or regular ghosts are incubated with these enzymes, but the glycoprotein is not digested when "inside-out" ghost vesicles are treated with the same agents (12). Similar kinds of experiments were performed with galactose oxidase as a probe for the presence of galactose residues. Essentially no galactose oxidase-sensitive sugars were detected when inside/out vesicles were used as the substrate, a result which confirms the earlier claims that lectin-ferritin conjugates do not bind to the inner surfaces or red cell ghosts (13).

GLYCOPROTEINS MAY EXIST IN THE MEMBRANE AS MACROMOLECULAR COMPLEXES

Lectin-ferritin conjugates have been used to map the distribution of glycoproteins (and probably also glycolipid) components of different cell membranes, and studies of such labeled red cell membranes by freeze-etching and electron microscopy indicate that glycophorin molecules are connected in some way with the intramembranous particles, which measure 60-80 Å in diameter (the exact size is difficult to determine with the carbon-platinum replica technique) have generated great interest in recent years because of the possibility that they may be sites for transport of materials across the lipid bilayer. If glycophorin does span the thickness of the membrane, as some recent evidence indicates, parts of the hydrophobic segments of this molecule would be part of these particles, although they could only contribute a relatively small fraction of the total mass of these units. Many investigators have suggested that other membrane proteins might also reside at these sites, however the composition of these particles is still open to question.

If we consider which of the major membrane proteins in addition to glycophorin are likely to be part of these particles, only the polypeptides of Band III (Figure 2A) seem to be likely candidates. Bands I-II (spectrin), V and VI are loosely bound to the membrane lipids, probably via electrostatic associations with polar groups, and they have been designated the class of "peripheral" membrane proteins. The spectrin polymers (Bands I and II) are associated together in the form of fibrous strands of filaments which are located along the inner surface of the membrane (15). Although spectrin is clearly not one of the proteins associated with the intramembranous particles, some recent evidence suggests that spectrin filaments may be close to or possibly attached to structures which are part of the cytoplasmic side of the particle. Nicolson has suggested the spectrin filaments interact with the C-terminal segment of the part of the glycophorin molecule which extends through the intramembranous particles (16). A very schematic representation of this possible association is illustrated in Figure 3.

Since the polypeptides which make up Bands III and glycophorin are not extractable from membranes by the usual solubilization methods they have been designated integral or tightly bound components. Membranes from which the peripheral proteins are extracted retain all of Band III and glycophorin and some additional uncharacterized bands as well (Figure 2B), and interestingly enough, they still contain intramembranous particles (17). The "reciprocal" properties of these two classes of membrane proteins has been elegantly shown by Steck and co-workers who found that all of the peripheral proteins could be selectively extracted from membranes with an alkaline (ph 11) wash, while all of the integral proteins were retained in the extracted membrane fragments (18). Treatment of ghost membranes with the non-ionic detergent Triton X-100 produced the opposite results: most of the integral proteins were released into the medium while the peripheral proteins remained with the membrane fragments (19).

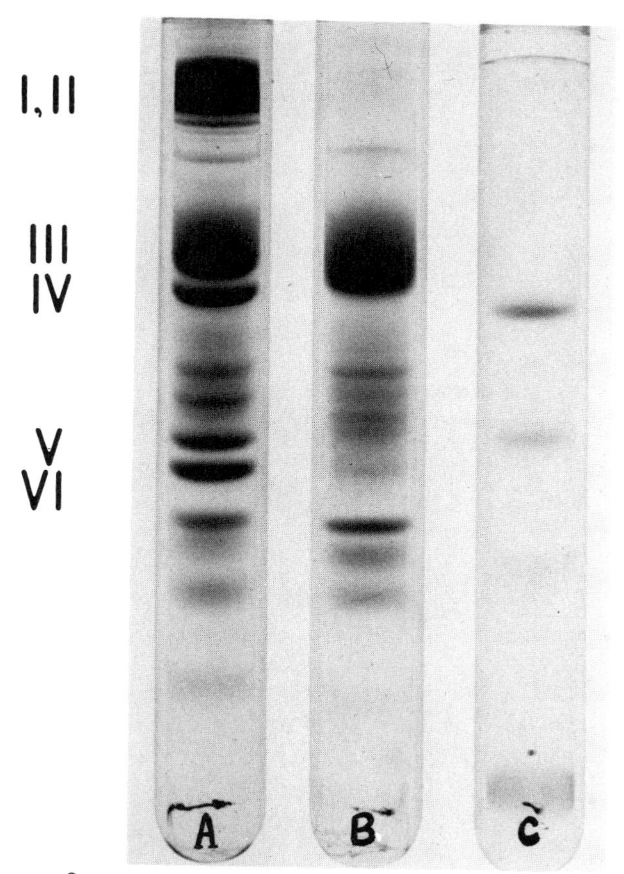

Figure 2.
SDS-acrylamide gels of complete human RBC ghost membrane (A) or partially extracted membrane fragments (B+C). The six major polypeptide species are labeled as suggested by Fairbanks et al (22). Membranes which have been treated with low concentrations of lithium diiodosalicylate (4) lack most of the so-called peripheral proteins but retain the major glycoproteins (B+C). A+B stained with Coomassee Blue, C stained with periodic acid-Schiff.

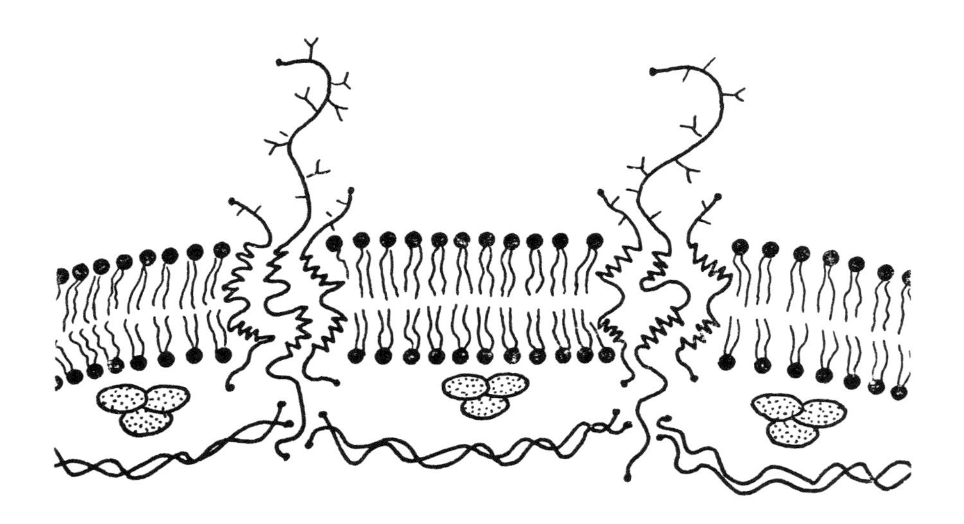

Figure 3

Glycophorin and the Band III polypeptides share another common feature in that they are both phosphorylated by endogenous protein kinases when red cell ghosts (20) or intact red cells (21) are incubated under appropriate conditions. It is perhaps significant that spectrin polymers are also phosphorylated under the same conditions.

Since glycophorin and the Band III polypeptides have common properties and seem to be retained or released from the membrane under identical conditions it is reasonable to consider the possibility that they are associated in the membrane in the form of a specific complex which may correspond to the intramembranous particle. Although the circumstantial evidence which supports this association is provocative, direct demonstration that this is so will ultimately depend upon the isolation and detailed characterization of these structural units.

REFERENCES

(1) S.J. Singer and G.L. Nicolson, Science, 175 (1972) 720.

(2) R.H. Kathan, R.J. Winzler and C.A. Johnson, J. Exp. Med., 113 (1961) 37.

(3) O.O. Blumenfeld, P.M. Gallop, C. Howe and L.T. Lee, Biochim. Biophys. Acta, 211 (1970) 109.

(4) V.T. Marchesi and E.P. Andrews, Science, 174 (1971) 1247

(5) H. Cleve, H. Hamaguchi and T. Hutteroth, J. Exp. Med., 136 (1972) 1140.

(6) V.T. Marchesi, T.W. Tillack, R.L. Jackson, J.P. Segrest and R.E. Scott, Proc. Nat. Acad. Sci. (USA), 69 (1972) 1445.

(7) A. Morawiecki, Biochim. Biophys. Acta, 83 (1964)339.

(8) R.J. Winzler, in: Red Cell Membrane, Structure and Function, eds. G.A. Jamieson and T.J. Greenwalt (Philadelphia J.B. Lippincott, 1969) p. 157.

(9) J.P. Segrest, R.L. Jackson, V.T. Marchesi, R.B. Guyer and W. Terry, Biochem. Biophys. Res. Commun., 49 (1972) 964.

(10) M.S. Bretscher, Nature New Biol., 231 (1971) 229.

(11) J.P. Segrest, I. Kahane, R.L. Jackson and V.T. Marchesi, Arch. Biochem. Biophys., 155 (1973) 167.

(12) T.L. Steck, in: Membrane Research, ed. C.F. Fox (Academic Press, New York, 1972) p. 71.

(13) G.L. Nicolson, (1972) Ibid.

(14) T.W. Tillack, R.E. Scott and V.T. Marchesi, J. Exp. Med., 135 (1972) 1209.

(15) G.L. Nicolson, V.T. Marchesi and S.J. Singer, J. Cell Biol., 51 (1971) 265.

(16) G.L. Nicolson and R.G. Painter, J. Cell Biol. 59 (1973) 395.

(17) H. Furthmayr and V.T. Marchesi, unpublished observations.

(18) T.L. Steck and J. Yu, J. Supramol. Struct. 1 (1973) 220.

(19) J. Yu, D.A. Fischman and T.L. Steck, Ibid, 1 (1972) 233.

(20) C.S. Rubin and O.M. Rosen, Biochem. Biophys. Res. Commun., 50 (1973) 421.

(21) D. Shapiro, P. Greengard and V.T. Marchesi, unpublished observations.

(22) G. Fairbanks, T.L. Steck, and D.F.H. Wallach, Biochem., 10 (1971) 2606.

DISCUSSION

H.G. WOOD: Before we start the discussion, Dr. Morrison has asked if he could make a few pertinent comments.

M. MORRISON: We have extended our investigations of the arrangement of the erythrocyte membrane, and I would like to tell you about the work (J. Biol. Chem., In press) of Drs. Mueller, Huber and myself on the glycoproteins of the red cell. We have employed our lactoperoxidase macromolecular probe and iodinated the intact normal human erythrocyte. The stroma was prepared from these cells and then the glycoproteins were purified by the procedure of Hamaguchi and Cleve (Biochem. Biophys. Acta *278*: 271 (1972)). The normal human erythrocyte has a number of glycoproteins which are readily separated on SDS acrylamide gel electrophoresis. The major glycoprotein has been labelled PAS I, and two readily detected minor glycoproteins, PAS II and III. As shown in Figure 1, not only is the major glycoprotein iodinated, but the two minor glycoproteins are also labeled. Our previous studies (Phillips, D. and Morrison, M., Biochemistry *10:* 1766 (1971), FEBS Letters *18*: 95 (1971)) have not pointed out the labeling which takes place on the minor glycoproteins since the amount of label incorporated into these proteins in the intact cell is small compared to the labeling in the major glycoprotein.

In order to determine whether the same protein exists on both sides of the erythrocyte membrane, a stroma preparation was resealed and labeled with lactoperoxidase system on the outside of this membrane. At the same time another aliquot of the stroma was resealed in the presence of lactoperoxidase. Resealed membranes were also prepared in which both lactoperoxidase and glucose oxidase were sealed inside the membrane. In the former case peroxide which is freely permeable to the membrane was added to the solution in order to initiate iodination. In the latter case, glucose was added to the solution. The glucose provided the substrate for glucose oxidase which generated peroxide. In order to be certain that the labeling was exclusively on the inside of the sealed ghost, sufficient catalase was added to the suspending medium to insure that little or no labeling took place

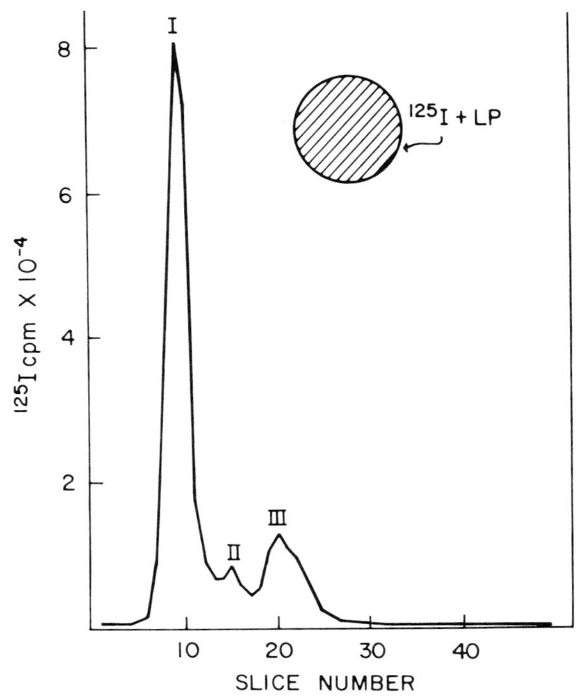

Figure 1. Lactoperoxidase catalyzed labeling of the glycoproteins extracted from intact human erythrocytes.

on the membranes exposed to the suspending medium. After labeling, the membrane glycoproteins were extracted and separated by SDS electrophoresis. The gel was sliced into 2 mm slices and counted.

The labeling pattern of the membrane proteins obtained with the enzyme on the inside of the membrane was very different from that obtained when the enzyme was outside of the membrane. In the case of the glycoproteins, however, the position of the enzyme with respect to sidedness of the membrane did not change the labeling pattern as dramatically. The results of labeling from the outside are shown in Figure 2 upper. The major glycoprotein and the two minor glycoproteins are all labeled. In Figure 2 lower, the results of labeling from the inside are shown and again the major glycoprotein and two of the

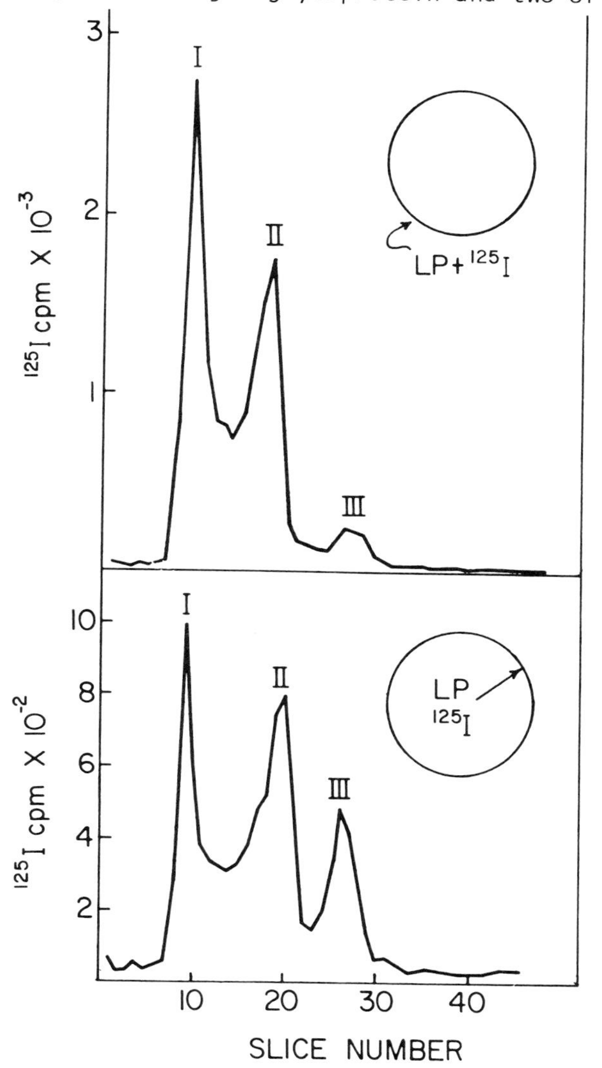

Figure 2. The glycoproteins extracted from resealed ghosts labeled externally with lactoperoxidase (upper figure); ghosts labeled internally with lactoperoxidase (lower figure).

minor glycoproteins are labeled. Essentially identical results were obtained on the inside labeling employing either peroxide or glucose to initiate iodination. The catalase on the outside of the resealed vesicles which contained glucose oxidase was in high enough concentration to inhibit iodination. Intact cells were not iodinated when similar concentrations of catalase were added to the reaction mixture. Thus, in this system little chance for erroneous labeling outside of the membrane is possible.

A comparison of Figures 1 and 2 shows that with the enzyme on the outside of the membrane, minor glycoproteins are labeled to a greater extent in the resealed membranes than in the intact cells. When the membranes are labeled on the inside, the minor glycoproteins are labeled more extensively when compared to the major glycoprotein.

These results have been interpreted as indicating that the lactoperoxidase system labeled (1) the major glycoprotein and two minor glycoproteins on the outside of both intact normal human erythrocytes and resealed erythrocyte membrane vesicles; (2) these proteins are also labeled when lactoperoxidase is sealed inside the resealed erythrocyte membrane vesicles; (3) it is concluded that all three glycoproteins penetrate the membrane of the normal human erythrocyte; (4) the differences in the extent of labeling of PAS I and II suggest that PAS I is not a simple dimer of PAS II.

A. KESTON: In 1944 I reported studies on iodination of proteins employing I^{131} iodide, a flavoprotein (xanthine oxidase) which generates hydrogen peroxide in the presence of xanthine, and lactoperoxidase which catalyzes the oxidation of iodide by peroxide. The protein, flavoprotein, and lactoperoxidase are present in milk in which these iodinations were performed. The system in milk was suggested to be a model for iodination reactions in the thyroid. The report also commented on the failure of catalase to inhibit the iodination reaction even when added in very substantial amounts, possibly because lactoperoxidase was more effective than catalase at low levels of H_2O_2. Another explanation of the lack of effect of catalase may be that H_2O_2 is bound to lactoperoxidase, as is iodine, and the bound H_2O_2 may not be easily accessible to catalase. I wonder whether the reason why the "inside" and "outside" in Dr. Morrison's experiments with membranes furnish similar products is that the catalase on the outside did not in fact destroy the peroxide on the outside. Were any control experiments carried out to elucidate this point?

M. MORRISON: Catalase would simply compete with lactoperoxidase for the available peroxide. In the presence of high concentrations of catalase and low concentrations of peroxidase, you get very little iodination. We did, in fact, do model experiments to show that under the conditions we employed there would be no iodination with the quantity of catalase we employed.

J. HOCHSTADT: Dr. Marchesi, I completely agree with you that Ted Steck has made vital contributions to this field. However, I believe it should be pointed out that he has recently reinterpreted some of his data on the vesicles. He previously interpreted his two vesicle bands to be both continuous, one inside-out, the other outside-out. He now believes as does Grant Fairbanks Laboratory, that those two populations of vesicles represent one population of continuous, intact vesicles and one population of non-continuous vesicles. This reinterpretation better explains the papain data you showed in an early slide, and I believe also fits in very nicely with the kind of rearrangements Dr. Morrison has just told us about.

M. MORRISON: No, that is not the answer. We are aware of the misinterpretation made in Dr. Steck's original paper and have thoroughly investigated the sidedness of our resealed membranes. They are sealed and they are outside-out.

S. HAKAMORI: Concerning Dr. Marchesi's speculation on the possibility of carbohydrate structures on the inside surface of the membrane, we described an "inside" carbohydrate in the June issue of JBC. Our conclusion was based on the comparison of the incorporation of radioactive label into membranes of intact red cells and into a previously extracted membrane fraction. We obtained extra label incorporated into the isolated membrane and this label was associated with a component of molecular weight near to 200,000.

V.T. MARCHESI: How can you be certain that the glycopeptide was really on the inside surface? Could it not have been in a masked form on the outer surface of the intact cell?

S. HAKOMORI: That is a possibility, of course. There is also a possibility that incorporation of label is dependent on the mobility and how excited the molecules are, so that some may be more reactive with the labeling reagent.

M. CZECH: We have some evidence that in fat cell plasma membranes the glycoproteins may also be present in large complexes. When the membranes are electrophoresed in the presence of SDS and mercaptoethanol, we detect two major glycoproteins of molecular weights 94,000 and 78,000. In the absence of reducing agents the 94,000 glycoprotein does not penetrate the gel and under these circumstances appears to be larger than 200,000.

I. KALOS: I would like to make a comment regarding the inside or outside localization of specific proteins in the intact cell. I have been attempting to locate the protein kinase which phosphorylates the membrane of 3T3 fibroblasts and I have shown that the protein kinase is able to phosphorylate substrate on the outside and able also to act on the inside. This was shown by the use of immobilized substrate (histone) which can be phosphorylated outside; this phosphorylation is also cAMP-dependent.

V.T. MARCHESI: How do you know that no kinase is being secreted outside?

I. KALOS: Control experiments show that only about 0.1% of the kinase is solubilized.

L. WARREN: Dr. Marchesi, how can you account for the fact that about five studies have shown that you can remove essentially 100% of the sialic acid of a red cell with neuraminidase and still keep the red cell intact? This is true of dogfish red blood cells, which I have studied. I don't think there are many glycoproteins in the red cell that do not contain sialic acid.

V.T. MARCHESI: But it is true that there doesn't seem to be any sialic acid on the band three polypeptides.

L. WARREN: Maybe that is the band you should look for as being inside.

MEMBRANE GLYCOPROTEINS AS PLANT LECTIN RECEPTORS

STUART KORNFELD, W. LEE ADAIR, CHARLENE GOTTLIEB
AND ROSALIND KORNFELD
Departments of Medicine and Biochemistry
Washington University School of Medicine

INTRODUCTION

The plant lectins are a group of proteins which are capable of binding to specific carbohydrate determinants on the surfaces of mammalian cells (for a recent review, see Ref. 1). Depending on the cell type involved, lectin binding can induce a variety of biologic effects as summarized in Table 1. In several of these instances, the lectins mimic the action of the known physiologic stimulator of the cell type involved. Thus the lectin may mimic the effect of insulin on fat cells (5), the effect of thrombin on platelets (6) and the effect of glucose on pancreatic beta cells (7). Two general points concerning these observations deserve special consideration. The first is that that these biological effects are not induced by all lectins. For example, the lectin from the mushroom _Agaricus bisporus_ is not mitogenic toward lymphocytes, yet it has insulin-like activity toward fat cells and is the only lectin so far tested which induces insulin release from islet cells (Table 2). Wheat germ agglutinin has insulin-like effects on fat cells but is not a mitogen even though it binds to lymphocytes. Concanavalin A, _Phaseolus vulgaris_ E-PHA and _Lens culinaris_ PHA are mitogenic toward lymphocytes, have insulin-like activity toward fat cells but do not induce insulin release from islet cells. The second noteworthy feature of these systems is that, in the few cases which have been examined up to now, it has been

TABLE 1

Biologic effects of various plant lectins

1. Agglutinate RBC's and other cell types
2. Agglutinate many tumor cells more readily than normal cells (2,3)
3. Stimulate resting lymphocytes to undergo mitosis (4)
4. Have insulin-like effects on fat cells (5)
5. Induce the platelet release reaction (6)
6. Induce insulin release from pancreatic islet cells (7)
7. Inhibit phagocytosis by granulocytes (8)
8. Inhibit fertilization of ovum by sperm (9)
9. Inhibit protein synthesis in most mammalian cell types (10-13)

The numbers in parenthesis refer to the appropriate reference.

demonstrated that the biologic effect exerted by the lectin results from the binding of the lectin to a specific cell membrane receptor.[1] Therefore, one explanation for the

[1] The one exception thus far noted concerns the inhibition of protein and DNA synthesis by the lectins of *Ricinus communis* (termed ricin and RCA I) and *Abrus precatorius*. In these cases the lectins first bind to membrane receptors and then most likely enter the cell where they directly inhibit protein synthesis (10-13). However, the possibility that even these lectins act at the cell surface has not been excluded.

TABLE 2

Specificity of the biologic effects of plant lectins

Lectin	Lymphocyte mitogen	Insulin-like activity with fat cells	Insulin release from pancreatic islets
P. vulgaris E-PHA	+	+	-
L. culinaris PHA	+	+	-
Concanavalin A	+	+	-
A. bisporus PHA	-	+	+
Wheat germ agglutinin	-	+	-
Ricin	(Toxic)	+	-

diversity of biologic effects induced by the plant lectins is that each lectin binds to a different membrane receptor with a distinctive carbohydrate structure. For this reason the work in our laboratory has been directed toward determining the structure of the cell membrane receptors for a number of plant lectins.

Quantitation of lectin receptors on various cell types- Most mammalian cells have relatively large numbers of lectin binding sites on their surface. The actual number of available receptors can be measured quantitatively by performing binding studies with purified lectins labeled with 125I. In most instances, double reciprocal plots of the binding curves are monophasic, but in the case of the Ricinus communis agglutinin (RCA I) and wheat germ agglutinin (WGA) the curves are biphasic. The results of a number of binding studies are summarized in Table 3.

TABLE 3

Plant lectin binding sites on various cell types

Lectin	Cell type				
	Human Erythrocytes	Normal human lymphocytes	Chronic lymphocytic leukemia lymphocytes	Mouse L1210 cells	Calf thymocytes
			Sites/cell		
P. vulgaris E-PHA	0.52×10^6	2.7×10^6	1.15×10^6	5×10^6	0.93×10^6
L. culinaris PHA	0.55×10^6	2.4×10^6	-	-	0.47×10^6
A. bisporus PHA	6.8×10^6	12.7×10^6	2.8×10^6	9.3×10^6	1.73×10^6
R. communis RCA I	1.2×10^6	3.6×10^6	-	3.5×10^6	1.66×10^6

The data in this table are compiled from References 13-17.

Of interest is the fact that each cell type tested has different numbers of binding sites for the various lectins and that the ratio of binding sites for the different lectins varies from one cell type to another. When human erythrocytes were treated with trypsin and then tested for lectin binding ability, there was a 40-50% decrease in the amount of P. vulgaris E-PHA, L. culinaris PHA, R. pseudoaccacia PHA and A. bisporus PHA bound. In contrast, R. communis lectin binding was not affected by this treatment indicating that its receptor was located on a trypsin-resistant molecule.

Characterization of trypsin-released glycopeptides with lectin binding sites - Trypsin treatment of intact human erythrocytes releases a glycopeptide with a molecular weight of 10,000 to 15,000 from the N-terminal portion of the major sialic acid-containing glycoprotein of the membrane (18). When the released glycopeptide was isolated, it was found to contain binding sites for the P. vulgaris, L. culinaris, R. pseudoaccacia and A. bisporus lectins (15,16,19,20). When the residual sialoglycoprotein material was solubilized from the erythrocyte membrane with lithium diiodo salicylate and isolated, it was also found to contain significant numbers of receptor sites for these lectins. Thus, receptors for these lectins appear to be distributed throughout the carbohydrate containing region of the molecule. In contrast, the purified intact sialoglycoprotein from erythrocytes showed poor haptene activity toward the R. communis lectins, indicating that these lectins bind to oligosaccharide structures on the other membrane glycoproteins (21).

A schematic diagram of the trypsin released glycopeptide is shown in Figure 1. It consists of 20% protein in the form of a peptide backbone and 80% sugar in the form of multiple oligosaccharide chains extending from the backbone. The carbohydrate chains are of two basic types. The Type I chain, shown in the lower right of Figure 1, is a branched structure containing sialic acid, galactose, mannose, and N-acetylglucosamine. This oligosaccharide chain is linked to the peptide backbone by an N-acetylglucosamine asparagine linkage. The Type II chain, shown in the lower left of Figure 1, contains sialic acid, galactose and N-acetylgalactosamine and is linked O-glycosidically to

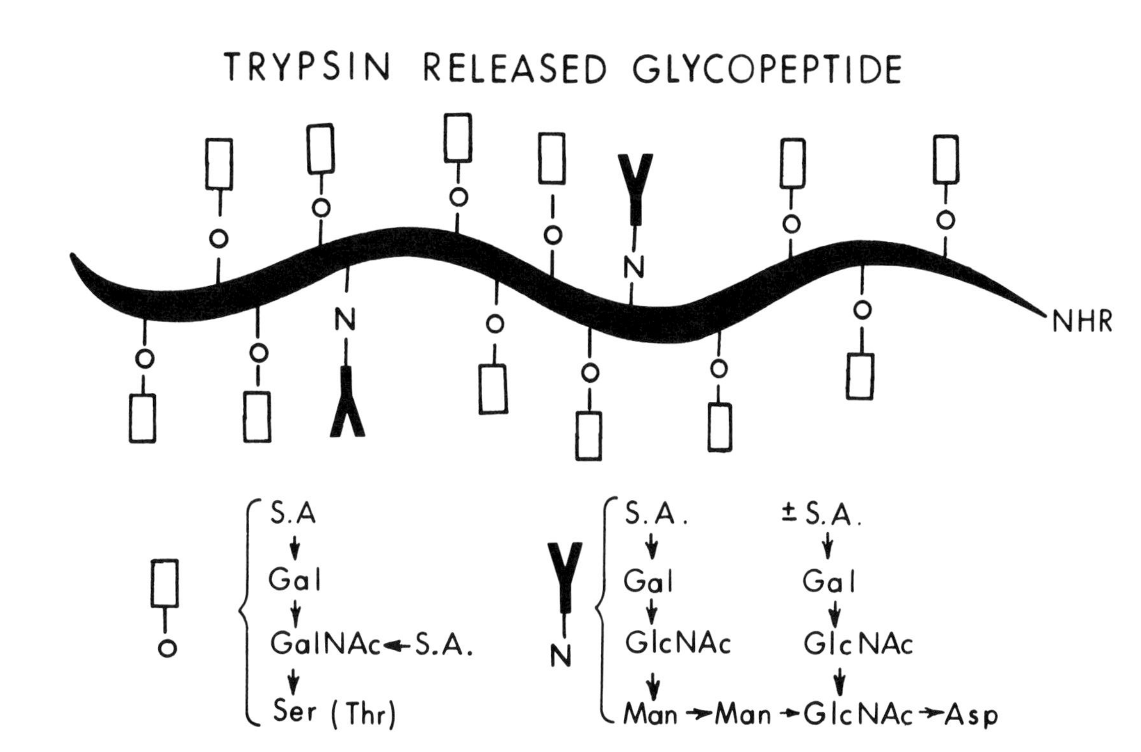

Fig. 1. Schematic diagram of the glycopeptide released from human erythrocytes by trypsin. The structure for the O-glycosidically linked oligosaccharide (lower left) is taken from Thomas and Winzler (22). The structure for the N-glycosidically linked oligosaccharide (lower right) is from Kornfeld and Kornfeld (19).

the hydroxyl groups of serine and threonine. There are about 6 Type II chains for each Type I chain. The Type II chains are alkali labile while the Type I chains are alkali stable. Using alkali treatment, pronase digestion of the glycopeptide and a variety of fractionation techniques, the glycopeptide I (GP I) and the glycopeptide II (GPII) molecules have been separated and tested for their lectin binding properties (15,16,19,20). The results of these experiments are summarized in Table 4. GP I is a potent haptene inhibitor of *P. vulgaris* E-PHA, *L. culinaris* PHA and *R. pseudoaccacia* PHA while GP II inhibits only *A. bisporus* PHA.

To discover which structural features of glycopeptides I and II determine their ability to bind each lectin, the glycopeptides were subjected to stepwise degradation with purified glycosidases and then tested for their lectin binding properties (15,16,19,20). Removal of the sialic acid from GP I did not diminish its ability to bind to *P. vulgaris* E-PHA. However, when the galactose residues were removed from GP I it lost 90% of its original inhibitory activity. Similar results were obtained with *R. pseudoaccacia* PHA. These results show the importance of the galactose residues in the binding of GP I to these lectins. The other sugars in the molecule, particularly the mannose residues, also influence lectin binding activity to a significant extent. This was best demonstrated by testing a variety of model compounds for lectin binding activity. The monosaccharide galactose, the disaccharide lactosamine (Gal β1,4 $\rightarrow$ GlcNAc) and other oligosaccharides devoid of mannose residues were completely inactive as haptene inhibitors of erythrocyte agglutination by *P. vulgaris*, *L. culinaris* and *R. pseudoaccacia* PHA. On the other hand, glycopeptides from fetuin and transferrin which have mannose residues in their cores as well as the sequence galactose $\rightarrow$ N-acetylglucosamine in their outer chains were found to have significant haptene activity. These data suggest that in addition to the outer chain galactose residues, the inner sugars of the oligosaccharide, particularly the mannose residues, are involved in the binding of *P. vulgaris* PHA and *R. pseudoaccacia* PHA to the glycopeptide. The ability of GP I to bind *L. culinaris* PHA also depends on the mannose residues in the core, but in this case it is the

TABLE 4

Haptene inhibitory activity of erythrocyte glycopeptides I and II

Lectin tested	GP I	GP II
	(inhibitory units*/mμm)	
P. vulgaris E-PHA	3.3	<0.01
L. culinaris PHA	4.4	<0.01
R. pseudoaccacia PHA	1.5	<0.06
A. bisporus PHA	<0.2	8.0

* One unit of inhibitory activity is defined as the amount of material which permits only 1^+ red cell agglutination in a standard hemagglutination inhibition system in 3 min.

N-acetylglucosamine residues in the outer branches which determine L. culinaris PHA binding and removal of the galactose residues has no effect (15,23).

The binding of A. bisporus PHA to GP II is determined by the galactose → N-acetylgalactosamine → Ser sequence of the Type II oligosaccharide (16). In this case the sialic acid residues actually hinder lectin binding. Thus removal of the sialic acid residues from GP II with neuraminidase results in an eight fold increase in haptene activity (Table 5). Subsequent removal of the galactose residue with β-galactosidase results in greater than 99% loss of haptene activity, demonstrating the importance of this sugar residue in determining A. bisporus PHA binding activity. Again, as in the studies with GP I, the simple sugars galactose and N-acetylgalactosamine are virtually devoid of activity. The disaccharide galactose → N-acetylgalactosamine has only 1/70 the activity of desialized GP II showing that the N-acetylgalactosamine

TABLE 5

Inhibitory activity of various compounds toward *A. bisporus* PHA

Compound	Inhibitory units/mμm
S.A. $\xrightarrow{\alpha 2,6}$ GalNAc S.A. $\xrightarrow{\alpha 2,3}$ Gal $\xrightarrow{\beta 1,3}$ GalNAc → Ser	8
Gal $\xrightarrow{\beta 1,3}$ GalNAc → Ser	66
GalNAc → Ser	< 1
Gal $\xrightarrow{\beta 1,3}$ GalNAc	1
Gal $\xrightarrow{\beta 1,3}$ N-acetylgalactosaminitol	< .01
Galactose	0.01
N-acetylgalactosamine	0.01

The data in this table are compiled from Reference 16.

→ Ser linkage is either a part of the lectin binding site itself or serves to orient the disaccharide in a particularly favorable conformation to bind the *A. bisporus* PHA.

These data demonstrate that various plant lectins bind to very different but specific sequences of sugars which are present on cell surfaces as distinct oligosaccharide units of membrane glycoproteins. The ability of the oligosaccharides to serve as lectin receptors is determined by a number of the sugar residues in the molecule. While certain sugar residues are essential for lectin binding capacity, they are not sufficient by themselves to allow lectin binding of a high affinity type. Finally, the sugar residues which determine lectin binding may be located in the interior of the oligosaccharide unit and, in some instances, the terminal sugar residues may actually hinder lectin binding.

Fractionation of membrane glycoproteins on lectin-Sepharose affinity columns - While the use of proteolytic enzymes to solubilize lectin receptors allows one to determine many of the characteristics of these molecules, the technique has several disadvantages. Firstly, it only releases receptors that are located on trypsin-sensitive glycoproteins. Secondly, it yields a mixture of glycopeptides which could either have been derived from a number of different membrane glycoproteins, or from different portions of the same glycoprotein. And thirdly, trypsin treatment degrades the intact membrane glycoproteins so that it is impossible to examine the lectin receptors in their native state. Therefore we sought a nonproteolytic method for solubilizing and fractionating intact membrane glycoproteins with lectin receptor activity. For this purpose we solubilized erythrocyte membrane glycoproteins with the non-ionic detergent Triton X-100 and determined that the solubilized material contained 70-95% of the binding sites for a number of lectins. The solubilized material was then passed through various lectin-Sepharose columns and the adsorbed glycoproteins eluted with the appropriate haptene sugar. Using this procedure we have been able to separate the major sialoglycoprotein of the erythrocyte membrane from another class of glycoproteins which have very different oligosaccharide units and lectin receptor activity (21).

The results of a typical experiment are summarized in Table 6. In this experiment Triton solubilized erythrocyte membrane glycoproteins were passed through a R. communis agglutinin I (RCA I)-Sepharose column and the adsorbed material eluted with 0.1 M lactose. The eluted material displayed extremely potent haptene activity toward RCA I

TABLE 6

Fractionation of erythrocyte glycoproteins on *R. communis* agglutinin-Sepharose

Erythrocyte membrane glycoproteins containing lectin receptors were solubilized with 7 volumes of 0.5% Triton X-100-56mM Na borate pH 8.0 using the procedure of Yu, *et al* (24). The soluble material from 77 ml of ghosts were passed over a column (2 x 8 cm) which contained 120 mg RCA I-Sepharose. The adsorbed material was eluted with 0.1 M lactose in 0.1% Triton X-100-56 mM Na borate pH 8.0. Fractions were assayed for protein, sialic acid and lectin receptor activity using a standardized haptene inhibitory assay system.

Fraction	Protein (mg)	Sialic acid (μmoles)	Total lectin inhibitory activity (I.U.) of each fraction toward		
			RCA I	*P. vulgaris* E-PHA	*A. bisporus* PHA
Triton extract	343	62	85,200	94,200	2,390,000
Lactose eluate	43	1.4	50,600	13,700	21,200
% in eluate	12.5%	2.3%	59%	14.5%	0.9%

and ricin, being 1200 times more potent than the haptene galactose. The material was also a strong haptene inhibitor of the A. precatorius lectin, which has a binding site which is similar to that of RCA I. The specificity of the column is demonstrated by the fact that the eluate contained 59% of the haptene inhibitory activity for RCA I but only 0.9% of the haptene activity for the A. bisporus PHA and 14.5% of the P. vulgaris E-PHA haptene inhibitory activity. Since the receptors for the A. bisporus PHA are located on the major sialoglycoprotein of the erythrocyte, these data demonstrate that the RCA I lectin receptors must be on glycoproteins which are distinct from the sialoglycoprotein. The finding that only 2.3% of the applied sialic acid was recovered in the lactose eluate supports this conclusion.

The strongest evidence that the R. communis lectin receptor glycoproteins are distinct from the sialoglycoprotein comes from the carbohydrate composition data (Table 7). The R. communis lectin receptor material is particularly rich in galactose and N-acetylglucosamine and almost devoid of sialic acid and N-acetylgalactosamine. This carbohydrate composition is strikingly different from that of the sialoglycoprotein. The R. communis receptor glycoproteins are major membrane components, containing approximately 19% of the total carbohydrate of the membrane.

When the R. communis lectin receptor glycoprotein material was subjected to electrophoresis in SDS on polyacrylamide gels, no carbohydrate containing bands could be detected with the periodic acid Schiff reagent (PAS). However, when the gels were sliced, hydrolyzed in acid and analyzed directly for aminosugar, the material could be detected as several glycoprotein peaks. Two of these glycoproteins may correspond to glycoproteins isolated by other investigators using different techniques (25). When Triton X-100 extracts were passed over ricin-Sepharose columns, very similar results were obtained. This finding is consistent with other observations which indicate that the binding site of ricin is very similar to the binding site of RCA I.

TABLE 7

Carbohydrate composition of human erythrocyte ghosts and glycoproteins

Carbohydrate analysis was performed by the GLC technique. The data are expressed as residues with the N-acetylglucosamine value being set at 3.0.

	Whole ghosts	Sialoglycoprotein*	WGA receptor	RCA I receptor
Sialic acid	2.5	8.1	9.9	0.6
Fucose	1.1	1.1	0.9	0.4
Galactose	7.8	8.1	6.9	2.5
Mannose	1.4	1.4	1.5	0.6
N-acetylglucosamine	3.0	3.0	3.0	3.0
N-acetylgalactosamine	3.9	7.1	7.5	0.6

* These values were taken from Ref. 25.

In contrast to these findings, the glycoprotein material eluted from a WGA-Sepharose column contained approximately an equal yield of receptor activity for the WGA, A. bisporus and P. vulgaris lectins and a much poorer yield of haptene activity for the R. communis lectins. Furthermore, the glycoprotein material isolated on the WGA-Sepharose column had a carbohydrate composition which was almost identical to that of the sialoglycoprotein (Table 7). On SDS polyacrylamide gels, the WGA receptor material gave a single band which was PAS positive and which had a mobility identical to that of the major sialoglycoprotein. These data indicate that the sialoglycoprotein is a receptor for the WGA. However, since the recovery of WGA receptor activity from the affinity column was only 6%, one cannot conclude that all of the WGA receptors are located on the sialoglycoprotein.

It is apparent from these data that affinity chromatography on lectin-Sepharose columns is an effective procedure for fractionating erythrocyte membrane glycoproteins. The technique offers a new approach to the fractionation of membrane glycoproteins since it is based upon the ability of the lectin to distinguish one oligosaccharide from another.

Relation of lectin receptor glycoproteins to the erythrocyte intramembranous particles - Tillack, et al (26), utilizing freeze-etch electron microscopy, have demonstrated an association between the receptor sites for P. vulgaris E-PHA and WGA and a 75A particulate constituent of the erythrocyte membrane, termed the intramembranous particle (IMP). These authors have suggested that the particles may represent an interaction between the hydrophobic portion of the sialoglycoprotein and the lipids of the membrane. In studies performed in collaboration with Drs. T. Triche and T. Tillack, we have found that the binding of A. bisporus PHA to erythrocyte ghosts shows an almost perfect correlation with the IMPs, confirming the previous observations that the sialoglycoprotein is associated with these particles. When ferritin conjugated RCA I was used, it was found that at low lectin concentrations there was a close correlation with the IMPs while at high lectin concentrations a poor correlation existed.

Thus high RCA I concentrations resulted in virtually complete coating of the surface of trypsinized ghosts which displayed marked aggregation of the IMPs. In all instances lectin binding was completely inhibited by the haptene galactose. These findings indicate that at least one but not all ofthe molecules which serve as receptors for the R. communis lectin is associated with the IMPs. We conclude that the IMPs contain at least two glycoproteins and that some membrane components which contain oligosaccharide receptors for RCA I (either glycoproteins or glycolipids) are not associated with the IMPs.

Use of lectins to isolate cell lines with altered membrane glycoproteins - Many questions concerning the structure, function and biosynthesis of cell membrane glycoproteins and their oligosaccharide units are difficult to approach using standard biochemical techniques. In considering this problem, it seemed to us that the availability of cell lines with altered surface carbohydrates might facilitate many studies of these membrane components. To isolate such cell lines, we have used the toxic lectin ricin as a selective agent. This lectin binds to a galactose-containing membrane receptor and subsequently causes an irreversible inhibition of protein and DNA synthesis leading to cell death (10-13). The experimental procedure was to add 0.1 μg/ml ricin to Chinese hamster ovary (CHO) cells growing in monolayer and test surviving cells for their ability to bind various lectins. Using this technique we have isolated a cell clone (15B) which is 80-fold less sensitive to the toxic action of ricin than the parent line and which displays major alterations in lectin binding properties (27). Fig. 2 illustrates the types of alterations seen, ranging from either marked or moderate decreases to an actual increase of different lectin binding sites. When double reciprocal plots were made using these data and the number of binding sites were calculated, it was found that the number of sites for ricin, RCA I, P. vulgaris E-PHA, A. precatorius PHA and soybean PHA on the 15B cells was decreased by 93% or more compared to the number on the parent cell line. The number of binding sites for L. culinaris PHA was decreased 83%, the number for WGA was decreased 60% and A. bisporus PHA binding was essentially unchanged. In contrast, the 15B cells bound 70% more ConA than the parent line.

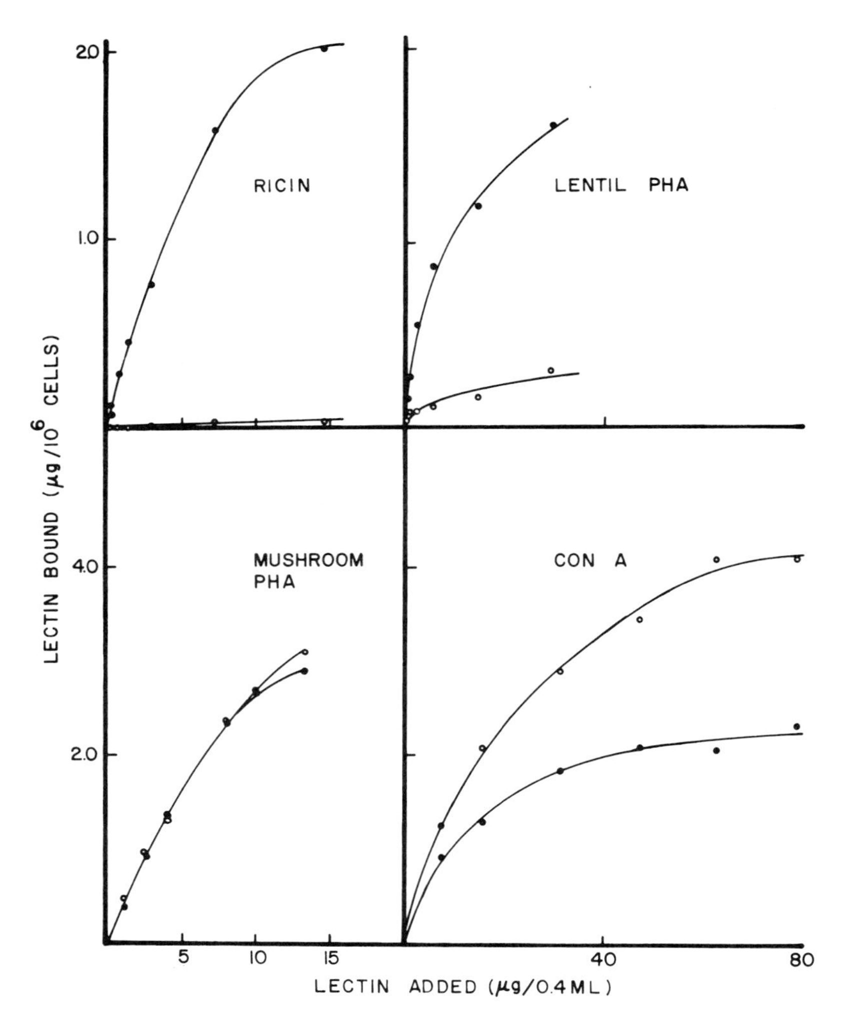

Fig. 2. Binding of ricin, L. culinaris PHA, A. bisporus PHA and ConA to CHO parent and 15B cells. 1-2 x 10^6 cells were incubated with 0.5 to 80 μg of 125I-lectin in a 0.4 ml reaction mixture. After 60 min the cells were washed twice and the amount of bound 125I-lectin determined. ●——●, CHO parent cells; o——o, 15B cells. (From Gottlieb and Kornfeld, Proc. Natl. Acad. Science - in press.)

The carbohydrate composition of crude membrane preparations of the 15B and parent CHO cells is shown in Table 8. The 15B membranes have 28% less sialic acid, 38% less N-acetylglucosamine, 49% less galactose, the same amount of N-acetylgalactosamine and 53% more mannose than the membranes of the parent cells.

The differences between the 15B and parent CHO cells in the patterns of lectin binding and membrane carbohydrate composition become more understandable when we relate them to the previously discussed data on the structure of lectin receptors on human erythrocyte membranes. In red cells P. vulgaris E-PHA, L. culinaris PHA and probably ConA bind to the Type I oligosaccharide while A. bisporus PHA binds to the Type II oligosaccharide (Fig. 1) The R. communis and A. precatorius lectins bind to a third type of oligosaccharide unit that contains predominantly galactose and N-acetylglucosamine. The finding that A. bisporus PHA binding to the 15B cells is essentially normal and that the N-acetylgalactosamine content of the membranes of these cells is about the same as the parent line suggests that the Type II-like oligosaccharide units of the 15B cells are intact. On the other hand, the loss of P. vulgaris E-PHA and L. culinaris PHA binding sites along with the preservation of ConA binding sites (α linked mannose residues {23,28}) is best explained by postulating a block in the synthesis of Type I-like units distal to the mannose residues. Such a block would also account for the decreased content of sialic acid, galactose and N-acetylglucosamine in the 15B membranes. Since galactose residues also serve as the determinant sugars for the R. communis and A. precatorius lectin receptors, we postulate that the 15B cells may have a defect in membrane oligosaccharide biosynthesis which affects the synthesis of at least two types of galactose and N-acetylglucosamine-containing units. The explanation for the increased mannose content of the 15B cells is not clear.

A different type of resistant cell has been isolated from a mouse L cell line. Six clones have been identified which possess 50-70% of the lectin binding capacity of the parental cell line yet are 10-100-fold less sensitive to the direct cytotoxic action of ricin. The mechanism of lectin resistance in these cell lines is currently under

TABLE 8

Carbohydrate composition of
parent CHO and clone 15B membranes

Crude membrane fractions were prepared by repeated freeze-thawing of the cells followed by sedimentation at 100,000 xg for 60 min. The membranes were washed once, resuspended in H_2O and analyzed for protein and carbohydrate content.

	CHO parent	15B clone	15B/parent
	mg/10^8 cells		%
Membrane protein	6.3	6.4	100
	nmole/mg protein		%
Sialic acid	5.3	3.8	72
Galactose	15	7.7	51
Mannose	19	29	153
N-acetylglucosamine	17	10.5	62
N-acetylgalactosamine	2.4	2.7	112

The data in this table are compiled from C. Gottlieb and S. Kornfeld, Proc. Natl. Acad. Sci. (1974), in press.

investigation. Hyman, et al (29) have recently employed immunoselection using lectin-antilectin and complement to isolate murine lymphoma cell variants which appear to be quite similar to our L cell variants.

CONCLUSION

Although plant lectins were discovered before the turn of the century, it is only within the past decade that their usefulness in the study of mammalian cell membranes has been appreciated. Investigations of the nature of membrane receptor sites for various lectins have served to enhance our knowledge of the structure and organization of complex carbohydrates present on mammalian cell surfaces. In addition, lectins can function as highly selective agents in the fractionation of membrane glycoproteins. Finally, certain lectins which are toxic toward mammalian cells can be used to select cell lines with altered membrane carbohydrates.

REFERENCES

(1) N. Sharon and H. Lis, Science, 177 (1972) 949-959.

(2) M. M. Burger and A. R. Goldberg, Proc. Natl. Acad. Sci., 57 (1967) 359-366.

(3) M. Inbar and L. Sachs, Proc. Natl. Acad. Sci., 63 (1969) 1418-1425.

(4) P. C. Nowell, Cancer Res., 20 (1960) 462-466.

(5) P. Cuatrecasas and G. P. Tell, Proc. Natl. Acad. Sci., 70 (1973) 485-489.

(6) D. Tollefsen, J. R. Feagler and P. W. Majerus, J. Clin. Invest., in press (1974).

(7) R. B. Lockhart-Ewart, S. Kornfeld and D. M. Kipnis, Clin. Res., 21 (1973) 622.

(8) R. D. Berlin, Nature New Biol., 235 (1972) 44-45.

(9) T. Oikawa, R. Yanagimachi and G. Nicolson, Nature, 241 (1973) 256-259.

(10) J.K. Lin, K. Liu, C. C. Chen and T. C. Tung, Cancer Res., 31 (1971) 921-923.

(11) S. Olsnes and A. Phil, FEBS Letters, 20 (1972) 327-329.

(12) K. Onozaki, M. Tomita, T. Sakurai and T. Ukita, Biochem. Biophys. Res. Comm., 48 (1972) 783-788.

(13) S. Kornfeld, W. Eider and W. Gregory, in: Control of Proliferation in Animal Cells (Cold Spring Harbor Press, 1974) in press.

(14) S. Kornfeld, Biochim. Biophys. Acta, 192 (1969) 542-545.

(15) S. Kornfeld, J. Rogers and W. Gregory, J. Biol. Chem., 246 (1971) 6581-6586.

(16) C. Presant and S. Kornfeld, J. Biol. Chem., 247 (1972) 6937-6945.

(17) R. Kornfeld and C. Siemers, J. Biol. Chem., 249 (1974) in press.

(18) R. J. Winzler, E. D. Harris, D. J. Pekas, C. A. Johnson and P. Weber, Biochemistry, 6 (1967) 2195-2202.

(19) R. Kornfeld and S. Kornfeld, J. Biol. Chem., 245 (1970) 2536-2545.

(20) A. M. Leseney, R. Bournillon and S. Kornfeld, Arch. Biochem. Biophys., 153 (1972) 831-836.

(21) L. Adair and S. Kornfeld, submitted for publication.

(22) D. B. Thomas and R. J. Winzler, J. Biol. Chem., 244 (1969) 5943-5946.

(23) S. Toyoshima, M. Fukuda and T. Osawa, Biochemistry (1972) 4000-4005.

(24) J. Yu, D. A. Fischman and T. L. Steck, J. Supramol. Structure, 1 (1973) 233-248.

(25) M. J. Tanner and D. H. Boxer, Biochem. J., 129 (1972) 333-347.

(26) T. W. Tillack, R. E. Scott and V. T. Marchesi, J. Expt. Med., 135 (1972) 1209-1227.

(27) C. Gottlieb and S. Kornfeld, Proc. Natl. Acad. Sci., in press (1974).

(28) L. L. So and I. J. Goldstein, J. Biol. Chem., 243 (1968) 2003-2007.

(29) R. Hyman, M. Lacorbiere, S. Stavarek and G. Nicolson, J. Natl. Cancer Inst., in press.

DISCUSSION

R. PORETZ: In an effort to examine the specificity of a number of lectins we have looked at the agglutinability of cells (erythrocytes and lymphocytes) which have been trypsinized and also treated with Ficin. It is interesting to note that one particular lectin, *Maclura pomifera* agglutinates cells treated with trypsin to the same extent as untreated cells. However, Ficin-treated cells are not agglutinated by this lectin. Looking at five other lectins (*S. japonica, P. lunatus, W. floribunda, L. alpinum, C. ensiformis*), trypsinization and ficinization each cause an increase in agglutinability of the cells. Have you noted any effect of other proteases on the *Ricinus communis* agglutinin; that is, on the binding of this agglutinin to cells or on the solubilization of its receptor?

S.A. KORNFELD: No, we haven't. I should point out that there is a significant difference between lectin binding and cell agglutinability. Most of our studies have been concerned with lectin binding and these may not relate directly to agglutinability as Dr. Nicolson discussed yesterday.

G.L. NICOLSON: I can give you a partial answer. We published a paper last year (G.L. Nicolson, J. Nat. Cancer Inst. *50* (1973) 1443) in which we looked at cell binding and agglutination with affinity purified RCA_I. In that study we used trypsin treatment of a variety of different cell types including erythrocytes, lymphomas, normal and transformed fibroblasts, etc. and found that there was no correlation between the number of RCA_I molecules bound and agglutinability. In all cases, a very brief trypsinization resulted in little or no change in the number of RCA_I sites but after very prolonged proteolytic digestion the number of sites actually went down. Brief trypsinization resulted in a very dramatic increase in agglutinability which was not greatly changed by further digestion. We have also looked at RCA_{II} binding and agglutination using trypsin and other proteolytic enzymes with similar results.

G. KOCH: I have several questions. First, concerning your ricin effect on DNA synthesis in PHA stimulated lymphocytes: David Webb *et al.* have shown that PHA exposure of lymphocytes induces drastic changes in cyclic AMP levels. Do Ricinus lectins alter cyclic AMP levels?

S.A. KORNFELD: We haven't examined the effect of ricin on cyclic AMP levels. Recently, it has been shown by other investigators that ricin inactivates the 60s subunits of ribosomes so that this seems to be at least one major site at which it works.

G. KOCH: This would indicate that ricin is taken up by cells and the inhibition of protein synthesis is due to an effect on ribosomes?

S.A. KORNFELD: Yes, I think the best postulate right now is that the ricin gets into the cell and has a direct effect on ribosomes. But we can't exclude the possibility that the ricin acts at the cell membrane.

G. KOCH: As soon as you see an effect of ricin on protein synthesis you cannot reverse the effect of ricin by exposure of cells to proteolytic enzymes. Can you at an earlier time?

S.A. KORNFELD: We have done experiments with human lymphocytes in which we allowed ricin to bind to cells and then at various times later we added lactose to take the lectin off the cell surface. The cells were then tested for DNA synthesis on day three. For the first two to three hours of ricin binding it was possible to reverse its effect, but longer binding caused an irreversible inhibition of DNA synthesis.

M. RIEBER: Do you have any information on how long it takes for the inhibitory effect of ricin on DNA synthesis to be revealed? Do you have any information with regard to the data shown yesterday and today that the lectin first interacts with the receptor outside the cell and later enters the cell? Also, do you have any information as to whether the interaction with the 60s ribosomes takes place preferentially on membrane bound ribosomes or on free ribosomes?

S.A. KORNFELD: With regard to the inhibition of DNA and protein synthesis by ricin, we have done a series of kinetic studies to examine this point. We found that maximal binding of ricin occurs in about 5 minutes, but there is no inhibition of protein synthesis for at least 20 minutes. It is only necessary to saturate about 0.1% of the receptor sites to completely inhibit protein synthesis. If one added enough ricin to saturate 100% of the binding sites, there is still a lag of at least 20 minutes before inhibition is observed. So there seems to be a fixed lag period. With regard to your second question, I don't think that it has been established whether or not ricin works preferentially on membrane-bound ribosomes or free ribosomal subunits. Some recent work on this has been published (Montanaro *et al.* Biochem. J. *136* (1973) 677).

G.L. NICHOLSON: Yesterday, I mentioned that in a cell-free system with a very low concentration of RCA_{II} ricin, protein synthesis was inhibited in a very short time, probably less than one or two minutes, but it took approximately 30 to 60 minutes in intact cells to show the inhibition. This seems to correlate with the time required for binding, clustering and subsequent endocytosis of RCA_{II} inside the cell and then release into the cytoplasm, presumably from the endocytotic vesicles. Several others including Drs. S. Olsnes and A. Pihl (FEBS Letters *20* (1972) 327, *28* (1972) 48; and J.Y. Lin et al. (Cancer Res. *31* (1971)) with polysomes and probably disrupts protein synthesis at that level. Inhibition of DNA synthesis lags behind this, that is at very low RCA_{II} concentrations it takes a few hours after inhibition of protein synthesis to see inhibition of DNA synthesis. It appears that inhibition of DNA synthesis is an indirect effect, as also concluded by Lin and his collaborators.

G. KOCH: I would like to make a comment on the regulation of protein synthesis in mammalian cells grown in suspension. Experiments performed by Soborio and Poug in my laboratory provided indirect evidence that one can drastically alter the rate of protein synthesis *in vivo* by events mediated by the cell membrane. Addition of DMSO or an increase in medium tonicity by addition or salt or sucrose blocks protein synthesis immediately and this effect is completely reversible on return to isotonicity. We think that these effects are mediated by the cell membrane because we don't observe comparable effects on protein synthesis by addition of DMSO or salt to an *in vitro* system.

S.A. KORNFELD: That is very interesting. I think that really it is still an open question as to whether ricin works at the membrane or must enter the cell to exert its toxic effect.

M. MORRISON: I wonder if we could interpret your data as suggesting that what we have all called lactoprotein I, that is the PAS-1 staining band, is heterogeneous and has two distinct chemical populations. Is that a correct interpretation of your data?

S.A. KORNFELD: No, I think the interpretation is that there are many glycoproteins and I don't think that bears on whether PAS-1 is one substrate or not. By that, I mean, I think the ricin receptors are clearly different glycoproteins than the major sialoglycoprotein.

M. MORRISON: I see. But didn't you isolate a pure PAS-1?

S.A. KORNFELD: You are referring to the wheat germ receptor experiments. It so happens that on the SDS gels we ran with this material, we only saw PAS-1 and not PAS-2. We really haven't systematically examined that point and tried to convert PAS-1 to PAS-2. So I would like to defer judgement as to whether or not PAS-1 contains more than one glycoprotein.

G.L. NICOLSON: I am a little disappointed with the heterogeneity of the product from ricin-Sepharose because the affinity column should be a very nice way to isolate band-3. Could this mean that the product is not band-3 or that some proteolytic degradation has occurred? Did you find band-3 coming out in the effluent?

S.A. KORNFELD: As you point out, the SDS gels of the affinity purified *R. communis* receptor reveal many bands. I think these represent many of the minor bands which are normally present on SDS gels of RBC ghosts. I suspect that the affinity column is pulling out a complex of glycoproteins and proteins, and that we are amplifying them so that they are now visible on SDS gels as a number of minor bands. I don't think we are getting significant proteolytic degradation, but we certainly can't exclude that possibility. We are currently trying to separate the affinity purified materials using other fractionation techniques.

W. FULLERTON: Dr. Kornfeld, regarding the glycoproteins which stain poorly with PAS, are these acidic glycoproteins? Have you looked to find whether they are sulfonated or carboxylated?

S.A. KORNFELD: No, we have only examined them on SDS gels.

S. ROSEMAN: Sharon made the point in his review article that because of the contaminating proteins found in many lectin preparations, hemagglutinating activity of these preparations isn't necessarily the normal lectin activity. I wish to raise the question concerning the purity of these reagents that are being used so ubiquitously. What kind of criteria do you use for purity? You remember that concanavalin A at one time was supposedly a pure reagent that everybody was using off the shelves, and now we know that the commercial stuff is not so pure - it consists of subunits cleaved to greater or lesser degrees.

S.A. KORNFELD: Yes, I think that's a valid point to consider. I would say that many of the lectins are pure proteins. They are homogeneous as examined by the standard physical techniques. In addition, in many instances, it is possible to completely block the biologic action with the appropriate haptene sugar, indicating that the biologic effect is really due to the carbohydrate-binding property of the lectin. But certainly not all lectins have been highly purified, and there could very well be many effects attributed to lectins which are really caused by contaminants in the lectin preparations.

G.L. NICOLSON: I direct Dr. Roseman to a recent article on RCA_I and RCA_{II}, which is concerned in detail with the points he has brought up (G.L. Nicolson, J. Blanstein, and M.E. Etzler, Biochemistry, Jan. 15 issue, 1974).

H.G. HEMPLING: Some effects of lectin have to do with aqueous pathways such as changes in conductance pathways and yet in the diagrams that you, Dr. Marchesi and Dr. Nicolson have shown, we don't see any water in cell membranes. What is the maximal amount of water that you are going to permit to be present in these membranes that will still maintain intact intrinsic proteins and a fluid lipid layer? What is the state of water in the membrane and how do you explain in Dr. Marchesi's favorite cell, the red cell, energies of activation of 4 kcal/mole of water which is classic for water movement in a beaker of water, whereas in your favorite cell, the lymphocyte, the energies of activation are about 18 kcal/mole which is comparable to an organized state with twice the structure of ice. Do you think it is important?

S.A. KORNFELD: It really wouldn't matter what I thought, because I know so little about it. But seriously, lectins do change membrane permeability. There are a number of studies which demonstrate that the transport of substances such as AIB and K^+ is enhanced by lectins. I suspect these changes in transport are a consequence of membrane alterations induced by the lectins. It might be very useful if someone studied the effect of lectins on the state of water in the membranes of the target cell.

COMPLEX CARBOHYDRATES AND INTERCELLULAR ADHESION

Saul Roseman

Introduction

The surfaces of eukaryotic cells play important, perhaps the central role in diverse biological phenomena including cell-cell recognition and intercellular adhesion. For this reason, the biochemistry of these cell surfaces has become an area of great interest within the past few years.

Much of the available information on the chemistry of eukayrotic cell surfaces is reviewed in this volume, and it is clear that we do not know the precise chemical structure of the surface of any given cell type. There is, however, some information available from immunological, histological and chemical studies, and these show that cell surfaces are rich in carbohydrates, i.e., the so-called glycocalyx. Since a given cell surface contains numerous antigens, including blood group, histocompatibility, tissue specific and others, a first approximation is that the eukaryotic cell surface is a mosaic comprised of different substances. Whether or not all the antigenic carbohydrate groups are attached to the same core protein is discussed elsewhere in this volume.

In some cell types, a fraction, perhaps the major fraction of the carbohydrate, consists of glycosaminoglycans, formerly called mucopolysaccharides. In the more general case, however, it appears that the major carbohydrate fraction consists of mixtures of glycoproteins and glycolipids. For present purposes, the latter two classes of substances are designated complex carbohydrates. In mammalian cells, the glycolipids are primarily glycosphingolipids, and the latter may be divided into two groups, one containing sialic acid (the gangliosides), and the other (neutral glycolipids) containing uncharged sugars; an additional minor class, the sulfolipids, contains sulfate bound to a hexose such as galactose. An example of a ganglioside, the most abundant ganglioside in human brain gray matter, is disialoganglioside whose structure is shown in Figure 1. The chemistry and biochemistry of the ganglio-

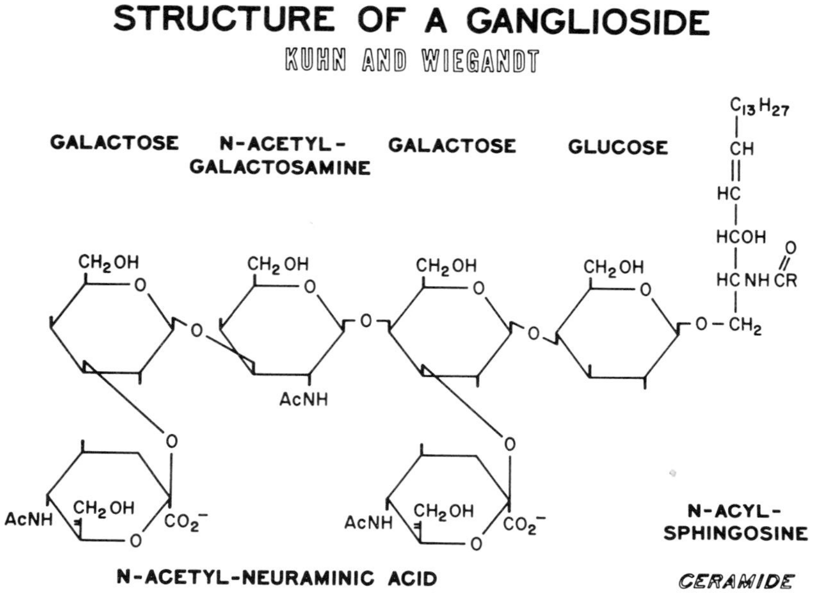

STRUCTURE OF A GANGLIOSIDE

KUHN AND WIEGANDT

GALACTOSE N-ACETYL-GALACTOSAMINE GALACTOSE GLUCOSE

N-ACETYL-NEURAMINIC ACID

N-ACYL-SPHINGOSINE

CERAMIDE

Fig. 1. Structure of disialoganglioside.

ides has been reviewed (1). A major variation in the glycosphingolipids is the length, structure and composition of the oligosaccharide chains attached to the ceramide moiety. Minor variations are also evident in that a given "pure" glycosphingolipid from a single source contains a mixture of sphingosines and fatty acyl residues.

The glycoproteins can also be divided into two major classes. The serum or blood type glycoproteins schematically illustrated in Figure 2 contain from one to a few, perhaps as many as 3, large, branched oligosaccharide chains attached to the protein core. The first sugar in the oligosaccharide chain is N-acetylglucosamine, glycosidically bound via the amide nitrogen atom of an asparagine residue in the protein (2). In general, the blood type glycoproteins contain oligosaccharide chains terminating in sialic acid or L-fucose linked to D-galactose which in turn is linked to N-acetyl-D-glucosamine; these trisaccharides are usually found linked to a "core oligosaccharide" containing D-mannose and N-acetylglucosamine.

GLYCOPROTEINS (SCHEMATIC)

BLOOD GLYCOPROTEINS

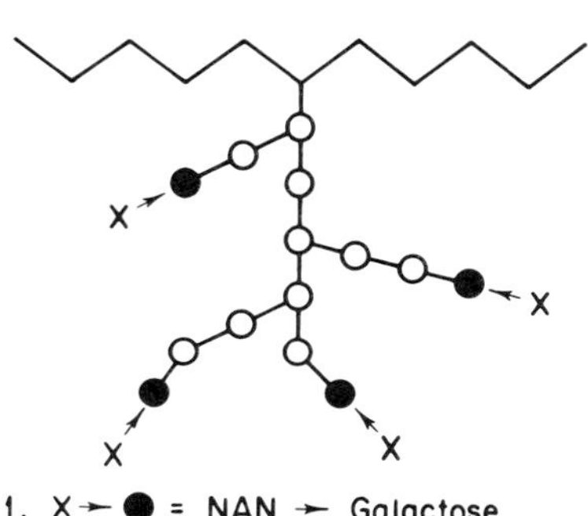

1. X → ● = NAN → Galactose
2. Few Polysaccharide Chains (branched)
3. OH^- Stable

SHEEP SUBMAXILLARY MUCIN (OSM)

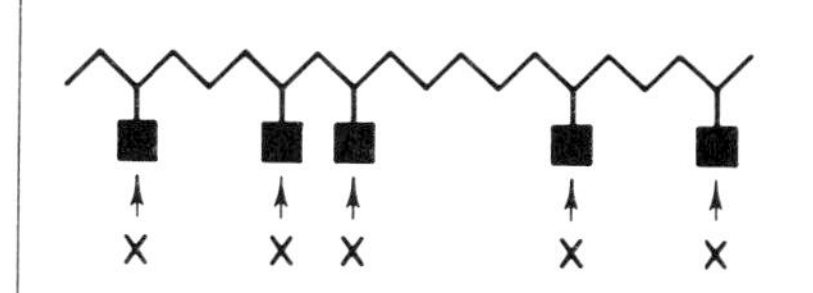

1. X → ■ = NAN → Acetylgalactosamine
2. Many Disaccharide Units
3. OH^- Labile

Fig. 2. Schematic structures of serum type and mucin type glycoproteins.

The second major class of glycoproteins are the mucins (Figure 2). Here, numerous short oligosaccharide chains, perhaps as many as 800, are found linked to an unusual polypeptide chain via O-glycosidic bonds to the hydroxyl groups in serine and threonine. The usual linking sugar appears to be N-acetyl-D-galactosamine, and the major amino acids in the mucin polypeptide are serine, threonine, glycine and proline; aromatic sulfur-containing amino acids are present in low to trace quantities (2).

Biosynthesis

The general pathways of synthesis of the complex carbohydrates may be visualized as occurring in three steps. All of the sugars found in these polymers are generated by the cell from glucose; the sugars are then "activated" by conversion to their corresponding sugar-nucleotide derivatives; at the last stage, the sugars are incorporated into the polymers. The pathways of synthesis of the monosaccharide-nucleotides are well established through the work of a number of laboratories (3,4) and in many cases the monosaccharides are synthesized at the level of the sugar-nucleotides. Some of these pathways can be quite complex, as illustrated in the metabolic pathway for the sialic acids (Figure 3). Figure 3 also shows a

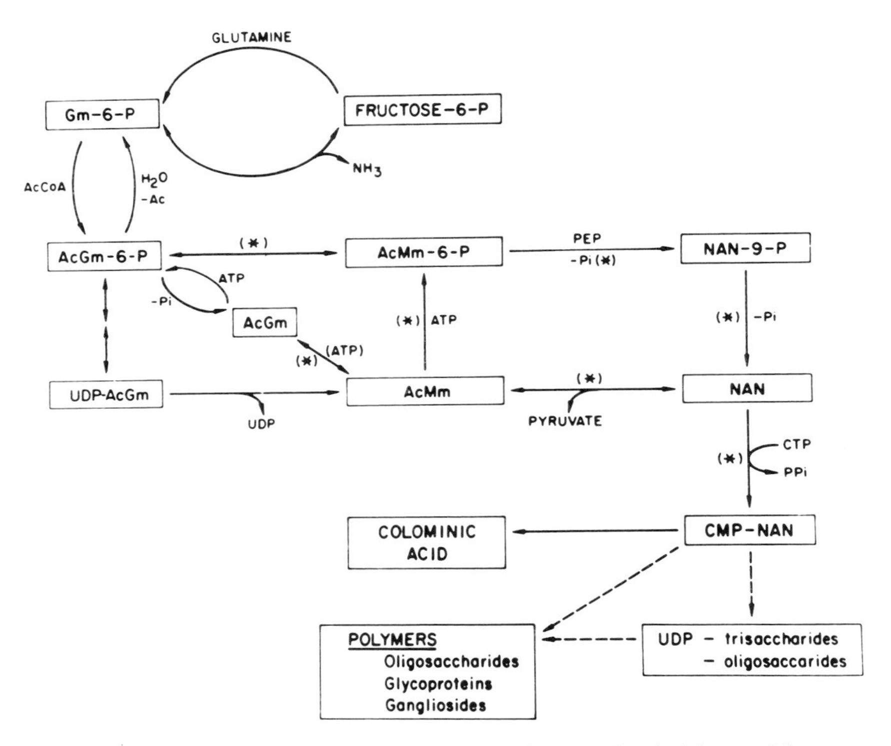

Fig. 3. Intermediary metabolism of sialic acid.

key reaction which will be discussed later in this review, that is, the introduction of the amino group into nitrogen containing sugars. Glucosamine-6-P is the key intermediate in the synthesis of derivatives of glucosamine, galactosamine, mannosamine, and sialic acid, and glucosamine-6-P is in turn derived from fructose-6-P by transfer of the amide group from L-glutamine; the enzyme is strongly inhibited by DON and other analogues of glutamine (5).

During the past few years, the area of major interest in the synthesis of complex carbohydrates has been in the last biosynthetic step, i.e., the incorporation of the sugars into the complex carbohydrates. From a theoretical point of view, one may consider this as two separate problems, the nature of the sugar donors and of the acceptors. The donors could be the monosaccharide-nucleotides, or oligosaccharide-nucleotides, or oligosaccharide-lipids of the type involved in the synthesis of bacterial cell walls and lipopolysaccharides (6). The acceptors could also be variable. In the glycosphingolipids, the initial acceptor could be sphingosine, with N-acylation occurring at some later stage or at the last stage after the oligosaccharide chain is completed, or, N-acylation could be the first in the sequence so that ceramide is the initial acceptor. All of the present data are in accord with ceramide being the acceptor.

The biosynthesis of the glycoproteins and mucins offer many more possibilities. Sugar incorporation could occur at the level of the amino acid, aminoacyl-t-RNA, the incomplete but growing polypeptide chain, or after the polypeptide chain is complete. Furthermore, in viewing the entire problem, there is no _a priori_ reason why oligosaccharide synthesis should necessarily be restricted to one of the stages in the formation of the polypeptide chain; some sugars could be incorporated early, while the outermost sugars in the oligosaccharide chains could be incorporated as the last steps in the biosynthetic pathway. In fact, present evidence suggests that the serum type of glycoproteins are synthesized in this manner.

As indicated above, the sugar-donors may not necessarily be monosaccharide-nucleotides. Drs. Behrens and Heath at this meeting review the evidence for the participation of dolicholo♦ligosaccharides. Another intriguing possibliity are oligosaccharide-nucleotides. Several such compounds have been isolated, and those studied in this laboratory (7) are shown in Figure 4. These intriguing compounds have the same structures as the terminal trisaccharide units in serum type glycoproteins such as orosomucoid (α_1-acid glycoprotein) and fetuin, and from

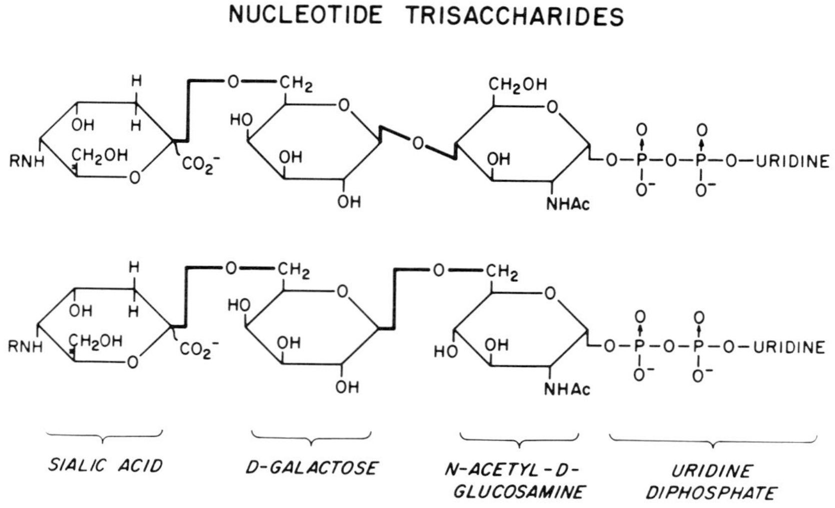
NUCLEOTIDE TRISACCHARIDES
RNH
OH
H
CH2OH
CO2-
O-CH2
HO
OH
CH2OH
NHAc
O-P-O-P-O-URIDINE
SIALIC ACID
D-GALACTOSE
N-ACETYL-D-GLUCOSAMINE
URIDINE DIPHOSPHATE

Fig. 4. Structures of goat colostrum nucleotide trisaccharides. R = acetyl or glycolyl.

this type of structural identity, it is reasonable to speculate that oligosaccharide-nucleotides serve as donors of "blocks" of sugar residues in the synthesis of the oligosaccharide chains of the complex carbohydrates. However, there is no evidence that these compounds participate in the biosynthesis of the complex carbohydrates, despite experiments designed to find such activities (8). In fact, the evidence is to the contrary in the sense that of some twenty or thirty enzymes involved in the biosynthetic pathways, all use the monosaccharide-nucleotides with the exception of the dolichol-oligosaccharide transferases described in this symposium. The enzymes that utilize monosaccharide-nucleotides are designated *glycosyltransferases*, and the general reaction catalyzed by this large group of enzymes is illustrated in Figure 5.

As indicated above, theoretical acceptors for the biosynthesis of complex carbohydrates such as the mucins or the blood type glycoproteins range from the amino acids *per se* to the complete polypeptide chain. The evidence from our own enzymatic studies, from early work on *in vivo* isotope incorporation, and from the work reviewed by

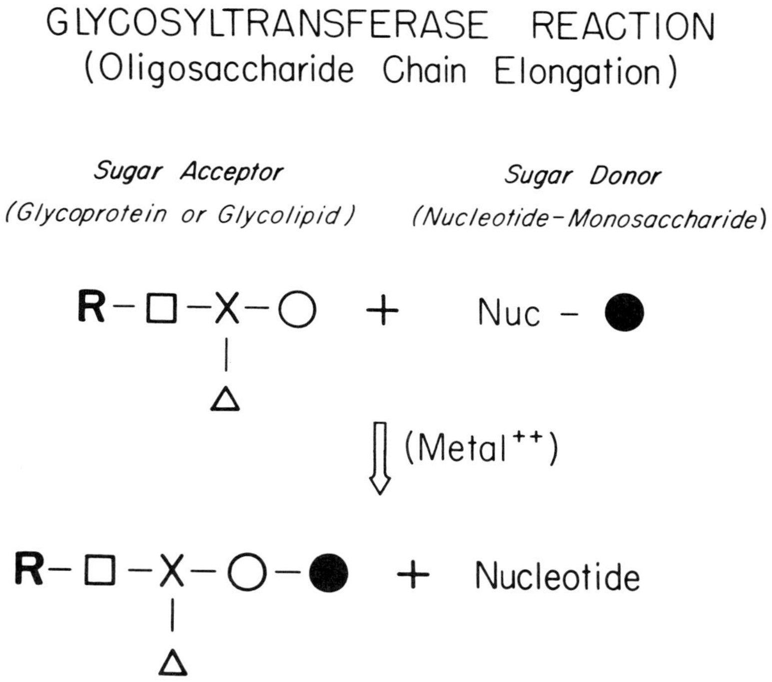

GLYCOSYLTRANSFERASE REACTION
(Oligosaccharide Chain Elongation)

Sugar Acceptor (*Glycoprotein or Glycolipid*)

Sugar Donor (*Nucleotide-Monosaccharide*)

R–□–X–○ + Nuc – ●
  |
  △

(Metal^{++})

R–□–X–○–● + Nucleotide
  |
  △

Fig. 5. The glycosyltransferase reaction. The sugar acceptor is a glycoprotein or glycolipid with an incomplete oligosaccharide chain; each of the symbols represents a different monosaccharide moiety.

Drs. Behrens and Heath at this meeting leads to the following conclusions: (a) Oligosaccharide chain synthesis in the mucins takes place on the complete polypeptide chain. (b) Serum type glycoproteins are processed in two stages. In the first stage, the core oligosaccharide units (consisging of N-acetylglucosamine and mannose) are incorporated from dolichol derivatives into the incomplete polypeptide in the rough endoplasmic reticulum. In the second stage, the oligosaccharide chains are completed by the action of a set of glycosyltransferases on the intact polypeptide containing putative sugar chains. Chain completion takes place in the smooth endoplasmic reticulum or more likely the Golgi complex, and possibly even at the cytoplasmic membrane.

The gangliosides can contain as many as eight glycose units comprising four different monosaccharides. Similarly, the oligosaccharide chains in the mucins that have been adequately characterized are sometimes pentasaccharides composed of four different monosaccharides, and the serum type glycoproteins can contain 16 or more glycose units composed of five different sugars. The soluble blood group glycoproteins are even more complex. If the biosynthetic pathways involve glycosyltransferases of the type illustrated in Figure 5, then how are "mistakes" avoided, and how can chain completion be effected? These problems are particularly interesting in light of the fact that the same monosaccharide unit can occur more than once in a single complex carbohydrate. Before answering these questions, it is important to emphasize that microheterogeneity is a general property of the complex carbohydrates. That is, any given preparation of a "homogeneous" mucin or serum type glycoprotein contains oligosaccharide chains of different lengths and complexity. For example, the pig submaxillary mucin that exhibits human blood type A reactivity contains oligosaccharide chains ranging from monosaccharides (N-acetylgalactosamine) to the antigenic pentasaccharide (9). This situation contrasts markedly with protein biosynthesis where it is almost invariable that a complete chain is synthesized. The contrast and reasons for the differences have been reviewed (10), and are briefly recapitulated here.

The concept is based on the properties and sub-cellular locations of the 15-20 glycosyltransferases studied in this laboratory. These enzymes consist of classes or families of enzymes, i.e., there are families of sialyltransferases, galactosyltransferases, N-acetylglucosaminyltransferases, and N-acetylgalactosaminyltransferases (10, 11). While each member of a given family utilizes the same sugar-nucleotide as the sugar donor, they show different specificities for the acceptor molecule. In the case of two sialyltransferases that utilize the same acceptor (12,13), the products are structural isomers of each other.

The importance of acceptor specificity is illustrated

in Figure 6 and 7. As a result of this specificity, the product of one reaction becomes the substrate for the next

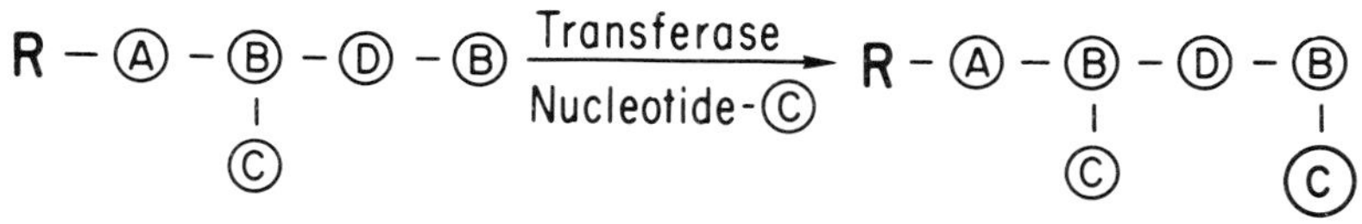

R – (A) – (B) – (D) – (B) —Transferase / Nucleotide-(C)→ R – (A) – (B) – (D) – (B)
with (C) attached to the first (B) in the substrate, and (C) attached to both (B) residues in the product

1. Sugar-nucleotides are glycose donors.

2. Sugars added as monosaccharides in a specific sequence.

3. Chain elongation at non-reducing ends or branch points.

R = protein or lipid

Fig. 6. Biosynthesis of oligosaccharide chains in glycoproteins and glycolipids by action of glycosyltransferases.

enzyme in the sequence, thereby defining both the composition of the final product and its monosaccharide sequence. This also means that where a sugar occurs more than once in an oligosaccharide chain, it is incorporated at each stage by a different glycosyltransferase of the same family. Further, this concept explains the microheterogeneity described above since chain completion requires optimal conditions for each enzyme in the sequence (sugar-nucleotide, metal ion concentration, etc.), and if a particular sugar could not be incorporated at the required step, then that chain could not be completed.

The concept of multiglycosyltransferases (10) comes from two considerations. First, enzyme specificity studies have shown that three separate sets of glycosyltransferase complexes are required for the synthesis of mucins, serum type glycoproteins, and gangliosides respectively. Second, the enzymes involved in a particular sequence appear to be located in the same sub-cellular fraction (11, 14). While a specific sub-cellular particle, containing only a single

The Multiglycosyltransferase Complex

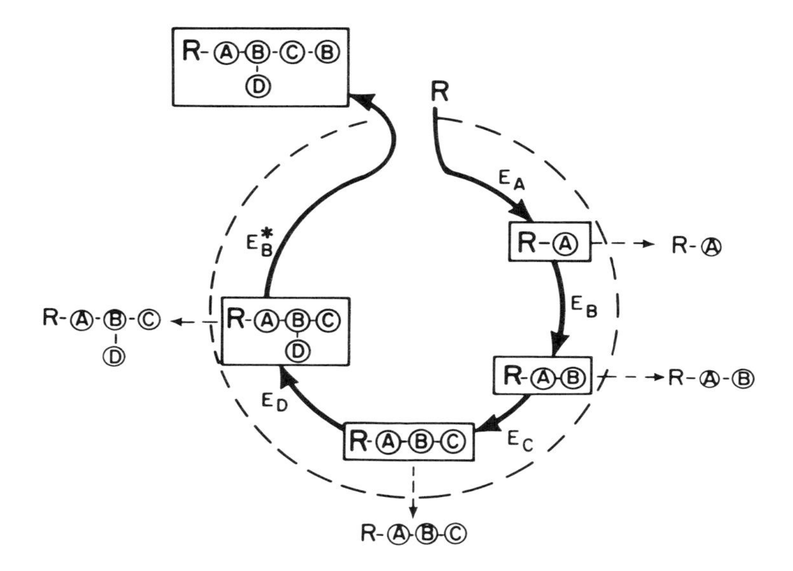

R = protein or lipid; different MGT Complexes for Glycoproteins, mucins, glycolipids

E_A, E_B, E_C, E_D = glycosyltransferases; $E_B \neq E_B^*$

Glycose donors are sugar nucleotides

Fig. 7. The multiglycosyltransferase system (or complex).

multiglycosyltransferase system, has not yet been isolated, we believe that such particles exist and function as schematically illustrated in Figure 7. According to this concept, all of the enzymes in the hypothetical particle would be required for complete oligosaccharide chain synthesis, whereas incomplete chains and microheterogeneity in glycoproteins would result as incomplete products were ejected or diffuse from the particle. In the case of the gangliosides, similar considerations explain why all of the expected intermediates in the biosynthesis of tetrasialoganglioside are found in brain extracts (1).

The entire concept of multiglycosyltransferases systems is predicated on a high degree of specificity of the glycosyltransferases for their acceptors. An example of this type of specificity is illustrated by one of the sialyltransferases which incorporates sialic acid from CMP-sialic acid into serum type glycoproteins (13). This enzyme utilizes a large number of glycoproteins provided that they are terminated by β-galactopyranosyl end groups. In these glycoproteins, after treatment with sialidase the terminal sugar, galactose, is linked to the penultimate sugar N-acetylglucosamine, most frequently via β, 1→4 bonds. The substrate specificity of this sialyltransferase has been examined in detail (13), and two examples of the kinetic data are shown in Figure 8 and 9. It should

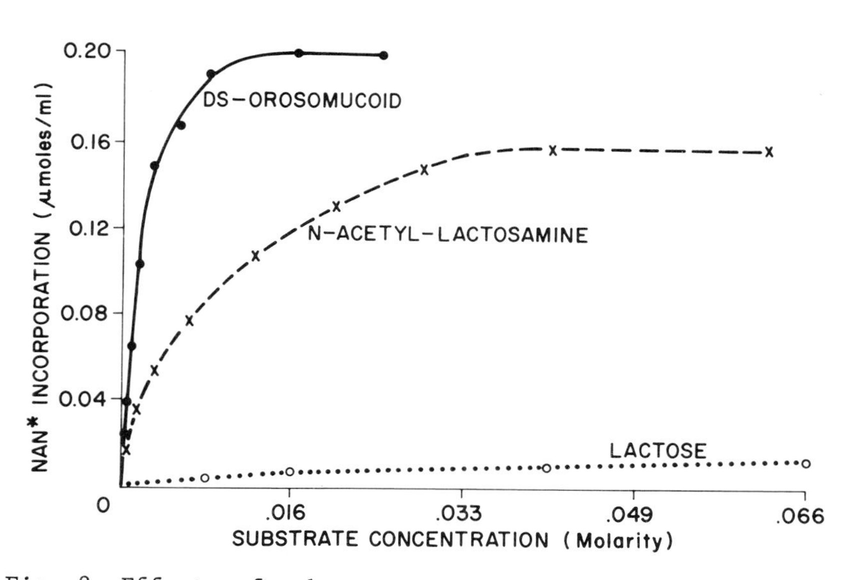

Fig. 8. Effects of substrate concentrations on rate of goat colostrum sialytransferase reaction. DS-orosomucoid represents sialidase-treated orosomucoid (α_1-acid glycoprotein).

GOAT COLOSTRUM SIALYL-TRANSFERASE(S): SUBSTRATE SPECIFICITY

GLYCOSIDES

SUBSTRATE	*RELATIVE ACTIVITY (%)*
Galactosyl – β 1,4 – N – Acetylglucosamine	100
Galactosyl – β 1,3 – N – Acetylglucosamine	19
Galactosyl – β 1,6 – N – Acetylglucosamine	3.5
Galactosyl – β 1,4 – Glucose [Lactose]	13
Aryl – β – Galactopyranosides	7 - 17

INACTIVE (<3%): *Cellobiose and other Disaccharides; α and β – Glucosides; α and β – N – Acetylgalactosamindes; α Galactosides; Pentosides; Monosaccharides.*

Fig. 9. Substrate specificity of goat colostrum sialyl-transferase. The relative activities represent relative V_{max} values.

first be emphasized that this enzyme shows a much greater degree of specificity for the glycoprotein than is apparent in Figure 8. That is, the concentration of the glycoprotein acceptor used in these experiments was calculated on the basis of its galactose content, not on the basis of the actual concentration of the glycprotein per se; the K_m based on the latter would be at least an order of magnitude less than that indicated by the results in Figure 8. Figure 8 and 9 show two more important parameters. The correct analogue of the glycoprotein is N-acetyllactosamine (Gal-β,1→4-N-acetylglucosamine), which is a good substrate for the enzyme. However, if the penultimate sugar N-acetylglucosamine is replaced by glucose (i.e., lactose), then the V_{max} drops by an order of magnitude, and it is not possible to saturate the enzyme with this substrate. The other parameter is shown in Figure 9. In this case, structural analogues of N-acetyllactosamine were used where the terminal β-galactopyranosyl group is linked to N-acetylglucosamine at the C-3 and C-6 positions. As can be seen, the V_{max} is then decreased from 5 to 25-fold. Thus, this sialyltransferase shows a high degree of specificity for glycoproteins bearing terminal β-galactosyl groups, but the nature of the penultimate sugar and the position of linkage of the β-galactosyl group to the penultimate sugar profoundly affect the rate of the reaction.

The specificities of the glycosyltransferases have led to the formulation of biosynthetic pathways for sheep and porcine submaxillary mucins (Figure 10), for the terminal trisaccharide units in serum type glycoproteins (Figure 11), and for the gangliosides (Figure 12). Some features of these pathways are of special interest. The N-acetylgalactosaminyltransferase that links N-acetylgalactosamine to the polypeptide chain of the mucin shows a remarkable specificity for this protein (15), despite the fact that it must incorporate a large number of N-acetyl-galactosamine residues into the chain. Even proteolytic digests of the protein are essentially inactive as accep-

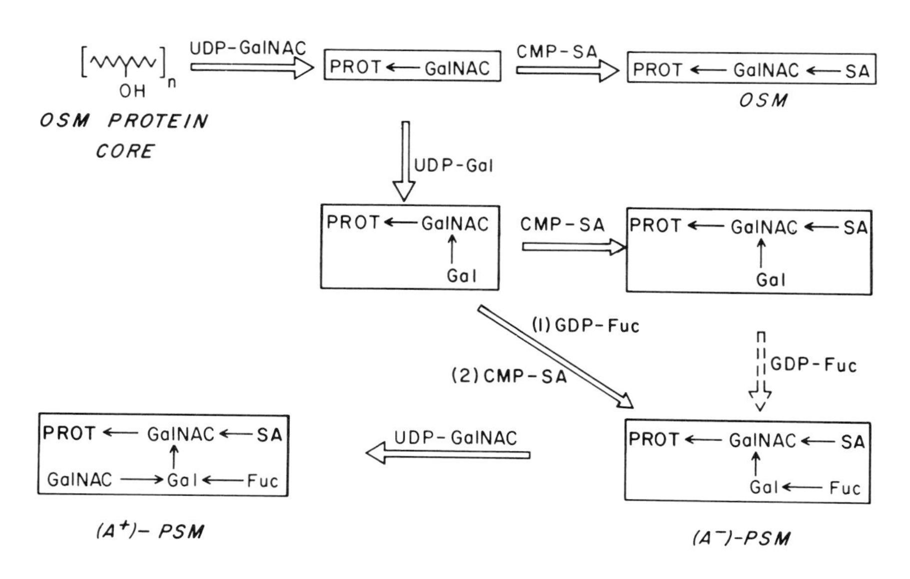

Fig. 10. Biosynthetic pathway for formation of oligosaccharide chains in sheep submaxillary mucin (OSM) and human blood group A reactive pig submizillary mucin (A^+-PSM).

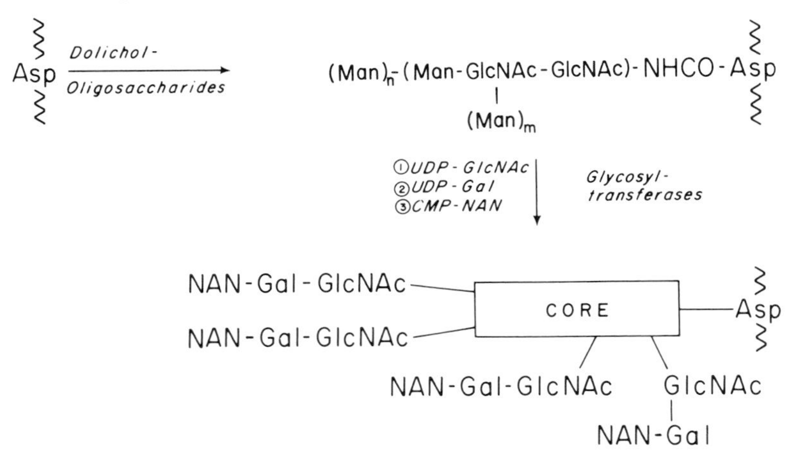

Fig. 11. Biosynthesis of serum type glycoprotein oligosaccharide chains.

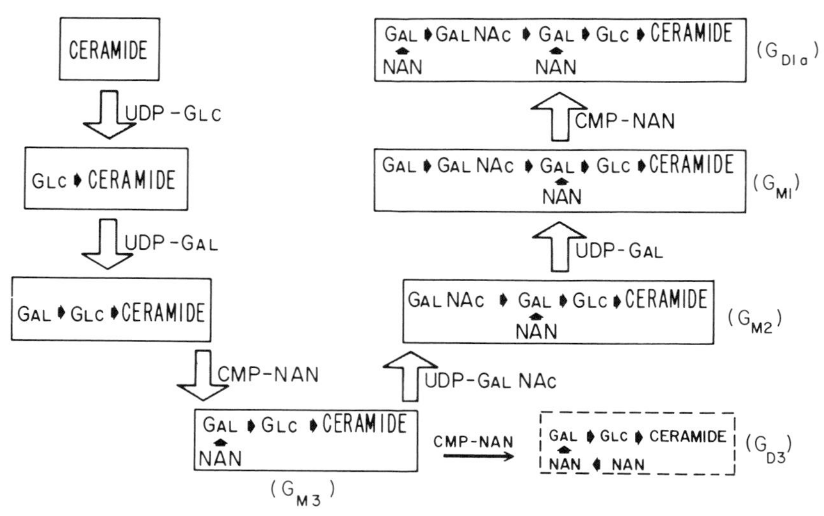

Fig. 12. Biosynthesis of gangliosides. G_{D3} is thought to be the precursor of trisialo- and tetrasialoganglioside.

tors. Another enzyme of interest in this pathway is the sialyltransferase (Figure 10). The reaction catalyzed by this enzyme may be the chain terminating reaction in the biosynthesis of the pig mucin oligosaccharides (16). In other words, if sialic acid is linked to a nascent chain, then chain growth is stopped. The evidence for the participation of dolichol in the biosynthesis of the core oligosaccharide units of the serum type glycoproteins is presented elsewhere in this symposium. Finally, the ganglioside pathway (Figure 12) represents an excellent example of a multiglycosyltransferase system. Here, there are two galactose and two sialic acid residues in the final product, disialoganglioside, and each of these sugars is added by a specific glycosyltransferase.

INTERCELLULAR ADHESION

Definitions

Selective intercellular adhesion was considered to be such an important process by embryologists early in this century that they postulated it to be the key to morphogenetic movements of cells and tissues, and to differentiation. An extensive literature has developed since the pioneering experiments of Holtfreter and his collaborators (17-19). However, the mechanism of the processes is unknown, and there is not even a generally accepted definition for the phenomenon. The literature is based on operational definitions. That is, each laboratory that studies intercellular adhesion defines it by the method used to observe and to sometimes quantitate it. While it is clear that each of these methods is being employed to study a phenomenon of fundamental biological importance, it is not at all clear that each is studying the same phenomenon, particularly at the molecular level. For example, there is no reason to believe that the "sorting out" of two cell types in an aggregate necessarily involves the same mechanism as those participating in the formation of aggregates from single cells, nor that the size of an aggregate is necessarily determined by the same events responsible for the formation of the aggregate in the first place. These considerations are underscored by the complexity of reactions known to occur when cells come into contact with each other (*viz*, formation of electrical junctions, desmosomes, etc.).

METHODS

In our own studies, we have elected to define cellular adhesion as the first measurable event, i.e., the formation of stable bonds between cells or between cells and a substratum. In other words, adhesion is the _rate_ of attachment of cells to other cells or to substrata such as glass, plastic, collagen, etc. _Homologous intercellular adhesion_ is the rate of attachment of like cells to each other. _Heterologous intercellular adhesion_ is the rate of attachment of unlike cells to each other. Cell-substratum adhesion is defined similarly.

Since it is clear that any studies on the adhesive mechanism are circumscribed by the methods used to define the process, this laboratory has paid particular attention to methodology. Three methods have been developed: (a) The Coulter Counter procedure takes advantage of the fact that the Coulter Electronic Particle Counter can discriminate between particles of different sizes, and can determine the number of particles of a given size in the suspension. The method (20) involves shaking a suspension of single cells under carefully defined conditons of temperature, shaker speed and geometry, and determines the number of single cells remaining in the suspension as a function of incubation time. The method is rapid, reproducible (Figure 13), simple, and measures the adhesive properties of large numbers of cells in the total population. Under carefully defined conditions, changes in adhesive rates of about 15% or more can be detected. The method suffers from a major disadvantage; it cannot discriminate between homologous and heterologous intercellular adhesion if mixed cell types are used. In addition, it measures by difference (i.e., the number of single cells remaining compared to the initial number), and it is therefore difficult to detect small classes of cells in the total population with different adhesive properties. (b) Roth and Weston (21) published a procedure which measured the adhesion of single radiolabelled cells to collecting aggregates. The latter consisted of cell aggregates added to the single cell suspension; aggregates were used composed both of the same and of a different cell type as

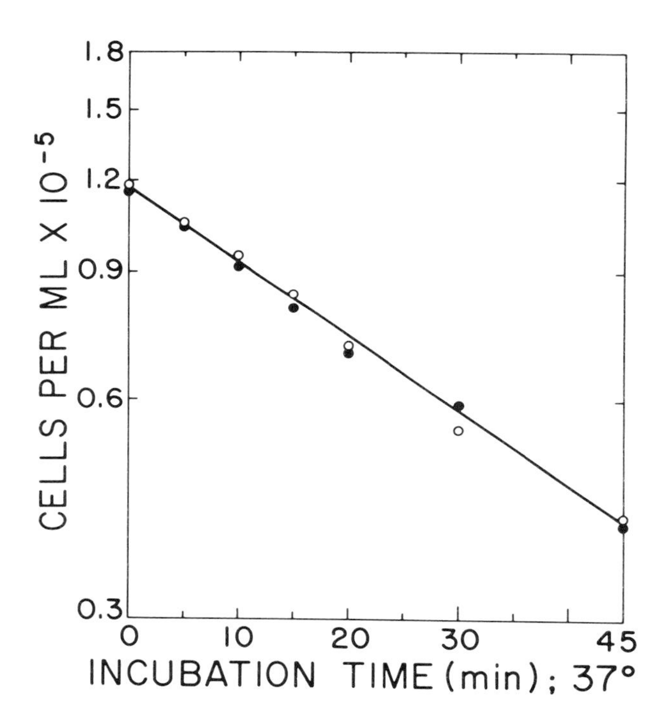

Fig. 13. Reproducibility of the Coulter Counter assay for intercellular adhesion. The closed and open circles represent the results, independently obtained, of two investigators, each of whom started with a neural retina from the same chicken embryo.

the single cells in the same flask. In the original method, quantitation was effected by histological and autoradiographic examination of the aggregates. This method has been modified in our laboratory (22). It offers the major advantage of simultaneously measuring homologous and heterologous intercellular adhesion. However, it suffers from serious disadvantages, the most important being that it is semi-quantitative at best, and that only a small fraction of the single cells adhere to the collecting aggregates under standard conditions. (c) The cell layer method (23) combines the advantages of both of the preceding methods. The principle of the assay is illustrated in Figure 14. A suspension of single, radiolabel-

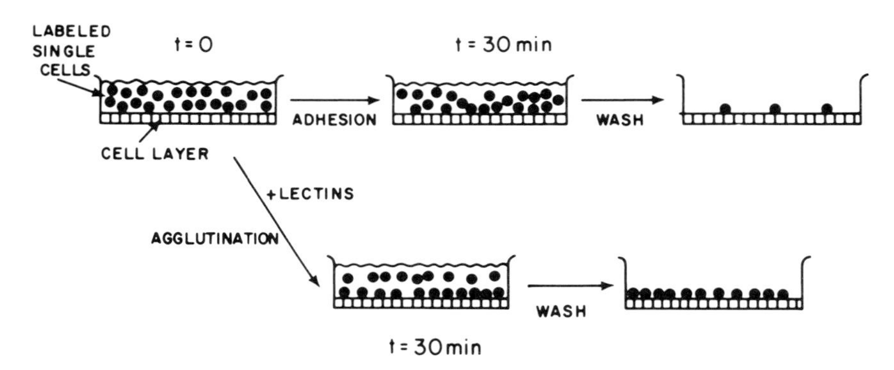

Fig. 14. Principle of cell layer assay. Radiolabeled single cells in suspension are shaken above a cell layer. The rate of adhesion is determined by incubating for various times, washing off the non-adhered cells, and counting those remaining attached to the cell layers. By incorporating a lectin, such as Concanavalin A, into the assay, the rate of attachment of the labeled cells to the layer is considerably enhanced. The difference in rates represents "agglutination".

ed cells is added to vessels containing monolayers of an homologous or heterologous cell type. The rate of adhesion of the single cells to the cell layer is then determined by counting the cells attached to the layer after washing off the unattached cells. The method is a direct procedure in contrast to the Coulter method, involving large numbers of cells (under selected conditions, more than 85% of the single cell population adheres to the layer), can detect classes of cells in the total population with different adhesive properties, permits separate treatment of the single cells and the cell layers to determine the effects of reagents on the adhesive rate, etc. The one major requirement for this assay is that stable cell layers must be employed.

Simple modifications of the cell layer method permit its application to different problems. Two examples will be given. Instead of measuring adhesion of single cells to cell layers, it is apparent that cell-substratum adhesion can also be measured by omitting the cell layer. Such studies have been conducted (23), and results on the kinetics of attachment, the effects of temperature, etc., lead to the conclusion that the mechanisms underlying cell-cell adhesion are different from those involved in cell-substratum adhesion. One such experiment is shown in Figure 15 (24). As can be seen, the addition of small

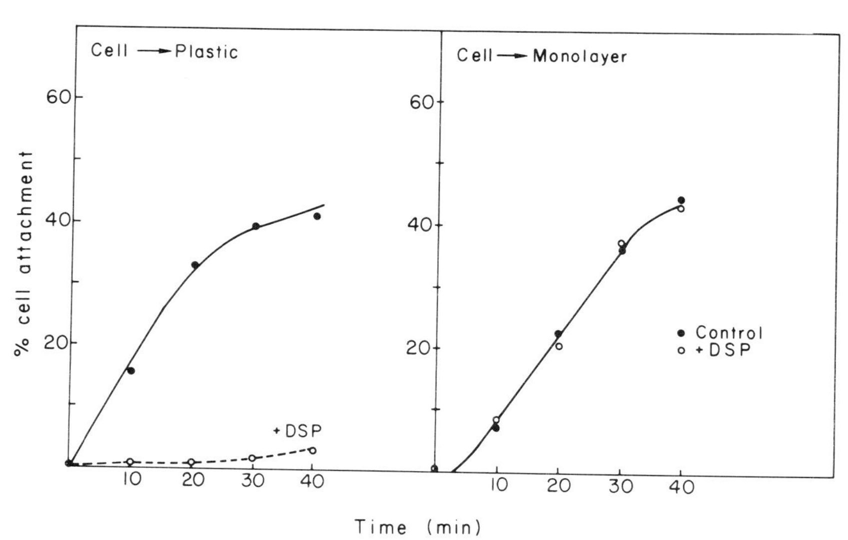

Fig. 15. Kinetics of adhesion of BHK cells to tissue culture plastic and to a cell layer, in the presence and absence of denatured serum protein.

quantities of denatured serum protein prevent the adhesion of BHK cells to tissue culture glass or plastic, whereas no effect is noted on the cell-cell adhesive process.

A second example of the applicability of the cell layer method is also illustrated in Figure 14. In the prolific studies conducted on the interactions between cells and plant lectins, there has been no acceptable quantitative method for measuring agglutination (25). As schematically indicated in Figure 14, the addition of a multivalent substance such as a lectin, which interacts with cell surfaces, promotes the formation of single cell attachment to the cell layer. This principle has therefore been developed to give a quantitative procedure for measuring lectin-promoted cell agglutination (26). These results also show that the cell layer assay should be capable of measuring multivalent molecules derived from the cells themselves, i.e., those presumed to be involved in the adhesive process. It may be noted here that the reverse effect would be expected if the reactive substance was monovalent. That is, a monovalent substance involved in adhesion should act as a "hapten", combine with the relevant surface molecules, and inhibit adhesion in the cell layer assay.

Possible Mechanisms

As noted above, despite extensive study of the process of intercellular adhesion, the molecular mechanism(s) is unknown. A mechanism was postulated in 1947 independently and simultaneously by two investigators, Tyler and Weiss (27,28). This hypothesis, illustrated in Figure 16, suggests that cell surfaces contain antibody-like and antigen-like molecules, and that complex formation or binding between these molecules on the surfaces of apposing cells results in intercellular adhesion.

In view of the high concentration of complex carbohydrates on cell surfaces, and the specificities of the glycosyltransferases described above, the Tyler-Weiss hypothesis was modified (11) as shown in Figure 16. Two alternate possibilities were suggested. In one, the surface molecules were the complex carbohydrates and the corresponding glycosyltransferases. In the other, the oligo-

I *Antigen – Antibody* (TYLER-WEISS-1947)

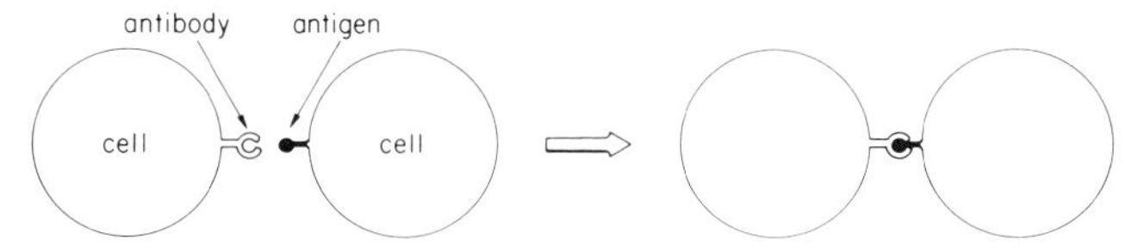

II *Enzyme – Substrate*

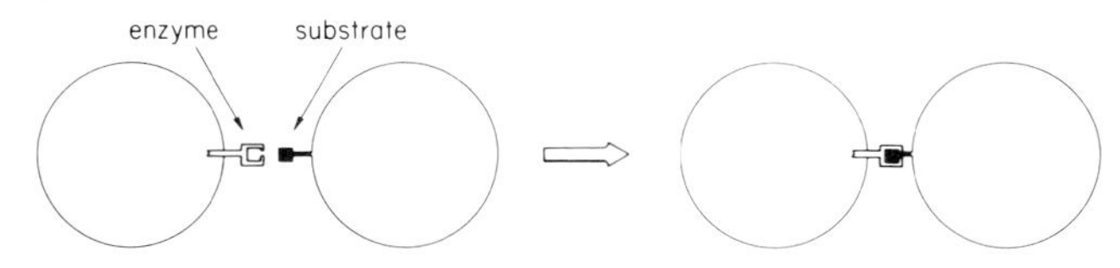

III *Hydrogen Bonds*

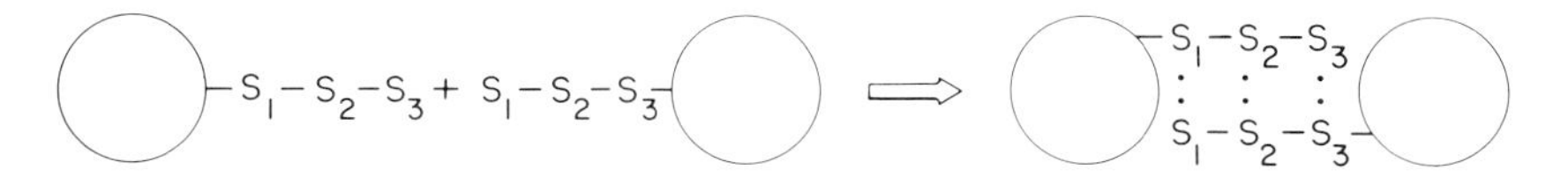

Fig. 16. Three possible mechanisms for intercellular adhesion. See text for details.

saccharide chains of the carbohydrates interacted by forming hydrogen bonds. Extremely stable hydrogen bonds are known to occur between neighboring monosaccharide units in polysaccharides such as cellulose and chitin.

In both hypotheses, surface carbohydrates are required for intercellular adhesion. The enzyme-substrate hypothesis further requires that glycosyltransferases be present on cell surfaces. Some of the evidence in support of these hypotheses will now be reviewed.

Requirement for Carbohydrates

The first results implicating sugars in intercellular adhesion came from studies with single cells obtained from a highly malignant mouse teratoma tumor (29). The ascites form of this tumor grows as small "embryoid bodies". When these bodies were dissociated, the resulting single cells were found to adhere to each other in a complex synthetic medium (Eagle's Medium 199), but not in a glucose balanced salts solution. The synthetic medium contained some 52 components in addition to the glucose and salts. Using the Coulter assay, we found that one compound, L-glutamine, was the active component in the synthetic medium. Typical results are shown in Figure 17. Since glutamine partici-

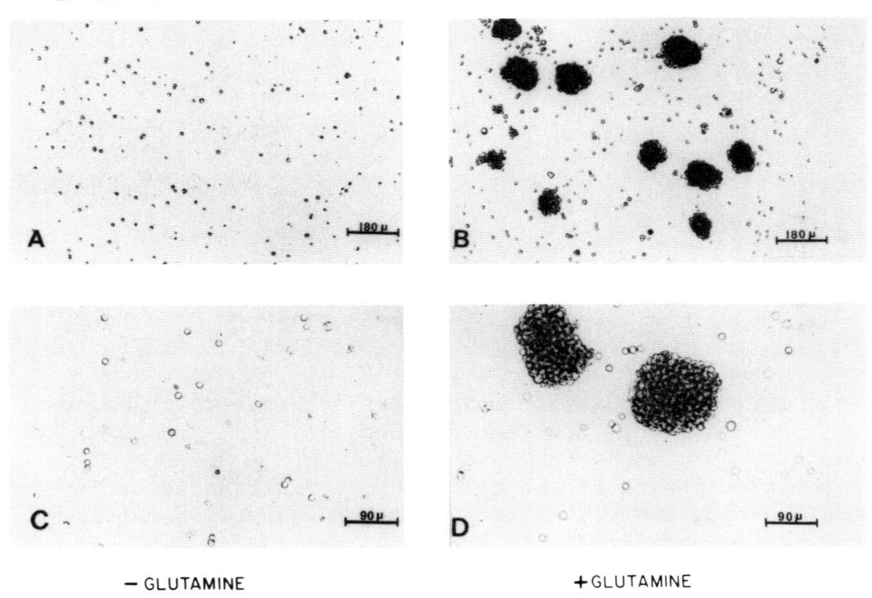

Fig. 17. Effect of L-glutamine on the intercellular adhesion of mouse teratoma cells. A and C represent the suspension when incubated in the absence, and B, D in the presence of glutamine.

pates in many biochemical reactions, we attempted to determine how it promoted adhesion by replacing it with a wide range of compounds. The only active compounds were found to be D-glucosamine and D-mannosamine. The hexosamines were active at concentrations comparable to those required for L-glutamine activity despite the fact that the experiments were conducted in the presence of the usual high concentrations of glucose, and the hexosamines are presumed to be taken up through the glucose transport system. These results are in complete accord with the known biosynthetic pathways for the amino sugars and the complex carbohydrates. That is, L-glutamine is the normal nitrogen donor for all nitrogen-containing sugars (Figure 3), the first product being glucosamine-6-P. However, the latter can be synthesized from glucosamine by the action of hexokinase and ATP, and similarly, mannosamine is an excellent substrate for hexokinase (30).

Recent experiments have provided more convincing and direct results on the role of carbohydrates in intercellular adhesion. The concept is as follows. If cells respond to carbohydrates on the surfaces of neighboring cells, then they may respond similarly by exposure to analogues of the complex carbohydrates linked to insoluble matrices. In the first series of such experiments (31), thioglycosides were linked via aminohexyl or other "linker arms" to Sephadex beads. The thioglycosides were used to avoid any potential problems with cell glycosidases. In addition, it was found that the usual methods for synthesizing affinity type beads gave undesirable by-products; the successful method involved synthesis of the complete ligands followed by coupling to the bead at the last step in the preparation. Various control beads were used in addition to beads covalently linked to galactose, glucose, or N-acetylglucosamine. A subjective assay was employed (Figure 18), light microscopic examination of a mixture of beads and cells after suitable periods of incubation above an agarose gel (to which neither the beads nor cells adhered). The microscope was focused just above the gel layer; where there was no effect, beads and a few cells were observed by virture of the fact that the beads settled more quickly than did the cells. Three tissue culture cell lines were used, BHK, 3T3, and SV40/3T3 cells. The results were essentially negative with all of the beads and cell types used with the exception of SV40/3T3 cells (and to a lesser extent, 3T3 cells) when these were mixed with the Gal-beads. In this case (Figure 19,20), large mixed aggregates of beads and cells were rapidly formed. The "kinetics" of the process are shown in Figure 20.

These rather dramatic results show an adhesive effect

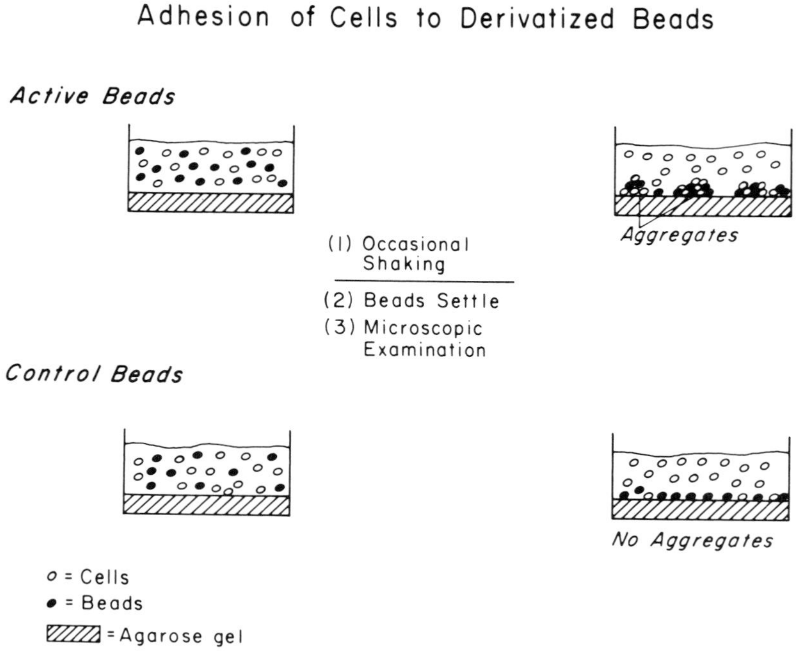

Adhesion of Cells to Derivatized Beads

Active Beads

Aggregates

(1) Occasional Shaking

(2) Beads Settle

(3) Microscopic Examination

Control Beads

No Aggregates

o = Cells

• = Beads

▨ = Agarose gel

Fig. 18. Assay for the effect of derivatized Sephadex beads on the adhesion of SV40/3T3 cells. See text for details.

that is both cell and sugar specific. It should be stressed that the cell density used in these experiments was low, so that cell-cell adhesion in the suspension was minimal. The fact that large clusters of cells and Gal-beads were formed leads to the conclusion that two processes occured. In the first, cells adhered to the Gal-beads. In the second, these cells became much more adhesive for the cells in suspension, and as the latter adhered to the cluster, they in turn became more adhesive for new single cells in the suspension. Proof of this interpretation and a much more extensive study of cell-sugar specificity and of the effects of more complex carbohydrates on adhesion requires development of a quantitative assay. Studies along these lines are now in pro-

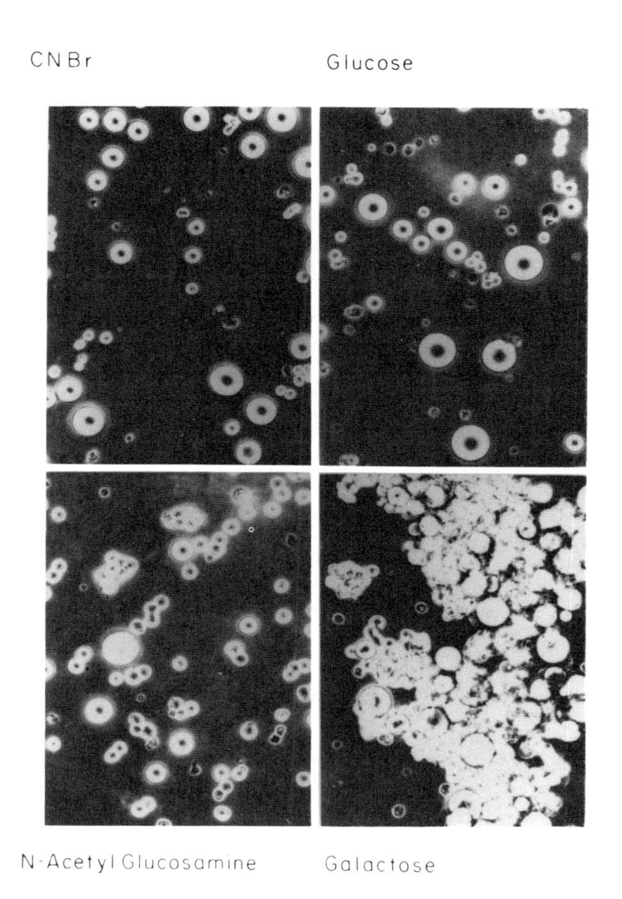

Fig. 19. Effect of attachment and adhesion of SV40/3T3 cells. See text for details.

Kinetics of Formation of Cell-Bead Aggregates
(SV40/3T3 Cells + Galactosyl-Beads, 37°)

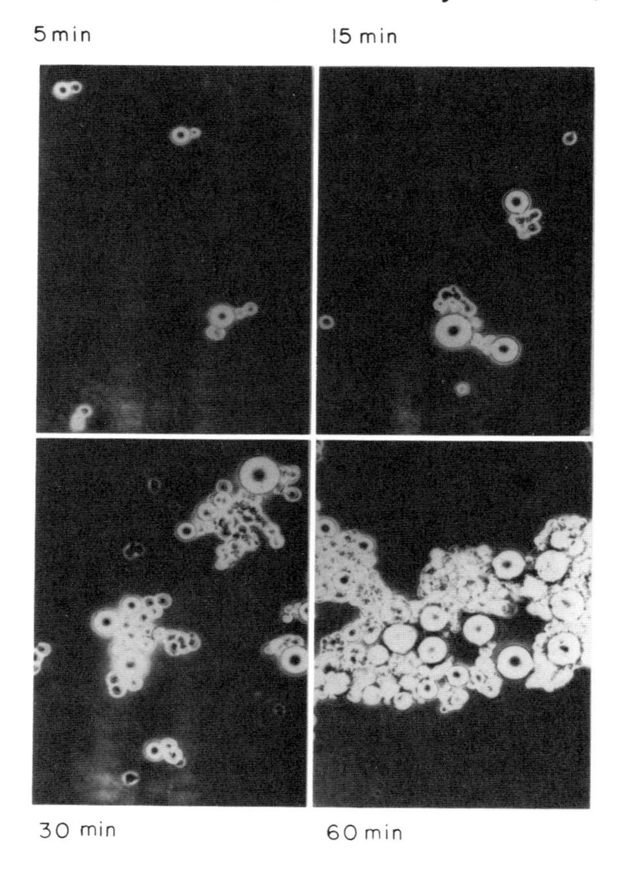

Fig. 20. Kinetics of attachment and adhesion of SV40/3T3 cells to galactose derivatized Sephadex beads.

gress, and should also permit us to examine the role of carbohydrates in other processes such as cell motility, and on "contact inhibition of growth".

If cells adhere to Gal-beads, then will Gal-beads adhere to a cell layer? Mr. Chipowsky has extended his

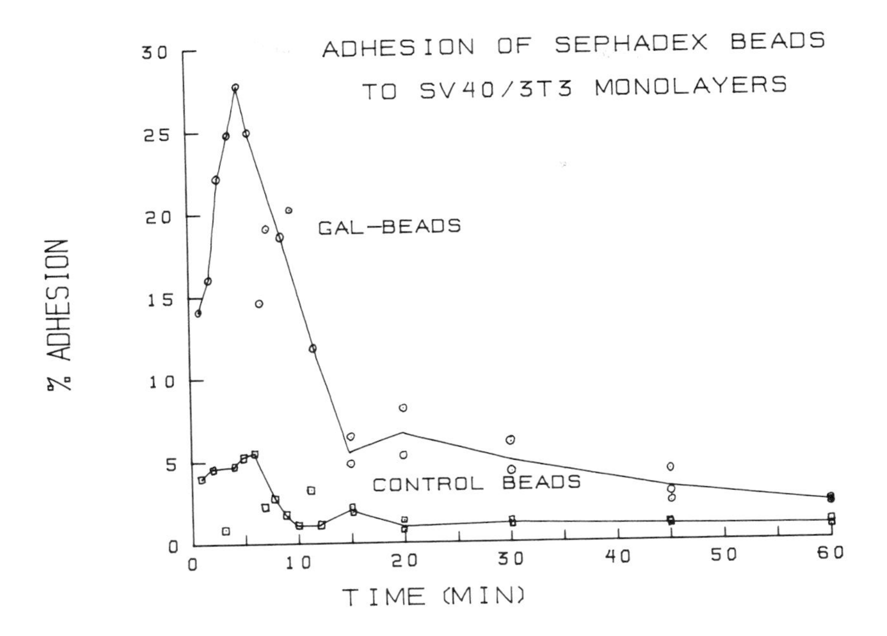

Fig. 21. Kinetics of attachment of beads to SV40/3T3 cell layers.

previous studies on the Gal-beads in an attempt to answer this question. Typical results are shown in Figure 21. Galactose-derivatized beads do adhere to SV40/3T3 cell layers, but show the surprising property of "deadhering" in a few minutes. The "deadhesion" may result from a change in the adhesive properites of the cells, or the cells may change the beads so that they become less adhesive, or both. In recent experiments, cell monolayers grown on uniformly labeled ^{14}C-glucose were exposed to either Sephadex or Gal-Sephadex beads. No radioactivity was found associated with the control beads, whereas the Gal-beads were significantly labeled. The labeled material appears to be covalently lined to the galactose residues on the beads, and if we can in addition show that it is not an artifact resulting from the action of glycosyltransferases liberated by cell lysis or other potential artifacts, then we may have finally obtained the first definitive evidence to support the contention that there are indeed cell surface glycosyltransferases and that they can modify the surfaces of neighboring cells using internally generated sugar nucleotides.

Cell Surface Glycosyltransferases

One of the hypotheses described above requires that glycosyltransferases be located on cell surfaces. The first experiments along these lines (32) were conducted with embryonic chicken neural retina cells. The latter were exposed to labeled UDP-galactose, and labeled Gal was found to be incorporated into endogenous acceptors. In these experiments, single cells were used at high density and in the presence of deleterious substances such as Mn^{++} which is required for most of the galactosyltransferases. Cell lysis and cell damage are therefore major problems. Some controls used to show that the intact cells and cell surface enzymes catalyzed the galactose transfer reactions were as follows: Exogenous acceptors were found to be very active in the system, and these included two high molecular weight substances, orosomucoid and a mucin preparation (different glactosyltransferases act on each of these polymers). Since it was possible that UDP-galactose was being taken up intact and then used, or was first hydrolyzed to labeled galactose or galactose-1-P and the latter taken up and utilized, large pools of unlabeled galactose, galactose-1-P or UDP-glucose were added to the incubation mixtures. None of these unlabeled compounds significantly affected the incorporation of the labeled galactose. In addition, the cell supernatant was examined for galactosyltransferase activity, but the activity was far too low to account for the observed incorporation.

Despite the results with the controls, we cannot and did not (32) conclude that we had unequivocally demonstrated the presence of cell surface glycosyltransferases. The work described above has been followed by a number of papers presenting similar types of results with a wide variety of cell types (33-36). All of these reports are consistent with the conclusion that cell surface glycosyltransferases exist, but it is our view that definitive evidence in support of this conclusion remains to be obtained.

The adhesion of blood platelets to collagen has been studied by several workers, particularly by Dr. Jamieson, who reviews this work elsewhere in this volume, and presents evidence for and against the idea that platelet-collagen adhesion occurs *via* surface glycosyltransferases and appropriate groups on collagen molecules.

Lipids

In addition to carbohydrates, lipids may be involved in the process of intercellular adhesion. This conclusion is based on work presented elsewhere in some detail (37). Only the major points will be summarized here.

The cell layer assay was used to directly measure the effects of cell fractions on the adhesion of 3T3 and SV40/3T3 cells to homologous layers. While the crude lipid fractions obtained from these cells were inactive, the lipids were purified by combinations of column and thin layer chromatography, yielding some 30-40 fractions, and three were found to be active at low concentrations. It should be emphasized that a large number of commercially available and other lipids were also tested, and were inactive. Assuming that the active fractions are homogeneous, then their properties may be summarized as follows: (a) A relatively non-polar, potent inhibitor of adhesion that appears to be a fatty acid, perhaps a unique fatty acid. This substance inhibits the adhesion of both cell types. (c) A polar phospholipid, representing less than 2% of the total phospholipid, stimulates (up to 6-fold) the adhesion of the 3T3 cells but not SV40/3T3 cells despite the fact that it was isolated from the latter.

Current studies are aimed at determining whether these substances are or are not homogeneous, and at establishing their structures by conventional methods and by GLC-mass spectrometry.

The Biological Process

To this point, intercellular adhesion has been discussed as if it were a static process with respect to the cell surface molecules responsible for the phenomenon. That is, as of a lock and key. This, in fact, is the basis of the hypotheses shown in Figure 16. However, we now believe that the process is a dynamic one.

Some of the results described earlier in this article suggest that the adhesive properties of cells can change very rapidly. The "triggering" effect of galactose-beads on the adhesive properties of SV40/3T3 cells is one example. Another, is the deadhesion of galactose-beads from the cell layer.

Another approach has been to study the morphology of cells shortly after they come into contact. Preliminary results (data not shown) indicate that there are demonstrable and probably important changes in morphology of these cells within three minutes after they come into contact. A more direct line of evidence to support the conclusion that the process is dynamic comes from experiments conducted by Dr. Jay Umbreit. One set of results is shown in Figure 22. Using the Coulter Counter assay, he studied the kinetics of adhesion of chicken embryonic neural retina and liver cells. As shown in the Figure, "loose" aggregates are rapidly formed, without a lag period. These aggregates are dissociated by several inversions of the suspension in which they form, or by simple dilution. Further, their formation is not inhibited by cyanide or other metabolic poisons. The second type of aggregate is "tight", requires a lag period of a few minutes at 33°, and its formation is strongly inhibited by metabolic poisons. The tight aggregates cannot be dissociated by subjecting them to simple shear forces or dilution, but are undoubtedly of the type previously reported (20) which could only be dissociated by vigorous treatment with proteases. The reversible and irreversible aggregates are separable by density gradient centrifugation. While we believe that the reversible aggregates represent intermediates in the formation of the reversible aggregates, this important point remains to be established. If it is correct, then these and perhaps the morphological results will undoubtedly revise our ideas on the nature of the biological process.

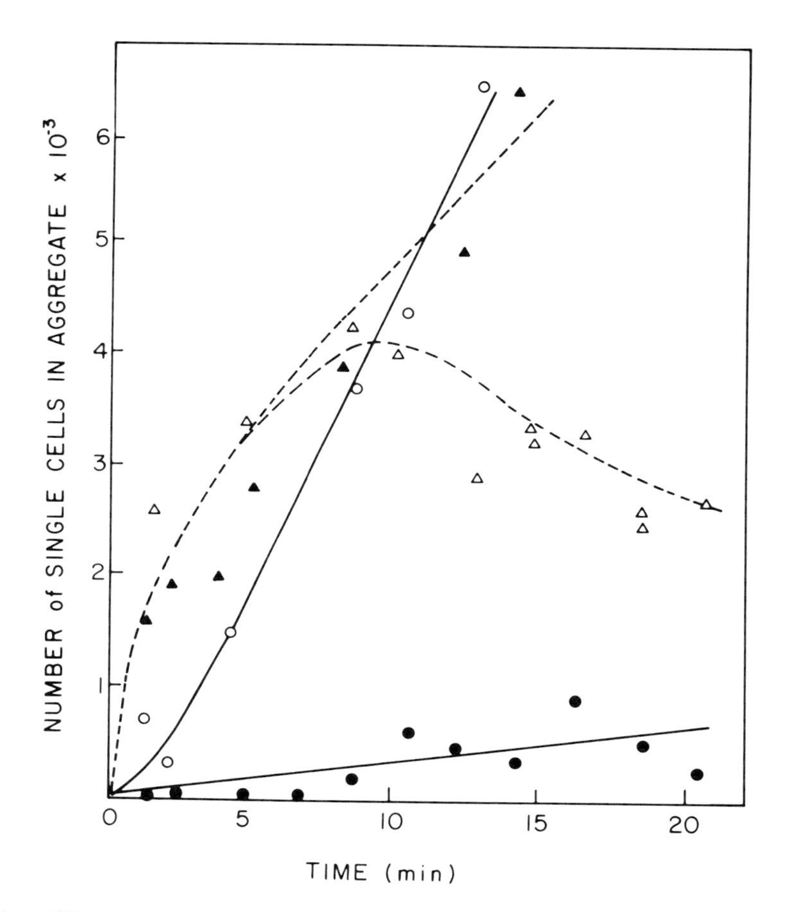

Fig. 22. Kinetics of formation of "tight" and "loose" aggregates by chicken embryonic neural retina cells. The Coulter Counter assay was used, and the results are presented as the number of cells in the aggregates (i.e., the number of cells disappearing from the single cell population). The dashed lines represent the "loose" or reversible aggregates, dissociable by inversion of the suspension. The solid lines represent "tight" or irreversible aggregates. The open symbols represent aggregate formation in the absence of cyanide, while the closed symbols show the formation of aggregates in the presence of cyanide.

Membrane Messengers

Cells show a remarkable ability to respond in different ways to diverse signals from the environment. If these signals cannot be taken up as solutes or by endocytosis, then how are the signals transmitted to the cytoplasm? We have recently considered this question (38) and have suggested a model whereby the membrane acts as a transducer *via* messengers built into the membrane (Figure 23). This is not a unique idea, but an extension of the second messenger hypothesis of Sutherland. The point of Figure 23 is to broaden our concepts of membrane messengers, or,

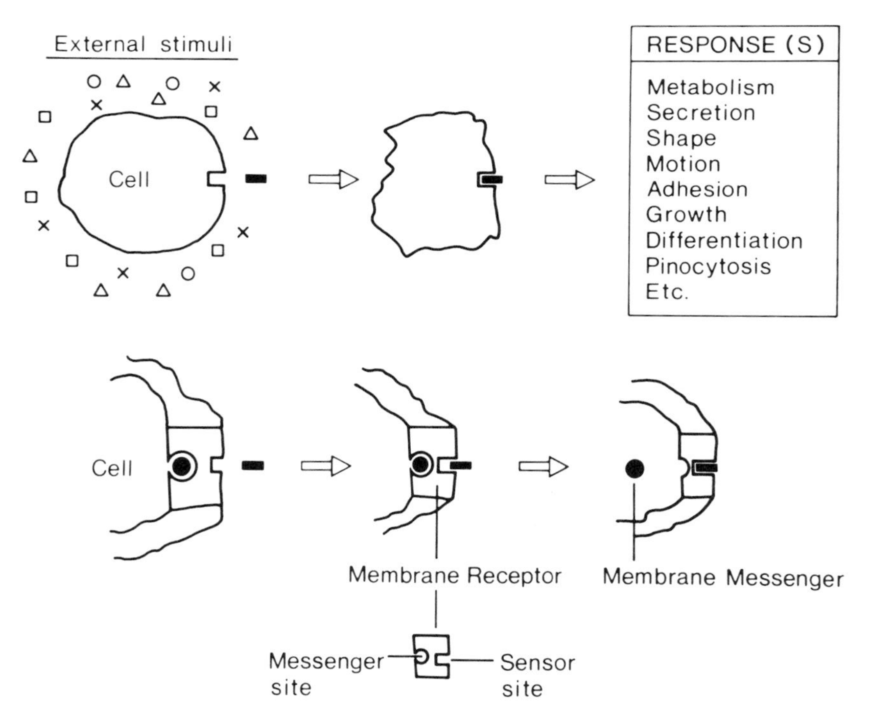

Fig. 23. Membrane Messenger Hypothesis. External stimuli represent different signals to the cell which cannot be taken up through the membrane or by endocytosis.

in short, that cyclic AMP, cyclic GMP, and Ca^{++} are insufficient to explain all of the responses that can be elicited from a single cell by diverse stimuli. The nature of the messengers, stoichiometry, and other considerations are discussed in the original article.

Discussion

This article briefly reviews the pathways of synthesis of the complex carbohydrates of the glycocalyx, and the evidence suggesting that these substances and possibly cell surface glycosyltransferases are involved in intercellular adhesion. It is clear that we are only at an early stage in obtaining a real understanding of the complex phenomena underlying intercellular adhesion and cell-cell recognition. Undoubtedly many of our current notions are simplistic and will require substantial modification as more is learned at the molecular level. Hopefully, the author has drawn clear boundaries between observation, interpretation and speculation, and that the speculation will serve some useful purpose in the design of experiments that will prove or disprove them.

Summary

The surface of eukaryotic cells called the glycocalyx is rich in glycolipids and glycoproteins. The pathways of synthesis of these complex carbohydrates are reviewed.

Intercellular adhesion, cell-cell recognition, contact inhibition of growth and motility, and other important physiological phenomena are mediated by the cell surface. Intercellular and cell-substratum adhesion is defined operationally as the first measurable event involving the formation of cell-cell or cell-substratum stable bonds, that is, as a rate phenomenon. Using this definition and the methods developed to quantitate the phenomena, the mechanism of intercellular adhesion was investigated. Evidence is presented indicating that carbohydrates are involved in the presence of glycosyltransferases on cell surfaces. These results are in accord with postulated mechanisms for intercellular adhesion. Additional experiments indicate that the process is more complex than that of a simple fit between a "lock and key".

A "membrane messenger" hypothesis is also offered to explain the diverse responses that can be elicited from a single cell by different signals from the environment.

REFERENCES

1. Svennerholm, L., J. Lipid Res., 5, (1964) 145.

2. Gottschalk, A., 1966. Glycoproteins. Elsevier Publishing Co., Amsterdam.

3. Roseman, S., Ann. Rev. Biochem., 28 (1958) 545.

4. Ginsburg, V., Adv. in Enzymology, 26 (1964) 35.

5. Ghosh, S., Blumenthal, H., Davidson, E. and Roseman, S. J. Biol. Chem., 235 (1960) 1265.

6. Scher, I. and Lennarz, W., Biochem. Biophys. Acta (Reviews), 265 (1972) 417.

7. Jourdian, G.W. and Roseman, S., Ann. N.Y. Acad. Sci., 106 (1963) 202.

8. Johnston, I.R., McGuire, E.J. and Roseman, S., J. Biol. Chem., 248 (1973) 7281.

9. Carlson, D.M., J. Biol. Chem. 243 (1968) 616.

10. Roseman, S., in Proceedings of the 4th International Conference of Cystic Fibrosis of the Pancreas (Mucoviscidosis). E. Rossi and E. Stoll, Eds. New York: S. Karger, pp. 244-268 (1968).

11. Roseman, S., Chem. Phys. Lipids, 5 (1970) 270.

12. Carlson, D.M., Jourdian, G.W. and Roseman, S., J. Biol. Chem., 248 (1973) 5742.

13. Bartholomew, B.A., Jourdian, G.W. and Roseman, S., J. Biol. Chem. 248 (1973) 5751.

14. Schachter, H., Jabbal, I., Hudgin, R., Pinteric, L.,

McGuire, E. and Roseman, S., J. Biol. Chem. 245 (1970) 1090.

15. McGuire, E. and Roseman, S., J. Biol. Chem. 242 (1967) 3745.

16. Schachter, H., McGuire, E.J. and Roseman, S., J. Biol. Chem., 246 (1971) 5321.

17. Curtis, A.S.G., Biol. Rev., 37 (1962) 82.

18. Manly, R.S., Adhesion in Biological Systems. Academic Press, New York, 1970.

19. Trinkaus, J.P., Cells into Organs. Prentice Hall, Englewood Cliffs, N.J., 1969.

20. Orr, C.W. and Roseman, S., J. Membrane Biol., 1 (1969) 109.

21. Roth, S. and Weston, J.A., Proc. Nat. Acad. Sci. U.S.A. 58 (1967) 974.

22. Roth, S., McGuire, E.J. and Roseman, S., J. Cell Biol. 51 (1971) 525.

23. Walther, B.T., Ohman, R. and Roseman, S., Proc. Nat. Acad. Sci. U.S.A., 70 (1973) 1569.

24. Umbreit, J. *et al*., Methods in Enzymology, in the press.

25. Lis, H. and Sharon, N., Ann. Rev. Biochem., 42 (1973) 541.

26. Rottmann, W.L., Walther, B.T., Hellerquist, C.G., Umbreit, J. and Roseman, S., J. Biol. Chem., 249 (1974) 373.

27. Tyler, A., Growth (Symposium), 10 (1946) 6.

28. Weiss, P., Yale J. Biol. Med., 19 (1947) 235.

29. Oppenheimer, S.B., Edidin, M., Orr, C.W. and Roseman, S., Proc. Nat. Acad. Sci. U.S.A., 63 (1969) 1395.

30. Jourdian, G.W. and Roseman, S., Biochemical Preparations, 9 (1962) 44.

31. Chipowski, S., Lee, Y.C. and Roseman, S., Proc. Nat. Acad. U.S.A., 70 (1973) 536.

32. Roth, S. and McGuire, E.J. and Roseman, S., J. Cell Biol., 51 (1971) 536.

33. Roth, S. and White, D., Proc. Nat. Acad. Sci. U.S.A., 69 (1972) 485.

34. Bosmann, H.B., Biochem. Biophys. Res. Commun. 48 (1972) 523.

35. Weiser, M.M., J. Biol. Chem., 248 (1973) 2542.

36. Shur, B. and Roth, S., Amer, Zool., 13 (1973) 1129.

37. Hellerquist, C., Rottmann, W.T., Walther, B. and Roseman, S. Cell Adhesion in Proceedings of the Los Almos (New Mexico) Symposium (1973) "Mammalian Cells: Probes and Problems", in the press.

38. Roseman, S., The Neurosciences: Third Study Program, MIT Press (1973).

DISCUSSION

W.J. WHELAN: In relation to the "lock and key" hypothesis, should you describe one component as a substrate, or should it more appropriately be termed a substrate analog or inhibitor? If it were a true substrate and a reaction took place, the enzyme and the decomposed substrate would pull apart.

S. ROSEMAN: I think the best answer to that is to consider whether glucose is a substrate of hexokinase in the absence of ATP.

E.C. HEATH: Couldn't you test the "lock and key" hypothesis, if you were able to make the oligosaccharide portions of the surface glycoprotein that are involved? Perhaps you can make a family of glycopeptides from surface constituents which could be used to compete with the interaction of cells. Have you tried to do that experiment?

S. ROSEMAN: The problem is not quite that simple. If you add a monovalent hapten to the cell layer assay that we use, you will get inhibition of adhesion. However, if you add a polyvalent hapten you may get stimulation. In our original work with Concanavalin A which is polyvalent, we got stimulation of adhesion with binding to both the cells and the cell layers. We have tested glycopeptide fractions in our assay and observe some stimulation of cell adhesion, possibly because these glycopeptides are polyvalent. We have also found considerable activity in the lipid fractions and depending on which fraction we tested, found potent inhibition or marked stimulation. Whether these are glycolipids or not we do not know as yet. I feel that adhesion is going to involve more than one step.

P. ROBBINS: I just wanted to make a comment about Ed Heath's question on glycopeptides. Dr. Teru Sasaki in our laboratory has isolated glycopeptides from chick cell surfaces and added them back to a confluent monolayer of chick cells. This caused some dissociation of the cells and another round of cell growth.

I. SCHENKEIN: In the experiment where you show adhesion and disadhesion to the Sephadex beads, can you utilize the released cells right away for another adhesion experiment or do you have to wait for recovery?

S. ROSEMAN: We have not done that experiment yet. The problem is just getting reproducible shear forces. You would be surprised how difficult it is to get absolutely reproducible results.

M. RIEBER: Do you think that the conditions in your specific adhesion assay, which measures the adhesion of cells to a confluent monolayer, truly represent the conditions that may exist in normal cells when they are proliferating?

S. ROSEMAN: I think our method measures the specific adhesion of normal cells. If you take normal cells and feed them under good conditions, they can be made to overgrow. What I have been talking about, that is the adhesion of one cell to another, does not necessarily relate to contact inhibition of growth. I think growth inhibition is much more complex. What you seem to be implying is that the adhesive sites may be located only around the sides of normal cells and not at the top. That is not true. In addition to tissue cultured fibroblasts, we have tested embryonic cells and these also, as one would expect, showed specific adhesion in this assay.

TRANSFER OF MANNOSE FROM DOLICHOL MONOPHOSPHATE MANNOSE TO YEAST GLYCOPROTEIN ACCEPTORS.

R. K. Bretthauer and S. Wu, Biochemistry and Biophysics Program, Department of Chemistry, University of Notre Dame, Notre Dame, Indiana

The mannosylation of dolichol monophosphate by GDP-mannose in yeast particulate enzyme preparations has been demonstrated by several investigators (see Ref. 1). Largely from results of kinetic studies, dolichol monophosphate mannose (DMP-mannose) has been implicated as an intermediate in the transfer of mannose residues from GDP-mannose to mannan-protein acceptors.

Utilizing DMP-[^{14}C]mannose as substrate with a *Hansenula holstii* particulate enzyme fraction (2), we have demonstrated the transfer of mannose to endogenous glycoprotein acceptors. The reaction is dependent on time and concentrations of enzyme, DMP-mannose, Triton X-100, and possibly $MgCl_2$. The product which is insoluble in chloroform:methanol or butanol can be solubilized with hot, neutral citrate buffer or with hot NaOH. The citrate-solubilized products which are excluded from Sephadex G-100 fractionate on DEAE-Sephadex A-25 in a manner similar to the radioactive glycoprotein products obtained from GDP-[^{14}C]mannose. Pronase digestion of the Sephadex G-100 excluded products results in radioactive products being retarded on Sephadex G-100 but excluded from Sephadex G-25. Treatment of these glycopeptides with NaOH:$NaBH_4$ releases radioactive oligosaccharides. These results imply that DMP-mannose is a mannosyl donor to oligosaccharides attached to protein through O-glycosyl linkages. (Supported by NSF Grant GB-38356.)

REFERENCES

(1) P.Jung and W. Tanner, Eur. J. Biochem. 73 (1973) 1.

(2) R. K. Bretthauer, S. Wu, and W. E. Irwin, Biochim. Biophys. Acta 304 (1973) 736.

STRUCTURAL ANALYSIS OF PURIFIED SINDBIS VIRUS GLYCOPEPTIDES

V.N. Reinhold and K. Keegstra

Biology Section, Arthur D. Little, Inc.

and

Biology Department, Massachusetts Institute of Technology

Cambridge, Massachusetts.

Glycopeptides were isolated from purified Sindbis virus by pronase digestion. The glycopeptides were purified by ion-exchange and gel-filtration chromatography. Each purified glycopeptide was submitted to methylation analysis. The resulting reduced alditol acetates were analyzed by gas-liquid chromatography-mass spectrometry with all data handled by an on-line IBM 1800 computer. The mass spectrum obtained for each repetitive scan was automatically compared with a stored library file of methylated alditol acetates.

Library search algorithms provide for a name, best-fit printout when spectral identity is approached, along with a correlation index of that identity. Structural analysis of complex mixtures with the described analytical system has facilitated what would otherwise be laborious and time-consuming efforts in carbohydrate sequence determination. Supported by grant NIH RR00317 and NIH 5F02CA51649-02.

STUDIES ON GLOBOSIDE BIOSYNTHESIS IN MOUSE ADRENAL TUMOR CELLS.

K-K. Yeung, J. R. Moskal, D. A. Gardner* and S. Basu, Department of Chemistry, Univ. of Notre Dame, Notre Dame, Indiana 46556 and *Molecular Biology Dept., Miles Laboratories, Inc., Elkhart, Indiana 46514.

The localization of globoside and globoside-related glycosphingolipids in plasma membranes of rat kidney, bovine kidney, human biliary adenocarcinoma and erythrocyte membrane has been demonstrated by various laboratories. The probable role of these glycosphingolipids as surface antigens[1] in normal, transformed and tumor cells has been of interest in recent years. The activities of the following three glycosyltransferases involved in globoside-related glycosphingolipid biosynthesis _in vitro_ were detected in the Y-1 mouse adrenal cortex tumor cell line[2] and in tumors derived from this cell line:

$$\text{Glc-cer} \xrightarrow[\text{(A)}]{\text{UDP-Gal}} \text{Gal-Glc-cer} \xrightarrow[\text{(B)}]{\text{UDP-Gal}} \text{Gal-Gal-Glc-cer} \xrightarrow[\text{(C)}]{\text{UDP-GalNAc}} \text{GalNAc-Gal-Gal-Glc-cer.}$$

The enzymatic activities were detected in a membrane fraction isolated by discontinuous sucrose gradient. Cells were homogenized with a Polytron 10ST homogenizer and the membranes were concentrated at the junction of 0.5 and 1.2 M sucrose at 52,000 x g. A correlation between globoside content and biosynthesis _in vitro_ in tumor, cultured Y-1, and normal adrenal cortex cells is under investigation. Increased activities of UDP-Gal:lactosylceramide α-galactosyltransferase[3], B, and UDP-GalNAc:GloboTri-cer β-N-acetylgalactosaminyltransferase[4], C, found in tumor cells, may be properties associated with neoplasia.

REFERENCES

1) C. G. Gahmberg and S. Hakomori, J. Biol. Chem., 248 (1973) 4311.
2) Y. Yasumura, V. Buonassisi and G. Sato, Cancer Res., 26 (1966) 529.
3) P. Stoffyn, A. Stoffyn and G. Hauser, J. Biol. Chem., 248 (1973) 1920.
4) J-L. Chien, T. Williams and S. Basu, J. Biol. Chem., 248 (1973) 1778.

CELL ENVELOPE GLYCOPROTEIN BIOSYNTHESIS IN FUNGI

M.K. Raizada, J.S. Schutzbach and H. Ankel, The Medical College of Wisconsin, Milwaukee, Wisconsin 53233

The major cell envelope glycoprotein of *Cryptococcus laurentii* contains mannose, galactose and xylose. A particulate enzyme fraction of the organism catalyzes transfer of mannosyl, galactosyl and xyosyl residues to an enzyme-bound acceptor from GDP-mannose, UDP-galactose and UDP-xylose. Using exogenous mono- and oligosaccharides as acceptors, this enzyme fraction was shown to contain at least 4 mannosyl,[1] 2-xylosyl[2] and 2 galactosyl transferases, each specific for the formation of a distinct linkage. Starting with free mannose as acceptor, several of these transferases can be utilized in stepwise *de-novo* synthesis of either α-Man-2-Man-6-Man(β-2-Xyl)-3-mannose(I) or α-(Gal-6)$_n$-Gal-β-mannose(II). Endogenous products obtained after transfer of mannose, galactose or xylose to enzyme-bound acceptor are solubilized with pronase, whereafter they are retained by columns of DEAE-cellulose and Sepharose-ConA. After treatment with $NaOH/NaBH_4$ none of these products binds to DEAE-cellulose and only the mannosyl product is completely retained by Sepharose-ConA. The major oligosaccharide released after $NaOH/NaBH_4$ treatment of the endogenous mannosyl product is a pentasaccharide, which has the same linkages as those formed during *de-novo* biosynthesis of oligosaccharide I. Similarly, the oligosaccharide released from the endogenous galactosyl product has properties comparable to oligosaccharide II. Our data suggest that all three monosaccharide residues are transferred directly from sugar nucleotide to a common glycoprotein acceptor, but that separate oligosaccharide side chains are formed, interconnected only by a common polypeptide core. The fact that these side chains can be synthesized separately from sugar nucleotides and the lack of formation of detectable lipid-bound oligosaccharides suggest biosynthesis of a major portion of the cell envelope glycoprotein directly from sugar nucleotides without involvement of intermediate lipid carriers. (Supported by NSF GB 18090.)

(1) Schutzbach, J.S., and Ankel, H., J. Biol. Chem. 246 2187 (1971).

(2) Schutzbach, J.S., and Ankel, H., J. Biol. Chem. 247 6574 (1972).

CELLULAR LECTIN RECEPTORS

R. D. Poretz, Department of Biochemistry, Rutgers University, New Brunswick, New Jersey

The finding that some lectins are capable of specifically agglutinating tumor cells, cause an immunosuppressed state in mice, and are mitogenic agents with respect to lymphocytes has caused a renaissance of interest in these plant proteins[1].

Analytical agglutination was conducted with the microtitration system employing purified lectins from P. lunatus, S. japonica, W. floribunda, and C. ensiformis, as well as partially purified agglutinins from E. europeaus and M. pomifera. Agglutination patterns using untreated and ficinized human erythrocytes demonstrated that though many of these lectins show similar specificities to simple sugars, each reacts with different cellular receptors. Every lectin except M. pomifera displayed up to 64 fold greater activity with ficinized cells as compared to untreated erythrocytes. However, the agglutinability of the ficinized cells by M. pomifera was greatly decreased. The agglutination activity of lectins which do not display human blood group specificity with murine erythrocytes was similar to those activities with human cells. Those lectins which are blood group specific are inactive or only slightly active with murine erythrocytes. The activity of concanavalin A with enzyme treated murine erythrocytes suggests that structures capable of reacting with this lectin are uncovered by the action of ficin, and these or other determinants are unblocked by removal of non-reducing terminal β-D-galactosyl residues. The activity patterns of the lectins differ with murine erythrocytes and lymphocytes.

REFERENCES

(1) N. Sharon and H. Lis, Science 177 (1972) 949.

TOPOGRAPHICAL ALTERATIONS OF THE FAT CELL SURFACE MEMBRANE ELICITED BY CONCANAVALIN A.

Michael P. Czech and W.S. Lynn, Duke University Medical Center, Durham, N.C. 27710

Our finding that concanavalin A mimics the stimulatory effect of insulin on fat cell D-glucose transport prompted us to study the interaction of this lectin with the fat cell plasma membrane.[1] Purified plasma membranes contain two major glycoproteins of 94,000 and 78,000 molecular weight as estimated by SDS-gel electrophoresis.[2] In the present studies gels were incubated with ^{125}I-concanavalin A subsequent to electrophoresis of membranes in SDS and extensive washing in phosphate buffer. After washing the gels free of unbound lectin, the label retained was exclusively associated with the 94,000 and 78,000 molecular weight regions, indicating these glycoproteins represent the major membrane receptors for this lectin. Carbohydrate and amino acid composition was determined for these membrane components.

Using the lactoperoxidase labelling technique (^{125}I) the 94,000 M.W. glycoprotein, and to a lesser extent the lighter glycoprotein, represented the major membrane components labelled and therefore exposed on the fat cell surface.[3] Addition of concanavalin A to intact cells greatly increased the incorporation of ^{125}I into these glycoproteins as well as other membrane components. Appropriate controls indicated the lectin does not increase the penetration of labelling reagents into fat cells. These data apparently reflect a dramatic alteration in membrane topography reflecting increased exposure of the major glycoproteins and other membrane proteins to the external labelling reagents following lectin binding to the cell surface.

[1] M.P. Czech and W.S. Lynn, Biochim.Biophys. Acta 297 (1973) 368.

[2] M.P. Czech and W.S. Lynn, J. Biol. Chem. 248 (1973) 5081.

[3] M.P. Czech and W.S. Lynn, Biochemistry 12 (1973) 3597.

Immunochemical evidence for putrescine sites on the membrane of mammalian cells.

by G. Quash, Dept. of Biochemistry, University of the West Indies, Mona, Kingston, Jamaica.

The *in vivo* cytolysis of tumours from any tissue could be attempted, if an antiserum directed against the antigen(s) common to all types of tumours were available. Such common antigens could be polyamines as their levels in human serum and urine rise in all patients with malignancy examined so far(1). These levels decrease on reduction of the tumour mass (2).

Antibodies to polyamines were prepared. By cell count, light and electron microscopy, their effect on the growth of chick embryo fibroblasts, mouse fibro - blasts, L cells and BHK (Py) was examined *in vitro*. With complement present, lysis of all four cell types occured with 8% antiserum - (the minimum threshold concentration for 4.5×10^5 cells); in its absence only L cells and BHK (Py) were affected. These showed morphological modifications with increases in lysozomes and lipid inclusions (3).

In the presence of complement, pre-cytolytic events include membrane distortion and cell enlargement. These were arrested and the cells recovered on adding putrescine to the medium. Spermine and spermidine did not inhibit cytolysis. The foregoing results are interpreted as evidence for the presence of putrescine on the cell membrane.

References

1. Russell D., Levy C, Schimpff S. and Hawk I. Cancer Res. 1971 31 1555.

2. Bachrach U. and Ben - Joseph M. Polyamines in Normal and Neoplastic Growth Raven Press, New York 1973, Pg. 15.

3. Quash G., Delain E. and Huppert J. Exptl. Cell Res. 1971 66 426.

ATTEMPTS OF PURIFICATION, SPECIFICITY AND INHIBITION OF THE LECTIN OF ERYTHRINA EDULIS.

V. MONTES de GOMEZ and G. PEREZ, Chemistry Department
National University of Colombia, Bogota, Colombia.

After proving the presence of lectin (PHA) using the method of SAGE and VASQUEZ (1), we try to purify it by fractional precipitation with $(NH_4)_2SO_4$ followed by chromatography on Sephadex G-100 and G-200 or on DEAE-Cellulose with discontinuous gradient of phosphate concentration within 0.05 M and 0.2 M plus NaCl 1.0 M, pH 7.0; the fractions with PHA (+) were examined by PAGE and it was observed that in neither treatment the purification was complete.

As to the specificity of PHA, it was observed that it can agglutinate all types of human erythrocytes, their membranes (2), rabbit, guinea pig and cow's erythrocytes but not those of dog, sheep or horse, nor those of horse trypsinized following the method of JAFFE (3).

Attempts to inhibit the PHA activity before and after forming the agglutinate gave the following results: agar arbinose, mannitol, mannose, pectin, Na_2EDTA, sucrose, sephadex, Na galacturonate, sorbose, xylose do not inhibit it, while galactose, galatosamine and lactose do destroy this activity. The treatment of the crude extract of PHA with KIO_4 0.06 M (4) destroy the carbohydrates present in it but does not destroy its PHA capacity; the same treatment used on erythrocytes destroy its capacity of being agglutinated. These experiences show that if the PHA is a glycoprotein, the glycosidic part is not essential for its activity and that the agglutination is caused by the union of the PHA with groups of the membrane sensible to KIO_4, possible derivatives of galactosamine which can act as inhibitors and are present in the membrane.

REFERENCES

(1) SAGE and VASQUEZ (1967), J. Biol. Chem., 242,120.
(2) DODGE et al., (1962), Arch. Biochem. Biophys.,100,119.
(3) JAFFE and BRUCHER (1972), Archivos latinoamericanos de nutricion, 22,267.
(4) BURGER and GOLDBERG (1967), Proc. Nat. Acad. Sci. USA, 57,359.

CELL SURFACE TOPOLOGY AND QUANTITY OF GENETICALLY DEFINED ANTIGENS

S. Ostrand-Rosenberg, Division of Biology, Caltech, Pasadena, California 91109.

Experiments were undertaken to investigate the relationship between 1) gene dosage and antigenic expression on the cell surface; and 2) the distribution of serological specificities on the cell surface and the genetic configurations that control them. Genetically and serologically defined bovine red cell specificities were studied.

Bovine red cells were labelled with bovine antibodies against the cell surface specificities of interest. The cell surface antigens were visualized for electron microscopy by either the hybrid antibody technique[1], or by an indirect labelling technique. Platinum shadowed, carbon reinforced replicas of the entire cell surface were prepared for transmission EM[2]. *Busycon caniculatum* hemocyanin and horse spleen ferritin served as the visual labels. Dosage effects, as detected serologically by time of complement-mediate lysis for cells negative, heterozygous, and homozygous for the Z cell surface specificity, were examined by the quantity of label per cell population. Homozygous Z cells were found to have approximately twice as much label as Z/- cells. Serologically, cells displaying the J antigen, a soluble serum substance which adsorbs to the red cell surface, demonstrate a broad spectrum of antibody-mediated lysis. Some cells require stronger anti-J reagents for lysis while other cells are lysed by minute quantities of anti-J antibody. By electron microscopy one can demonstrate whether those J positive cells lysed by smaller amounts of anti-J reagent have larger quantities of J label than those J positive cells which require stronger anti-J antibody for lysis. Studies were also undertaken to examine the red cell surface topological distribution of antigens whose genes reflect a trans configuration of the serological factors as compared to those in cis configuration, e.g., B_1/G cf. B_1G, and R_2/X_2 cf. R_2X_2.

REFERENCES:

(1) T. Aoki, U. Hämmerling, E. de Harven, E. Boyse, and L. Old, J. Exp. Med. 130 (1969) 979.

(2) S. Smith and J. Revel, Devel. Biol. 27 (1972) 434.

SUPPORT:

USPHS, Training Grant #GM-86; AEC, AT043767

Thanks to Dr. C. Stormont for supplying the cells and reagents.

THE FRACTIONATION AND BIOSYNTHESIS OF MEMBRANE COMPONENTS IN ERYTHROID CELLS

B. Brennessel and J. Goldstein, N.Y. Blood Center and Biochemistry Dept. Cornell U. Medical College, N.Y., N.Y.

Glycoprotein preparations (1,2) from O erythrocyte membranes have been further fractionated in terms of blood group activity by using Phytohemagglutinin from *Phaseolus vulgaris* (PHA) coupled to Sepharose 4B. The one major and two minor glycoproteins of the erythrocyte membrane, as resolved by SDS-polyacrylamide gel electrophoresis, were shown to bind specifically to PHA and could be eluted with 5M $MgCl_2$ at 4°. The recovered glycoproteins possessed M-or N-activity but were devoid of H-activity. H-activity, which could be detected in concentrated preparations of the starting glycoproteins, was recovered in the fraction which did not bind to PHA-Sepharose. It was found to contain no detectable glycoproteins or MN-activity, and could, in turn bind to *Ulex europeus* lectin attached to Sepharose. Preliminary evidence indicates that the H-activity thus separated from glycoproteins is associated with another class of molecules most probably glycolipid.

The above separation procedures, as well as preparative SDS-polyacrylamide gel electrophoresis are being used to study the incorporation of ^{14}C-labeled sugar precursors into membrane components of human reticulocytes. Although starting glycoprotein preparations from reticulocytes of mixed cell types usually contain an average of 10% of the radioactivity of the membranes, the use of our separation procedures reveals that the radioactivity is not associated with the purified glycoproteins. The nature of the labeled material which does not bind to PHA-Sepharose is being investigated. Also, since rabbit membrane glycoproteins have been found to contain receptors for PHA and thus can be fractionated, experiments are in progress to compare the biosynthesis of human and rabbit erythroid membrane constituents.

REFERENCES

1. H. Hamaguchi and H. Cleve, Biochim.Biophys.Acta 278, (1972) 271.
2. V.T.Marchesi and E.P.Andrews, Science 174 (1971) 1247.

PYRUVATE INDUCED CELLULAR FLATTENING AND PSEUDOPODIA FORMATION BLOCKED BY CYTOCHALASIN B

L.R.Pickart and M.M.Thaler, Dept.of Pediatrics, University of California, San Francisco

The phases of the cell cycle are accompanied by characteristic alterations in surface morphology. Under usual culture conditions, the cells flatten and increase their surface area at the time of DNA replication, then assume a spherical form prior to cell division.[1,2] We have found that pyruvate, in concentrations which occur in many cell culture media, causes cellular flattening, the development of pseudopodia and the stimulation of DNA synthesis in a monolayer culture of hepatoma cells.

When the hepatoma cells were cultured in a minimal growth medium containing 0.5% calf serum, the majority of cells were spherical and grew in clusters.[3] The addition of pyruvate (100 to 500 μg/ml) caused cellular flattening and the development of long pseudopodia on a majority of the cells within 48 hours at both high and low cell densities. At high cell densities, the pseudopodia form interconnecting matrices between cells. DNA synthesis went up 4.6 fold while RNA and protein synthesis was only slightly stimulated. Cytochalasin B (0.5-2.0 μg/ml) blocked the development of flattening and pseudopodia. Removal of pyruvate from medium reversed the effects. EDTA (1 mg/ml) caused the retraction of pseudopodia into the cells in 1 to 3 minutes.

REFERENCES

1. H. Dalen and P. Scheie, Exp. Cell Res. 57 (1969) 351.
2. K. Porter, D. Prescott, and J. Frye, J. Cell. Biol. 57 (1973) 815.
3. L. Pickart, and M.M. Thaler, Nature New Biol., 243 (1973) 85.

POSSIBLE ROLE OF ALVEOLAR HYDROXYLATED GLYCOPEPTIDES IN STORAGE AND SECRETION.

S. Bhattacharyya, M. Passero and W.S. Lynn, Duke University Medical Center, Durham, N.C. 27710

Two glycopeptides (molecular weight 62,000 and 36,000) which contain 8% heterosaccharides, 15% glycine, 1.2% hydroxyproline and 0.3% hydroxylysine have been found as the major proteins (80%) obtained by alveolar lavage of man, dog, rat, and rabbits. The 36,000 molecular weight peptide is the major one. These peptides also accumulate excessively in rats fed with cortisone or in patients with alveolar proteinosis (M. Passero et al P.N.A.S. 70, 973, 1973). In addition the major peptides found in isolated lamellar bodies of Type II alveolar cells of rabbits and rats are identical in composition to the two unique glycopeptides isolated from lavage material. Isolation and characterization of three large fragments obtained from cyanogen bromide treatment of the human 36,000 M.W. peptide indicate that the monosaccharides are distributed unhomogenously along the peptide.

It appears that these unusual glycoproteins with high glycine content are used for packaging, storage in lamellar bodies and secretion of surfactant by alveolar Type II cells. Whether other secretory cells, e.g. submaxillary, mammary, or pancreatic, use similar peptides for packaging and secretion is under study.